Biology and Ecology of Venomous Marine Scorpionfishes

Biology and Ecology of Venomous Marine Scorpionfishes

Ramasamy Santhanam
Former Dean
Fisheries College and Research Institute,
Tamil Nadu Veterinary and Animal Sciences University,
Thoothukudi, Tamil Nadu, India

ELSEVIER
ACADEMIC PRESS
An imprint of Elsevier

Academic Press is an imprint of Elsevier
125 London Wall, London EC2Y 5AS, United Kingdom
525 B Street, Suite 1650, San Diego, CA 92101, United States
50 Hampshire Street, 5th Floor, Cambridge, MA 02139, United States
The Boulevard, Langford Lane, Kidlington, Oxford OX5 1GB, United Kingdom

British Library Cataloguing-in-Publication Data
A catalogue record for this book is available from the British Library

Library of Congress Cataloging-in-Publication Data
A catalog record for this book is available from the Library of Congress

ISBN: 978-0-12-815475-5

For Information on all Academic Press publications
visit our website at https://www.elsevier.com/books-and-journals

Publisher: Charlotte Cockle
Acquisition Editor: Anna Valutkevich
Editorial Project Manager: Tasha Frank
Production Project Manager: Sreejith Viswanathan
Cover Designer: Christian Bilbow

Typeset by MPS Limited, Chennai, India

Contents

Foreword

I am delighted to write the Foreword for this book titled "Biology and Ecology of Venomous Marine Scorpionfishes". The scorpaenid fishes (also called rockfishes) of the family Scorpaenidae include scorpionfish, lionfish or turkeyfish, and rockcods. They are the world's most *venomous* species and are widespread in tropical and temperate seas but mostly found in the *Indo-Pacific*. While the *lionfish* (*Pterois volitans*) are extensively harvested from the wild for the live aquarium trade, rockcods (all members of genus *Sebastes*) are considered good eating and are caught by sport fishermen as well as commercially for as food fish.

It is the most comprehensive book about venomous scorpionfishes ever produced. This title represents an incredible resource, and its content is excellent. The author gives the biology and ecology of scorpionfishes in a general account followed by species accounts comprising common name, distribution, habitat, and biology. The chapter "Venomology of Scorpionfishes" which deals with the venom apparatus, characteristics of venoms, envenomation, and treatment and management of envenomations would be of great use for the sea-goers in general and scuba divers in particular. A general introduction about the scorpionfishes provides an important perspective.

Prof. R. Santhanam has succeeded superbly in creating a most useful overview of the hitherto-untouched marine scorpionfishes, a treatise that will be most welcomed by ichthyologists, marine biologists, fisheries researchers, and marine biology students worldwide. It will also be a welcome reference work for biologists and ecologists, and environmental groups, as well as commercial and amateur fishers of world seas.

Libraries, and any institution, including colleges and universities, with patrons interested in the lives of scorpionfish will find the reasonably priced volume worth having.

It is my hope and expectation that this book will provide an effective learning experience and referenced resource for all fisheries professionals. I congratulate the author for this timely and for the rich contribution to the science of Fisheries.

(S. FELIX)

Preface

The scorpaenid fishes (Family Scorpaenidae) include many of the world's most venomous species of scorpionfish, rockfish, and lionfish. Most species of this family are confined to the Indo-Pacific Oceans. Two species of Indo-Pacific lionfish (*Pterois volitans* and *Pterois miles*) were inadvertently introduced into the Eastern Atlantic in the 1990s and have invaded and overpopulated shallow reef systems in the Gulf of Mexico and Caribbean Sea. They are the masters of camouflage living on or near the bottom at all depths from shore to about 2200 m. Almost all the species of Scorpaenidae are considered venomous as their fin spines are capable of inflicting serious or fatal wounds in humans.

The scorpaenid fishes are not commercially fished because of their venom, but some species are caught as incidental bycatch for human consumption. California scorpionfish, *Scorpaena guttata*, has an important part of the commercial and sport fishery in southern California, especially within the Los Angeles port complex. In the recreational fishery, they are taken primarily aboard commercial passenger fishing vessels and private vessels, and occasionally from piers and jetties. Several brightly colored species of scorpaenids are of aquarium value. Species such as *Scorpaena brasiliensis*, *Taenianotus triacanthus*, *Scorpaenopsis macrochir*, *Rhinopias frondosa*, and *Rhinopias eschmeyeri* command a high price in the aquarium trade. The venoms of certain species of scorpionfishes, however, have been reported to possess pharmaceutical properties. While the toxic extracts of *Scorpaenopsis venosa* have shown antimicrobial activity, the venom of *Scorpaena plumieri* possesses antitumoral effects.

Though a few books are presently available on the Biology of marine fishes, a comprehensive book on the "Biology and Ecology of Venomous Marine Scorpionfishes" has not so far been published. This publication would therefore fill up this long felt need. The aspects covered in this publication include general biological characteristics of scorpaenid fishes; profile (taxonomy, common name, global distribution, habitat, diagnostic features, biology, and fisheries, pharmaceutical and aquarium values) of 250 species of venomous marine scorpionfishes; and venomology of scorpaenid fishes with information on venom apparatus of scorpaenid fishes; characteristics of scorpaenid fish venoms; envenomation of scorpaenid fishes; and treatment and management of scorpaenid fish envenomations.

It is hoped that the present publication when brought out would be of great use as a standard reference for researchers, teachers, and students of various disciplines such as fisheries science, marine biology, zoology, life sciences, and environmental sciences; libraries of colleges and universities; and as a valuable guide for sea-goers.

Suggestions from the users are welcome.

The Author

Acknowledgments

The author is very grateful to Prof. Dr. S. Felix, Vice-Chancellor, Tamil Nadu Dr. J. Jayalalithaa Fisheries University (TNJFU), Nagapattinam, India, for his kind Foreword. He sincerely thanks Prof. Dr. B. Ahilan, Dean in-Charge, Fisheries College & Research Institute, TNJFU, Chennai, for his valuable suggestions. Special thanks must be expressed to the international agencies Wikipedia, Smithsonian's National Museum of Natural History, Smithsonian Tropical Research Institute, Australian National Fish Collection (CSIRO), FishBase, Discoverlife, Western Australian Museum, National Museum of Natural History (France), Solitary Islands Underwater Research Group (SURG), St Helena Government, Reef Builders, Flickr, Archipelago Marine Research Ltd., Fish Biosystem, National Ocean Service (NOAA), and Shutterstock, and internatinal friends John E. Randall, Kirk Kilfoyle, Dulcic, Jug-Dujakovi, Bartulovic, Glamuzina, Haskovic, Skaramuca, Dimitri A. Pavlov, Guillermo Moreno, H. Geoffrey Moser, Elberth Ahlstrom, Thomas E. Laidig, Keith M. Sakuma, Mary M. Nishimoto, Greg Hix, Renee Coles-Hix, James A. Morris, Jr., C. Calloway, Yabe, Naranji Muddula Krishna, Kwang-Tsao Shao, Thomas Gloerfelt-Tarp, Sujatha, N. M. Krishna, Mizuki Matsunuma, Hiroyuki Motomura, Thorke A. S. Østergaard, Minoru Ishida, Kunio Amaoka, Keith Sainsbury, Pr. G. Duhamel, R. Causse, P. Béarez, S. S.Mishra, Cambraia Duarte, Pedro Miguel Niny, Denis Tweddle, M. Eric Anderson, Raju Saravan, Naoko Kanehira, Hisashi Imamura, Kunto Wibowo, Luciano Gomes Fischer, Jimenez Prado, Ross Robertson, Keiichi Matsuura, Fernando Bento, Olivier Bodere, Teodoro Vaske Junior, Michael N Trevor, Béarez, Philippe Béarez, John Snow, Andreas Völckers, Louis H. DiSalvo, Serge Abourjeily, Hiroyuki Tanaka, A. M. Arias, Patrick O'Donnell, Milton Love, Mark Allen, Gerald R. Allen, Ian Shaw, Theo Modder, Xisco Ordinas, Robert Patzner, J. Brown, Toshiyuki Suzuki, Amezcua Linares, Felipe, Info Nero di Sole, Philip Good, Benjamin Victor, M. Erdmann, Bruce B Collette, Osmar Luiz Jr., Hamid Osmany, David Greenfield, Alfredo Carvalho Filho, David Catania, David C. Cook, Hermosa, Jr., V. Gregorio, Richard Winterbottom, Patrick Durville, Hsuan-Ching Ho, Ken Clifton, Mary Jane Adams, Díaz de Astarloa, Juan Martín, Joi Ito, Tetsuji Nakabo, Umeyoshi Yamada, and Stephen Vives who were very kind enough to provide images of certain species of venomous scorpionfishes for the present purpose.

The secretarial assistance provided by Mrs. Albin Panimalar Ramesh is also gratefully acknowledged.

1 Introduction

The oceans and seas also form the most terrifying and unforgiving habitats for many hazardous animals. While their poisonous animals transfer their poisons when touched or eaten, the venomous creatures actively inject their venoms. Among these two categories, venomous animals are the most important as they are often fatal to the victims.

Categories of marine hazards: The categories of marine hazards and the animals responsible are given below.

Category	Source animals
Contact irritants	Sea anemones, sea cucumbers, stinging corals, crown-of-thorns starfish, hydroids, box jellyfish, sea urchins
Ingested toxins	Shellfish toxin, scombroid toxin, tetrodotoxin, ciguatoxin
Injected toxins	Cone shells, blue-ringed octopus, stingrays, scorpionfish, stonefish, sea snakes
Predators	Barracuda: Moray eels, large grouper, saltwater crocodile, sharks

Source: Anon. http://web.utah.edu/umed/students/clubs/international/presentations/dangers.html.

Facts on marine hazards: It has been reported that the box jellyfish have been known to kill people within 3 minutes, blue-ringed octopus in 30 minutes, and pufferfish (eaten) in 17 minutes. At least 65 people have been killed by box jellyfish in the last century, over 30 of them on beaches between Mackay and Cairns. While aboriginal people long knew about box jellyfish, it was not until after the death of a 5-year-old boy at Cardwell, in 1955, that *Chironex fleckeri* was identified by scientists. The Irukandji jellyfish (*Carukia barnesi*) was first scientifically identified in 1961 by Cairns doctor, John Barnes. He named it after the local Irukandji aboriginal people (Anon. http://web.utah.edu/umed/students/clubs/international/presentations/dangers.html).

The geographical locations of human fatalities from marine animal envenomation are given below.

Venomous marine animal	Affected area
Blue-ringed octopus	Australia, Singapore
Cone shell	Australia, Fiji, India (Banda), New Caledonia, Japan (Okinawa), Vanuatu
Sea snake	Burma, Malaysia, India (Madras), Indonesia (Java), Japan (Okinawa), Oman, Vietnam
Stingray	California, Colombia, Fiji, New Zealand, Surinam, Texas
Stonefish	Australia (Thursday Island), East Africa, Japan, Seychelles

Source: Fenner, 2004.

Venomous marine animals: Venomous, stinging invertebrate animals include sponges, coelenterates (jellyfish, hydroids, corals, and sea anemones), echinoderms (sea urchins,

Biology and Ecology of Venomous Marine Scorpionfishes. DOI: https://doi.org/10.1016/B978-0-12-815475-5.00001-7

starfish, and sea cucumbers), annelid worms (bristleworm), and mollusks (cone shells, octopi, and nudibranchs). Stinging vertebrates discussed include stingrays, catfish, scorpionfish, and leatherjacks (Schwartz and Meinking, 1997). Although the venomous marine animals are present worldwide, they are mainly confined to the Indo-Pacific region.

Hazardous fishes: There are more than 500 species of fishes that are poisonous to man, while another 250 species have venomous organs (Angelini and Bonamonte, 2016). Among the marine fishes, the scorpionfishes (rock fishes) of the family Scorpaenidae assume greater significance as they have been found responsible for severe injuries and occasionally deaths in humans around the world. The more venomous scorpionfishes have been observed on the Brazilian coast, East Atlantic Ocean, and in the Southwestern Atlantic region.

Diversity of scorpaenid fishes (family Scorpaenidae): This family enjoys the dubious distinction of containing more venomous species than any other family of bony fishes. As per Food and Agriculture Organization of the United Nations (FAO), this family includes scorpionfishes, rockfishes, rosefishes, stingfishes, stonefishes, turkeyfishes, and waspfishes with nine subfamilies, viz., Apistinae, Plectrogeninae, Pteroinae, Scorpaeninae, Neosebastinae, Sebastinae, Setarchinae, Synanceiinae, and Tetraroginae (Poss, http://www.fao.org/tempref/docrep/fao/009/x2400e/x2400e20.pdf; Anon., 1983). This family originally had 56 plus genera and 418 species (Anon. http://www.wetwebmedia.com/scorps_dang_mars_art.htm; Bonecker et al., 2014). Among them, 57 species from the 23 genera (*Apistus, Brachirus, Centropogon, Choridactylus, Erosa, Gymnapistes, Helicolenus, Hypodytes, Inimicus, Leptosynanceja, Minous, Notesthes, Pterois, Ruboralga, Scorpaena, Scorpaenodes, Scorpaenopsis, Sebastapistes, Sebastes, Sebastodes, Sebastolobus, Snyderina*, and *Synanceja*) were found to be possessing severe venomous properties (Halstead, 1970; Roche and Halstead, 1972). As the classification and relationships of this family are in controversy and some of the fishes that used to be in family Scorpaenidae are now placed under several different families including family Apistidae (wasp scorpionfishes), family Tetrarogidae (waspfishes), and family Synanceiidae (stonefishes). The present publication which is based on the Taxonomy of Eschmeyer (1998) deals with about 250 species of venomous species of the family Scorpaenidae (scorpionfishes or rockfishes) with the major categories, viz., scorpionfishes (subfamily Scorpaeninae), rockfishes (subfamily Sebastolobinae), and lionfishes (turkeyfishes) (subfamily Pteroinae).

Distribution of scorpionfishes: Scorpaenidae are well represented throughout the world's oceans. However, they mostly occur in all tropical and temperate seas. Most scorpionfishes are found distributed in Red Sea, Pacific Ocean to Australia, and Hawaii, and a few scorpionfishes (no lionfishes or stonefishes) live in the Caribbean.

Habitat: Most scorpion fishes live on or near the bottom (benthic or benthopelagic). They lie in crevices, in caves, and under overhangs.

Envenomation and injuries: Though most of the fishes of this family are venomous and are capable of inflicting serious and potentially fatal wounds, less than 5% of the fishes of this family have been studied even in the most preliminary manner and, to date, the basic chemical structure of even a single fish venom has not been elucidated. It is evident that the major reason for this apparent lack of knowledge is the inherent and extreme instability of fish venoms (Saunders and Taylor, 1959).

Impact of lionfish on native fish: Lionfish are an invasive species from the Indo-Pacific that were introduced to the Atlantic through the aquarium trade. They consume vast quantities of more than 50 native fish and invertebrate species some of which are ecologically and economically important. In the Caribbean and Western Atlantic, the spread of lionfish has created a great concern for the health of coral reefs already threatened by other stresses. For example, manipulative field studies in the Bahamas have shown that lionfish can reduce recruitment of native reef fish by up to 79%, with predation being the likely mechanism. After lionfish invaded a mesophotic reef, a shift of native reef fish has also been reported from coral and sponge communities to algal-dominated communities. Increased predation on herbivorous fish was implicated as the cause of this shift (Gardner et al., 2015).

1.1 Human Uses of Scorpaenid Fishes

Fisheries: Due to the venomous spines, very few species of Scorpaenidae are commercially exploited. Although all are edible, most species are small and relatively dangerous to handle, or occur on hard bottoms, and thus do not form the basis of large fisheries, despite being at times common in bycatch and their flesh generally of excellent quality for food. Because most species are too small, not abundant enough, or found on rocky bottoms, they are presently of little commercial importance in the Western Central Atlantic. However, larger species are often encountered in local fresh-fish markets and highly valued as food. A few shallow-water species are caught by sport fishermen on hook-and-line, some taken by the handline fishery, and others caught incidental to trawl fisheries. *Helicolenus dactylopterus* is commonly taken by trawlers working moderate depths (400–600 m) for shrimp. Some scorpionfishes are of considerable commercial importance outside Area 31, especially in cold water regions (Poss, http://www.fao.org/tempref/docrep/fao/009/x2400e/x2400e20.pdf). One such commercially exploited species is the Eastern red scorpionfish (rock cod), *Scorpaena cardinalis*. In New South Wales, this species is exploited as a recreational catch which amounted to 20–50 t/annum (three times more than commercial catch, 5–10 t/annum). As far as its exploitation status, it is fully fished (Anon. https://www.dpi.nsw.gov.au/__data/assets/pdf_file/0007/375883/Eastern-Red-Scorpionfish.pdf). Rockfishes such as *Sebastes mystinus*, *Sebastes melanops*, and *Sebastes flavidus* are also important to the commercial and recreational fisheries of the northeastern Pacific. In 1993, they accounted for approximately 18% of the commercial groundfish catch in weight as per Pacific Fishery Management Council (PFMC) and 19% of all species in the recreational landings off California, Oregon, and Washington. Among the rockfishes, the genus *Sebastes* is a very important part of the commercial and sport fishery catches off California, Oregon, Washington, Canada, and Alaska. These rockfishes comprise one third of the recreational fish catch in California, with species such as *S. mystinus*, *S. melanops*, and *S. flavidus* comprising up to 30% of this proportion. In addition, *S. mystinus* and *S. melanops* are also important components of the commercial rockfish catch.

Aquarium values: There are more than 200 recognized species of scorpionfishes in artificial aquariums around the world. Owing to their vibrant colorations (in shades of red, orange, yellow, tan, brown, gray, and white) and complex markings and behavior, scorpionfishes are kept in tanks by numerous people. Further, they are extremely hardy fishes. They adapt very well to life in the aquariums and are extremely disease resistant (Anon. https://a-z-animals.com/animals/scorpion-fish/). Tropical scorpionfishes of various kinds are extensively harvested from the wild for the live aquarium trade. The lionfish (*Pterois volitans*) is particularly popular. The popular aquarium varieties of Scorpaenidae are given below.

Reef scorpionfish, *Scorpaenodes caribbaeus*
Cheekspot scorpionfish, *Scorpaenodes evides*
Merlet's scorpionfish, *Rhinopias aphanes*
Easter Island scorpionfish, *Rhinopias cea*
Japanese scorpionfish, *Rhinopias argoliba*
Eschmeyer's scorpionfish, *Rhinopias eschmeyeri*
Weedy scorpionfish, *Rhinopias frondosa*
Strange-eyed scorpionfish, *Rhinopias xenops*
Shortfin lionfish, *Dendrochirus brachypterus*
Zebra lionfish, *Dendrochirus zebra*
Bluefin lionfish, *Parapterois heterura*
Spotfin lionfish, *Pterois antennata*
Common lionfish, *Pterois volitans*
Miles's lionfish, *Pterois miles*
Leaf scorpionfish, *Taenianotus triacanthus*

Food values: Scorpionfishes are venomous but not poisonous. Further, all these fishes are edible, but most are small and dangerous to handle and therefore are not valuable food fish. The larger and more abundant species, caught primarily as a by-catch of deepwater trawlers, are sold commercially in some regions of the world. The flesh is white and considered an excellent food (Anon. https://www.mexican-fish.com/scorpionfish-family/). The California scorpionfish, *Scorpaena guttata,* is considered to be a very good food fish once the spines have been avoided.

Bourbetto (scorpionfish stew): This is the most famous seafood dish on the Greek island Corfu. It is simple and delicious and is prepared from the Red Scorpion Fish, *Scorpaena scrofa,* which is a close edible relative to other species of scorpion fish distributed around the globe, including the lion fish, flatheads, etc. (Field, 2014).

Pharmaceutical values: The crude and fractions of venom gland and gonad of *Scorpaenopsis venosa* have been reported to show antimicrobial activity against bacterial strains, viz., *Pseudomonas* sp., *Staphylococcus aureus, Vibrio harveyi, Vibrio parahaemolyticus, Proteus* sp., and *Escherichia coli;* and three fungal strains, viz., *Rhizoctonia solani, Pythium proliferum,* and *Aspergillus niger* (Sasikala and Ravindran, 2013).

Development of new drugs from scorpaenid fish venom: Earlier studies made on lionfish venom showed that it contained targeting bioactive substances, viz., peptides and proteases which could be of more use in the development of new drugs in future. Overall, research on fish venom is slow and still in the early stages, but it is a field that is actively being explored and is sparking the interest of current and rising scientists (Brown, 2012).

1.2 Status and Threats

1. Like other creatures of the intertidal zone, the scorpionfishes are also affected by human activities such as reclamation and pollution. Trampling by careless visitors and overcollection also has an impact on local populations.
2. Some species may be caught as incidental bycatch when lines and hooks are set near reefs. Their coral reef habitat may also be in jeopardy due to warming oceans and global warming
3. These slow-growing fishes are often threatened by overfishing and coral reef habitat destruction by trawl nets.
4. Sport fishing puts some pressure on this species.
5. Some species are considered to be very good food fish once their spines have been avoided. Such species are sought after in some Asian communities in Southern California. There is a small commercial fishery for such species, and they are caught by angling, gillnetting, and otter trawling. The extent of the impact of fishing on these populations is unknown.
6. Many species—especially lionfish and some of the rarer scorpionfish—are highly prized in the aquarium trade. As it is impossible to breed them in captivity, all individuals must be captured from the wild, putting pressure on wild populations in certain areas. Harvesting tropical scorpionfishes for the live aquarium trade may involve the use of cyanide or blasting, which damage the habitat and kill many other creatures.

Conservation measures: Conservation measures are needed to protect the coral reef habitats of scorpaenid fishes from further loss and destruction. Common approaches for coral reef conservation include establishing protected areas or no-take zones that exclude anthropogenic disturbance of reef ecosystems. Conserving existing areas of coral reef often results in improved ecosystem service provision. For coral reef restoration, rearing, transplanting, and monitoring of coral reef fragments are possible approaches.

2 Biology of Marine Scorpionfishes

Scorpaeniformes (scorpionfishes and flatheads) is the fourth largest order of fishes encompassing about 20 families, 250 genera, and 1000 species. The classification of this order is controversial, not only in terms of monophyly (descendants of a common ancestor) but also at the subordinal (rank) and familial level (Washington et al., https://swfsc.noaa.gov/publications/CR/1984/8499.PDF). Among the different families of Scorpaeniformes, the family Scorpaenidae assumes greater significance due to its constituents, viz., venomous scorpionfishes (rockfishes) or stingfishes which comprise three subfamilies, viz., Scorpaeninae (scorpionfishes); Sebastinae (rockfishes); and Pteroinae (lionfishes or turkeyfishes).

Biology of the Scorpionfishes (Rockfishes) (Family Scorpaenidae)

Characteristics of the family Scorpaenidae: The scorpaenid fishes are very similar to the sea perches, but with a bony stay across the cheek. The common name of these fishes comes from the stinging pain that they can inflict. All the species of this family are considered to be venomous mainly due to their dorsal fins. Some species also have venomous anal and pelvic fins. Their pectoral and caudal fins are not venomous as they do not possess stinging spines. The rockfishes are not so toxic as scorpionfishes and lionfishes venom, but still they are capable of inflicting a painful sting. These fishes remain mostly stationary during daylight, but are active at night.

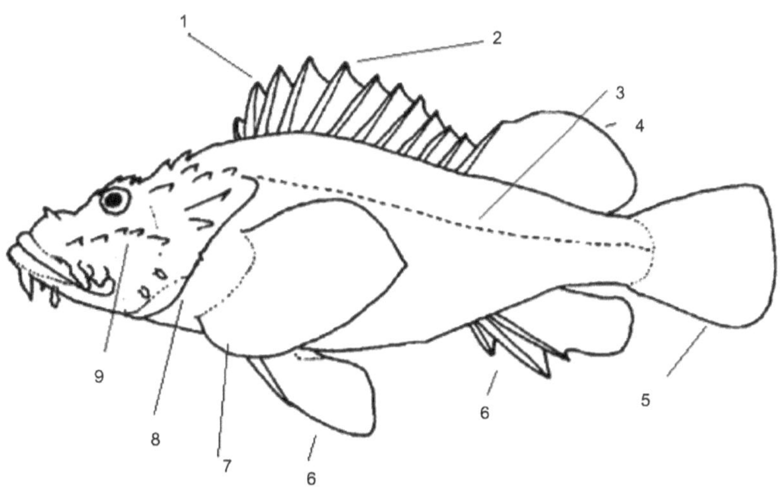

External anatomy of a scorpaenid fish: (1) Spinous dorsal fin, (2) spine of dorsal fin, (3) lateral line, (4) soft-rayed dorsal fin, (5) caudal fin, (6) spinous anal fin, (7) soft-rayed pectoral fin, (8) gill cover, (9) suborbital ridge. Size: Many species of the family Scorpaenidae are relatively small, typically under 200 mm SL and their biology is poorly studied. *Author.*

Biology and Ecology of Venomous Marine Scorpionfishes. DOI: https://doi.org/10.1016/B978-0-12-815475-5.00002-9

Coloration of body: The body of many fishes of this family is red, reddish brown, or brown in color, and it may have barred or mottled color patterns that are typically darker dorsally than ventrally. Further, fishes caught below about 50 m are mostly red and white, often with spots of darker red, brown, or black. Like many venomous animals, the scorpionfishes use their brilliant color as a warning. However, unlike many venomous animals, these colors are not on permanent display. Using a behavior known as "flashing," many scorpionfishes have the ability to display their colors when needed, such as when a predator is nearby. Their colors are often found on the inner surface of their pectoral fins. Some species also have brilliant colors on the skin within their jaw structure that can be exposed when the fish opens its mouth. This serves as an effective warning defense for approaching conspecifics. For example, in *Scorpaena plumieri*, the coloration is quite variable as the bright white spots on the pectoral fins get extended when the fish is disturbed.

Coloration in *Scorpaena plumieri*. *Photo courtesy: Kirk Kilfoyle.*

Camouflaging: Most species of *Scorpaenidae* are strongly camouflaged, and this remarkable ability helps the fishes blend into their surroundings. This camouflage largely assists these fishes in prey capture. Often, herbivorous prey will be attracted to the scorpionfishes and their encrusting algae. Other times, the encrusting algae can appear to be a "safe-haven" to passing-by fish and crustaceans. Further, the habitat of scorpionfishes varies greatly amongst species according to their camouflage. For example, some members are best suited to mud bottoms, while others have camouflage that renders them unnoticeable amongst corals belonging to *Pocillopora* spp. corals.

Behavior: The scorpaenid fishes are generally solitary in nature, except during courting and mating when a pair may be located. Groups larger than a pair are rarely found. An exception to this is found in the leaf scorpionfish, which commonly occurs in groups of three. Fighting does occur between conspecifics to defend their territory or their mates. The red lionfish, *Pterois volitans* displays a peculiar defensive behavior toward potential enemies, in which it swims aggressively toward the enemy with the dorsal fin in front in order to stab the other fish. This type of active attacking behavior might also exist in other lionfishes.

Food and feeding: The scorpaenid fishes feed primarily on large and small crustaceans including decapods, amphipods, and isopods. Many feed on small fishes as they attain larger sizes. Other food items of these fishes include polychaetes and small cephalopods. Most species are ambush predators. Gut contents of some of the scorpionfishes analyses showed the following:

Larger species—*Scorpaenopsis gibbosa*—piscivores
Smaller species—*Iracundus signifier* and *Dendrochirus brachypterus*—piscivores
Smaller species—*Scorpaena maderensis* and *Scorpaenodes guamensis*—zoobenthivores

Reproduction: Most fishes of the family Scorpaenidae are known to lead solitary lives and evidently aggregate only for reproduction. The young of most species are planktonic, with many settling out of the plankton relatively quickly. The general characteristics of the eggs and larvae of the fishes of the family Scorpaenidae are given below.

Eggs: Most fishes of the family Scorpaenidae are ovoviviparous, producing between a few hundred and a few thousand eggs. In the fishes of the subfamilies, viz., Scorpaeninae and Pteroinae, the eggs are extruded in bilobed gelatinous egg masses which float at the surface. The eggs are slightly elliptical and have homogeneous yolk, a narrow perivitelline space, and a smooth chorion. A single oil globule is present in *Pterois* spp. (0.16–0.17 mm), and *Scorpaena* spp. lack an oil globule.

Larvae: In the fishes of the subfamily Scorpaeninae, the larval stages are known for only 3 of the 15 genera. Hatching of eggs occurs at about 2.0 mm or less, and newly hatched larvae have a large elliptical yolk sac, unpigmented eyes, pectoral fin buds, and lack a mouth. Finfold is inflated and, along with the body skin, it forms a balloon-like envelope that is attached principally at the snout and pectoral regions. Larvae are relatively deep-bodied. Head becomes massive. Snout has a steep profile. Pectoral fins are generally well developed and deep-based. While these pectoral fins are fan-shaped and enlarged in the genus Scorpaenodes, they are smaller but distinctively shaped in Scorpaena (fan-shaped with scalloped margin) and Pontinus (slightly wing-shaped). Ossification of fin rays occurs in early larvae (4–5 mm). Caudal, pectoral, and pelvic rays get ossified almost simultaneously, followed immediately by the dorsal and anal fins. The most prominent pigment is on the pectoral fins. Typical patterns are a concentration at the distal margin (*Scorpaenodes* and in some species of *Pontinus* and *Scorpaena*); a solid covering over most of the fin as in some species of *Scorpaena* and a diagonal bar as in some species of *Pontinus*. A melanistic sheath develops over the dorsal surface of the gut and gas bladder in most species of *Scorpaena*, whereas in *Scorpaenodes* and *Pontinus*, only the gas bladder is pigmented. In the subfamily Pteroinae, newly hatched larvae are small (I. 1–1.6 mm) and are similar in morphology to those of Scorpaeninae (Washington et al., https://swfsc.noaa.gov/publications/CR/1984/8499.PDF).

Skin shedding: Many scorpaenidae have the ability to shed the outermost skin layer, viz., the cuticle. Because of their sedentary lifestyle, it is not uncommon for algae or other organisms to grow on these fish. The skin shedding benefits the fish by removing any encrusted, attached algae, cyanobacteria, and even parasites. Some scorpaenidae shed their skin more often than others.

Defenses: All species of this family have very effective defenses, and most species have powerful stinging (venomous) spines in their dorsal fins which are capable of bringing instant paralysis. Many scorpionfishes have brightly colored pectoral fins which they flash at predators, warning them of their venom. When these fishes are away from their predators, the fins can be tucked away so as not to ruin their camouflage. It is also reported that while scorpionfishes rely on their amazing camouflage to escape predators, lionfishes generally use their intimidating fins as protection.

Conservation Status

Detailed studies have not been made on the stock assessment, habitat loss due to environmental factors, and other conservation issues of scorpaenid fishes.

Conservation of scorpionfish: In 2005, three species, viz., California scorpionfish (*Scorpaena guttata*), gopher rockfish (*Sebastes carnatus*), and cabezon (*Scorpaenichthys marmoratus*) were assessed under the Marine Life Management Act, Nearshore Fishery Management Plan. The Assessment results indicated that scorpionfish biomass declined before 1980 and has since increased (healthy state) mainly due to changes in recruitment (Anon. https://www.wildlife.ca.gov/Conservation/Marine/Scorpionfish).

Conservation of rockfish: Rockfish caught in deep water often sustain injuries—referred to as barotrauma—caused by rapid decompression and expansion of gases in the swim bladder. Fish that are released with inflated swim bladders cannot resubmerge and will die. Because of high release mortality, intentional catch-and-release fishing is greatly discouraged, particularly in depths of 20 m or greater. Alaska anglers can best prevent wasteful rockfish mortality by avoiding waters where unwanted catches are likely. When rockfish are caught incidentally despite avoidance efforts, proper deepwater release techniques can reduce mortality. Anglers can help conserve Alaska's valuable rockfish stocks by following some commonsense guidelines and employing deepwater release techniques.

Conservation of lionfish: Lionfish are not currently listed as threatened or endangered in their native range. However, the increase in pollution in coral reefs may negatively affect the lionfish's primary food sources (crustaceans and fish). If lionfish are unable to adapt to declines in their prey species, their numbers may decrease considerably.

2.1 Biology of the Scorpionfishes (Subfamily Scorpaeninae)

Characteristics of subfamily Scorpaeninae: Body of the species of this subfamily is weakly, rather than strongly, compressed. Body depth is about 20%–50% of standard length (SL). Head is moderate to large (37%–50% of SL) and is often depressed with cirri, particularly above eye. Eye is small to relatively large (4%–14% of SL). Snout is short to long (6%–20% of SL) and is often prominent. Mouth is often large and upturned. Numerous small conical teeth are present on upper and lower jaws. Vomer and palatine teeth may be present or absent. Branchiostegal rays are typically 7. Gill rakers are 1–9 in upper arch, and 4–20 in lower arch. Most species are with numerous head spines. Dorsal fin is with strong venomous spinous part bearing 8–18 spines connected to soft-rayed part posteriorly, with 7 1/2 to 13 1/2 soft rays, the last typically split to its base and counted as 1 1/2. There are three anal-fin spines. Caudal fin is either rounded or truncate and is never forked. Pectoral fins are usually large, with 11–24 rays; rays of larger individuals of most species are branched. Pelvic fins which are normally thoracic in position are with one strong spine and four to five branched rays. Scales in most species are relatively small and either ctenoid (rough to touch) or cycloid (smooth to touch) and entirely absent in others or present only as deeply embedded scale rudiments. Lateral-line is present, with 12–54 pored or tubed scales; when present, 4–8 scales above lateral line; and 10–19 scales below lateral line. All species of this subfamily possess striated swim bladder musculature which is extrinsic in nearly all species. Species without swim bladder also possess musculature. Pyloric caeca are 1–16. Vertebrae are 24–29.

The available information on the biology and ecology of the species of this subfamily are given elsewhere.

2.1.1 Reproduction

2.1.1.1 Scorpaena notatae

Until recently, most aspects of reproduction of this species were unknown although a few studies on the ovarian structure and process of oogenesis are available. Based on the individual fishes of this species collected from northwest Mediterranean, the males were dominant than females, and the sex ratio was 1.5. The spawning period of this species was found to be May against the common spawning period (July–October) of scorpaenid fishes (Munoz et al., 2005). The fecundity of this species was found to be 6000–33,000 eggs per female. This is an ovuliparous (external fertilization; intermediate between the most basic oviparity and the first transitional stages toward viviparity) species.

2.1.1.2 Scorpaena scrofa

Egg characteristics and embryonic development: Fertilized eggs of this species are pelagic and ellipsoidal with a homogeneous and unsegmented yolk surrounded with transparent gelatinous layer. Eggs range in diameter from 0.8 to 1.0 mm. There are no oil globule(s) in the eggs. The first cleavage occurs at about 1 hour after fertilization, the second after 2 hours 15 minutes, and the third at 3 hour 40 minutes. At 5 hours 15 minutes, the blastoderm is in advanced stages of cleavage, and this stage is called "mulberry stage." Gastrulation starts 11 hours 10 minutes after fertilization. Neurula stage starts 12 hours 45 minutes after fertilization. Formation of the embryo begins after 16 hours, and somatic segmentation after 18 hours 15 minutes. The Kupffer cells make their appearance 17 hours 10 minutes after fertilization. Between 20 and 23 hours, optic vesicles are formed; olfactory lobes get differentiated and the pericardial cavity develops. After 26 hours, the somite differentiation is completed, while optic vesicles and olfactory lobules are clearly visible. The heart beating begins after 26 hours 15 minutes. Hatching starts after 30 hours 25 minutes, and after 31 hours 45 minutes, the all yolk-sac larvae hatch (Dulcic et al., 2007).

Larval development: The newly hatched yolk-sac larvae (about 2 mm length) are transparent and pelagic; and they float at the surface. The fin fold invades much of the body. The eyes are unpigmented. Though the mouth is undeveloped, a short and simple gut is seen in this stage. The pectoral fins are visible. At the beginning of day 2 (i.e., about 25 hours after fertilization), pigmentation develops in the eyes. Now the two-thirds of the yolk-sac get absorbed. At the end of the second day, granular pigmentation develops in the eyes and formation of the maxillaries and lower jaw starts. After 46 hours of hatching, the mouth becomes functional. By the end of day 3, the foregut and hindgut become functional. The anus is s open. Pectoral fins (0.25 mm length) show three marginal rows of melanophores. The eye is completely pigmented. By the beginning of day 4, yolk-sac is completely absorbed. Now pectoral fins show four marginal rows of melanophores. The body and notochord are completely straight. Now the larvae are able to swim at the surface. One row of melanophores is visible on the ventral contour of the body from anus to the end of notochord. On the end of day 5, pigmentation is clearly and strongly visible from the stomach to the end of notochord. Few melanophores are visible on the peritoneal and anus region also. About one-third of pectoral fin is covered with 5–6 rows of melanophores. At this stage, formation of caudal fin starts, and several rows of melanophores are seen on the dorsal and ventral contour of the body. Pectoral fins have 18 rays and 5–6 marginal rows of melanophores. Formation of rays in caudal fin starts. By the end of day 6, head and opercular spines are clearly visible (Dulcic et al., 2007).

2.1.1.3 Scorpaenopsis possi

Adult individuals (about 13–20 cm total length (TL)) of this species possess various intensities of coloration (with a light orange or bright crimson ventral part of the body and pectoral fins in different fishes) irrespective of the sex. Sexual dimorphism is almost totally absent. These fishes are oviparous with females producing eggs. The eggs embedded in a floating gelatinous matrix (about 2.5 cm size) are distributed mainly within a single layer. Surface of the matrix consists of compartments, each of polygonal shape 9–13 μm in size. First clutch used for the incubation contains 1260 eggs. Before insemination, ovulated eggs (about 700–800 μm diameter) possess oval shapes, and they are separated from the gelatinous matrix. However, these eggs which are transparent with homogenous yolk without oil droplets are still floating at the water surface.

Laboratory experiments showed that following the first cleavage, subsequent cleavage cycles occurred with a periodicity of about 7 minutes. The blastoderm consisted of deeper cells and a layer of periderm located at the surface. During organogenesis (when the number of muscular segments exceeds 20), the end of the caudal part of the embryo was found separated from the yolk sac. The finfold appeared in the caudal portion of the embryo. The Kupffer's vesicle was located under the posterior area of the caudal part of the embryo. A zone of small vesicles connected with the embryo was seen in the area where its body was separated from the yolk sac. In the subsequent stages of embryonic development, the perivitelline space became extremely narrow. Weak constrictions of the heart tube were observed when the number of muscular segments in the body of the embryo reached 25. During hatching, 13 preanal and 12 postanal segments were seen in the embryo of 1.6 mm in length (Pavlov and Emel'yanova, 2007).

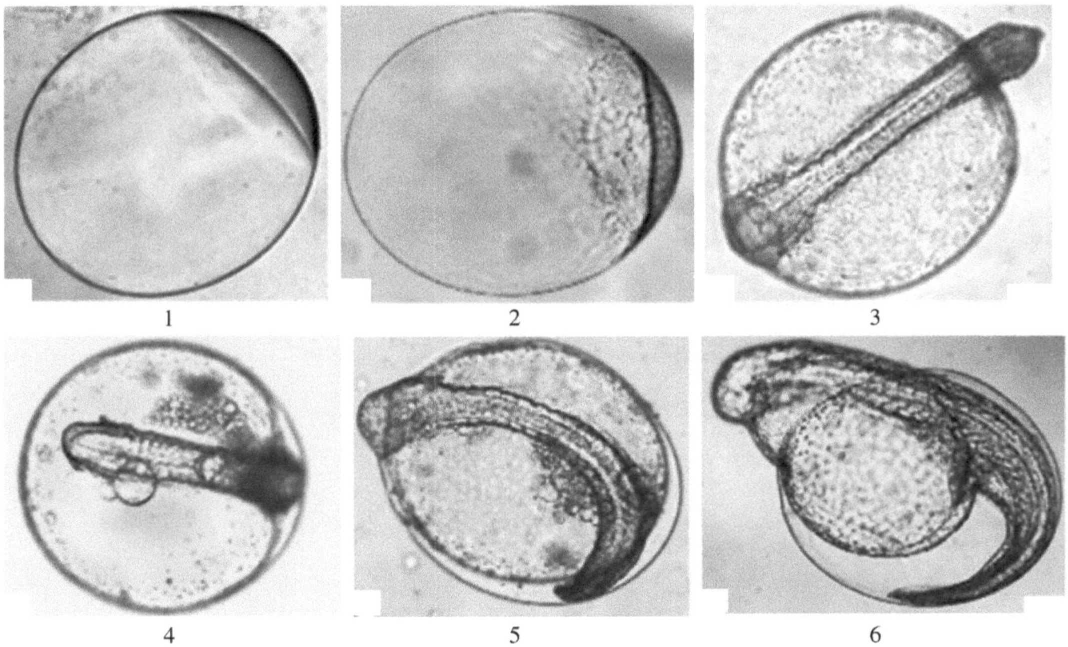

Embryonic development of *Scorpaenopsis possi*: (1) Egg before insemination, (2) Blastulation (4 h), (3) 20 segmented embryo (23 h), (4) 20 segmented embryo showing Kupffer's vesicle and finfold formation, (5) 25 segmented embryo (27 h), (6) hatching of the embryo (35 h). *Photo courtesy: Dimitri A. Pavlov.*

2.1.1.4 Sebastapistes cyanostigma

Little information is available about the detailed reproduction of this species. Adult individuals of this species (6–8 cm TL) possess an orange coloration. Sexual dimorphism is absent. Based on general morphology of the urogenital system in the individuals of both sexes, the difference between *Scorpaenopsis possi* and *Sebastapistes cyanostigma* is absent. As other scorpionfishes, these fishes are oviparous with females producing eggs. Eggs (about 10 mm in diameter) of this species are normally seen in the floating gelatinous matrix. After insemination, the egg enveloped attaches tightly to the surface of the yolk, and the perivitelline space is not seen. The first cleavage is seen within 65 minutes after insemination. The Kupffer's vesicle appears when eight somites appear in the body of the embryo. First movements of the embryo appear several hours before hatching, and weak constrictions of the heart are seen at the same time. At hatching, 12 preanal and 15 postanal segments can be seen in the embryo of 1.6 mm in length. At the surface of the yolk sac, opaque glandular cells are seen below large polygonal cells of epidermis. The cerebellum appears within 16 hours after hatching. When the embryo reaches 2 mm in length, the yolk is resorbed (Pavlov and Emel'yanova, 2007).

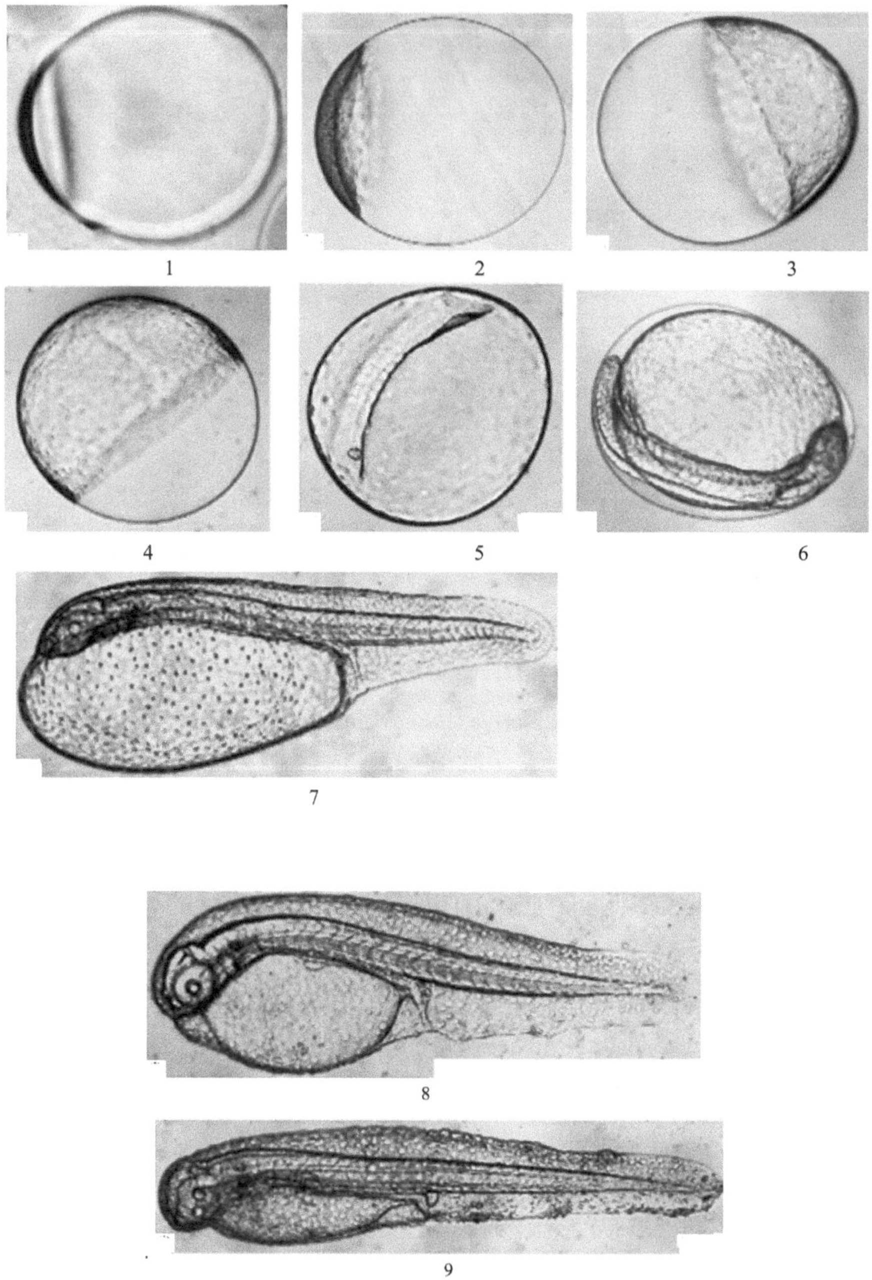

Embryonic development of *Sebastapistes cyanostigma*: (1) Egg before insemination, (2) end of blastulation (11 h), (3) end of gastrulation (13 h), (4) beginning of organogenesis (19 h), (5) formation of the Kupffer's vesicle (23 h), (6) embryo before hatching (35 h), (7) hatching embryo (42 h), (8) free embryo (58 h), (9) free embryo (69 h). *Photo courtesy: Dimitri A. Pavlov.*

2.2 Biology of the Rockfishes (Subfamily Sebastinae)

Characteristics of the subfamily Sebastinae: The species of this subfamily are found in all tropical and temperate seas with a few species entering freshwater. Most are found in the Pacific and Indian Oceans. Body of the species is somewhat compressed. Spines with venom glans are present in dorsal, anal, and pelvic fins, and spines are also seen on opercle and preopercle. The largest species reach a length of about 1 m. The venom of some tropical species is quite dangerous, but that of local species usually causes only moderate irritation. These scorpaenidae have internal fertilization, and most species (those in the genus *Sebastes*) are ovoviviparous. Some scorpionfishes (including our Shortspine Thornyhead, *Sebastolobus alascanus*) are oviparous, laying eggs in a gelatinous balloon. These species are major predators of other fishes and are important for sport and commercial fishing industries. Though this subfamily includes more than 120 species, only the venomous species of this subfamily are included in this publication (Anon. http://www.burkemuseum.org/static/FishKey/scorp.html).

2.2.1 Reproduction and Life History in Rockfishes

Life history of Pacific rockfish (*Sebastes* spp.): The species of the genus *Sebastes* are some of the longest lived fishes known, with maximum ages for many species spanning more than 50 years, and some approaching 150 years. The age of 50% maturity is variable among different species and it is reported as 5–20 years. In these species, delayed maturity, a long reproductive life span, and reproducing multiple times are adaptive responses to a low probability of successful reproduction in any given year. Although rockfish are sometimes reported as being ovoviviparous (bearing live young without maternal nutrition), many studies have confirmed that they are a primitive viviparous group and supply nutrients to developing embryos. The number of eggs produced at the 50% maturity age commonly range from 2000 to 500,000 eggs per year. However, individuals of some species have been reported to produce several million eggs with maximum length. Most species copulate in the fall, but sperm may be stored and fertilization may not take place until several weeks later. An exception is in the *Sebastolobus* spp. where buoyant, egg masses are released and fertilization is believed to be external. Copulation has been observed in very few species of Sebastes, so little is known about courtship or mating behavior in this genus. The female retains the embryos for 4–5 weeks until hatching, and parturition generally occurs in winter or spring for more southerly populations and spring or summer in northerly populations. In most species, reproduction occurs annually, but the production of multiple broods occurs in some species, and this may be related to the prevailing environmental conditions. Generally, the larvae are found in the upper mixed zone of the ocean for a variable period of time, before metamorphosing into juveniles. The young larvae move closer to shore as they grow during the spring months. After several months, juveniles of many species move to deeper adult habitats. Relatively poor ocean conditions for the past two decades may have reduced recruitment success in some California–Oregon species. In contrast, in waters off Alaska and northern

British Columbia, above average recruitment has occurred in recent years for some rockfish species (e.g., Pacific ocean perch, *Sebastes alutus*). A dominant feature of rockfish reproduction is a pattern of infrequent and irregular years with successful recruitment during periods with favorable environmental conditions, and many years with poor recruitment. Recent research has shown that older black rockfish (*S. melanops*) release larvae earlier in the season followed by progressively younger fish, and that successful recruitment may come from a relatively restricted time period within the spawning season. This pattern has also been observed in darkblotched rockfish (*Sebastes crameri*) and yellowtail rockfish (*Suillus flavidus*). In another study, it was found that recruits of shortbelly rockfish (*Sebastes jordani*) exhibited reduced genetic variability compared to the adult population, suggesting that surviving young of the year are the products of reproduction by only a small fraction of the adult population. Reproductive success appears then to be restricted to narrow spatial and temporal windows when conditions are favorable for larval survival (Parker et al., 2000). The reproduction of certain species of the subfamily Sebastinae studied already is given below.

2.2.1.1 *Helicolenus dactylopterus*

In this species, collected from Azores Archipelago, the females are internally inseminated between July and November and spawn between January and March. There is a separation between female and male reproductive cycles. The overall sex ratio observed was 1:1, but seasonal and length-related differences were observed. This suggests differences and changes in availability, probably related to the annual reproductive cycle and changes in feeding behavior. The reproductive mode is a zygoparous form of oviparity, intermediate between oviparity and viviparity. Eggs covered with gelatinous material are fertilized in the ovary. Eggs are released into the seabed at various stages of development, and larvae assume a planktonic existence once the gelatinous covering dissolves (Mendonça et al., 2006; Muñoz et al., 2010).

2.2.1.2 *Sebastiscus marmoratus*

In this viviparous species, no report is available on its life history and reproduction. However, the role of serum steroid hormones during its reproductive cycle has been studied in this species. Serum levels of estradiol-17β (E2) and testosterone (T) were moderately high throughout the spawning period from December until February (E2), and until postspawning in April (T). Serum progesterone (prog) fluctuated but remained low throughout the annual reproductive cycle; 17α, 20β-dihydroxy-4-pregnen-3-one (17α, 20β-diOHprog), on the other hand, was relatively high during the spawning period. During the spawning period, 7 of 12 females reared under laboratory conditions spawned twice at 10–16-day intervals. Histological observations indicated that oocytes developed gradually during gestation of the preceding brood and, after parturition, developed more quickly toward the end of vitellogenesis and subsequent fertilization. In repeat spawners, E2 and female-specific serum proteins remained high several days after the first parturition, then gradually decreased. Serum progesterone showed no significant changes over the period. The 17α, 20β-diOHprog, however, was low immediately after parturition, then rapidly increased, remained elevated during the middle of the period, and then decreased. These results indicate that E2 is involved in vitellogenesis, and 17α, 20β-diOHprog may have some important roles in gestation in the multiple spawner (Takano et al., 1991).

2.2.1.3 Sebastolobus alascanus

Reproduction: Unlike other representatives of this group, this species is oviparous with internal fertilization. Females possess specialized ovarian structures including stalk-like ovigerous lamellae and secretory epithelia. This species spawns pelagic bilobed, floating, gelatinous hollow egg masses, 15–61 cm long between December and May off the West Coast. These fishes are single spawners with determinate annual fecundity (3000–106,000 eggs per year). The length at which 50% of this species is mature is 18.2 cm for West Coast populations and is 21.5 cm for Alaskan populations. Off California, they begin to mature at 5 years, 50% are mature by 12–13 years, and all are mature by 28 years. However, Love (1996) reported that they all are mature by age 16, and they can live more than 62 years (McCain et al., 2005).

Growth and development: Spawning occurs at 600–1000 m. Eggs rise to the surface to develop and hatch. Larvae are pelagic for about 12–15 months. Juvenile of this species spend 14–15 months in midwater. Juveniles transforming to a benthic stage are larger than 50 mm TL. These fishes grow throughout their life. Males and females are of similar size. Although it is difficult to determine the age of older individuals, off California, they have been estimated to live to over 100 years of age, with a maximum size of 80 cm. The mean size of these fishes increases with depth and is greatest at 1000–1400 m. Further, they have a nonasymptotic growth pattern with relatively high growth rates after sexual maturity (McCain et al., 2005). The values of natural mortality (M) [estimated through gonadosomatic index (GSI)] were found to range between 0.013 and 0.017 for this species from both Alaska and the West Coast. Such a low estimate of mortality suggests that this species is an extremely long-lived one (Pearson and Gunderson, 2003).

2.2.1.4 Sebastes atrovirens

The laboratory-produced larvae had a size between 4.0 mm at birth and 6.5 mm at day 30. The mean length at birth was 4.3 mm and at 30th day, it was 5.3 mm. The pectoral fins were found extended to just in front of the anus by the time it reached a mean length of 4.5 mm (day 10). At birth, *Sebastes atrovirens* larvae were similar to *S. carnatus* with pigmentation on the ventral and dorsal aspects of the gut, the anus, posterior aspect of the hindgut, posterior dorsal midline of the trunk, and a complete ventral midline series. With further development, the larvae showed pigmentation on the lower jaw, forebrain, nape, anterior dorsal midline, and otic capsule. With continued growth, new pigmentation appeared on the posterior and posterioventral surfaces of the gut, cleithrum, and lower jaw of these larvae. The pigmentation patterns of *S. atrovirens* and *S. carnatus* remained extremely similar throughout all stages. In these larvae, there was no head spine development during the 28 days (Moreno, 1990).

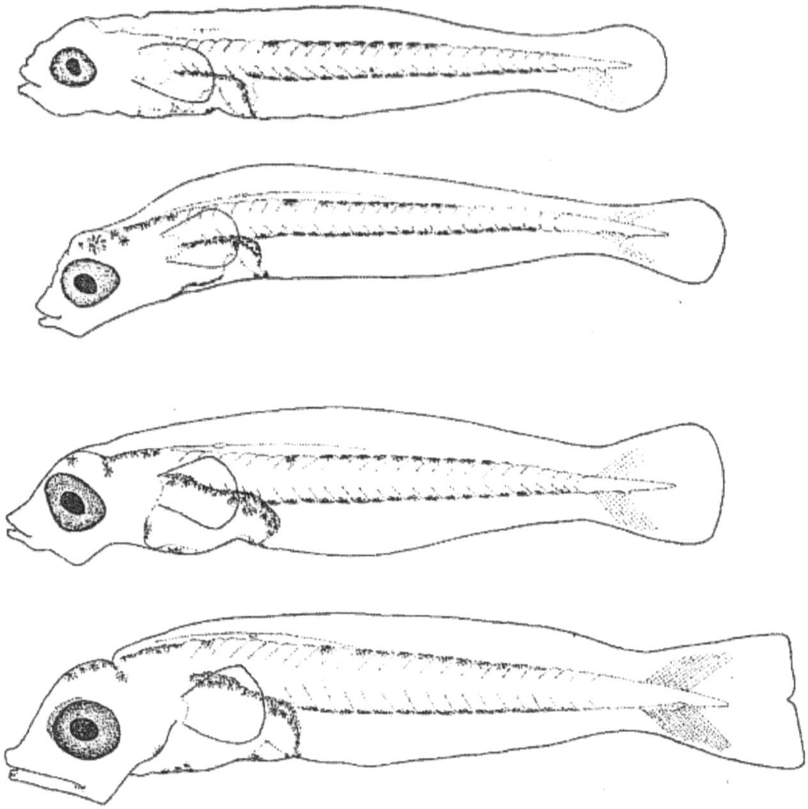

Top: 2-day old larva (4.4 mm), bottom: 10-day old (4.6 mm), top: 19-day old larva (5.4 mm), bottom: 28-day old (5.9 mm). *Image courtesy: Guillermo Moreno.*

2.2.1.5 *Sebastes auriculatus*

Reproduction: This species is viviparous, with planktonic larvae and pelagic juveniles. Brown rockfish mate in March and April in Puget Sound where they spawn once per year and probably give birth in June. Off Oregon, spawning occurs in May and June; the spawning season is longer off central California, at least from December to July. Also, off California females spawn more than once per season. A 31-cm female brown rockfish produces approximately 52,000 young and a 48-cm female produces 339,000 (McCain et al., 2005).

Growth and development: Brown rockfish are 5–6 mm in length at birth. They can grow to a length of 55 cm. They are 52 cm in length at the age of 18 years. Males and females probably grow at the same rate and mature at similar ages and lengths. However, other evidence indicates that females grow larger than males, and both species may live as long as 34 years. Off Oregon, 50% of brown rockfish mature at 31 cm (5 years), and all are mature at 38 cm (10 years) (McCain et al., 2005).

2.2.1.6 Sebastes carnatus

The yolk sac larvae produced in the laboratory showed little pigmentation. However, in this stage, complete ventral and dorsal midline pigment series were present along with a few melanophores scattered laterally on the yolk sac and a prominent anteriolateral trunk melanophore. Melanophores on the dorsal surface of the gut, anus, and the anterio- and posterio-dorsal aspects of the trunk were also quite frequent in this 1-day old larva. All other larvae in the 5.0–5.4-mm size class (day 13) showed pigments on the jaw, forebrain, nape, anus, and ventrally on the gut. The 6.0–6.4-mm size class (day 25) showed pigment on the maxilla, base, and border of the pectoral fins, cleithrum, and otic capsule. Formation of spines was first observed on larvae of 6.1 mm mean length (25 days) with the appearance of the first and second posterior preoperculars, pterotic, and parietal (Moreno, 1990).

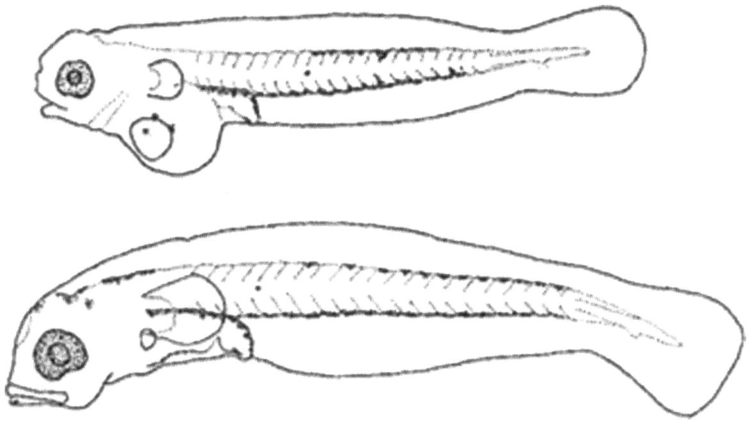

Top: 1-day old larva (3.5 mm), bottom: 13-day old (4.3 mm). *Image courtesy: Guillermo Moreno.*

2.2.1.7 Sebastes caurinus

Reproduction: Off central California, male copper rockfish may be sexually mature at 3 years of age (30 cm); all are mature by 7 years (40 cm). All females are mature off central California by 8 years (41 cm). In Puget Sound, Washington, sexual maturity occurs at age 4, but occasionally at age 3. These fishes spawn once per year. In Puget Sound, Washington, eggs mature by February. Fertilization occurs from March to May off Washington. Egg production ranges from 15,000 eggs in a 24-cm female to 640,000 in a female 47 cm long. Parturition occurs from April to June in Puget Sound, from February to April south of British Columbia, and from March to July in southeastern Alaska. Gravid females were observed in February and March off the central California coast (McCain et al., 2005).

Growth and development: The young pelagic larvae measure 5–6 mm in length at birth; they remain pelagic until 40–50 mm (standard deviation, SD). These fishes are slow-growing and are living up to 55 years with 66 cm in maximum length. In Humboldt Bay, California, fishes are 110–155 mm long as underyearlings, 138–196 mm at age 1 year, 172–231 mm at age 2, and 220–300 mm at age 3. Growth rates are found to be highest during the summer, coinciding with high feeding rates and upwelling (McCain et al., 2005).

2.2.1.8 *Sebastes constellatus*

Reproduction: As with other kinds of rockfish, fertilization, in this species is internal. It is viviparous (live young are born) with planktonic larvae and pelagic juveniles. Starry rockfish spawn from February to July in southern California (peaking in May) and April to May off central California. A 24-cm female spawns 33,000 eggs and a 3-cm female spawns 228,000 eggs. These fishes are multiple brooders (McCain et al., 2005).

Growth and development: Starry rockfish grow to 46 cm and live at least 32 years. Males and females grow at about the same rates, but males mature at a slightly smaller size than females. Males first begin to mature at 18 cm; 50% of males mature at 19 cm (6–7 years); and all mature by 27 cm. Females begin to mature at 21 cm, 50% are mature at 22 cm, and all are mature by 29 cm (McCain et al., 2005).

2.2.1.9 *Sebastes crameri*

Reproduction and life history: Like other rockfish species, this species is viviparous (it has internal fertilization of eggs, maternal nourishment of developing embryos, and gives birth to live young). The reproductive period is protracted, with insemination occurring from August to December, fertilization is from October to March, and parturition from November to June. In before caudal (BC), the release of larvae peaks in February and has been recorded even in June. The exact gestation period for this species is unknown; however, in most rockfish species, it lasts 1–2 months. This species has one brood per year and releases larvae all at once. Fecundity per female ranges from 20,000 to 610,000 larvae in BC and 19,000 to 500,000 larvae in Oregon. Fecundity increases with female age and size. The older the female the earlier larvae are released. Successful recruitment may occur within the spawning period. The larvae measure 4–9 mm in length at parturition and are fairly well developed. However, they are weak swimmers and their survival is strongly linked to environmental factors, such as ocean currents and upwelling events. Mortality is high during the early life stages. These fishes have a long pelagic larval duration, extending on average from late December through August on the US west coast. The pelagic phase varies from several months to a year during which time the larvae transform into juveniles, which then settle on the ocean floor. Fishes of BC mature from May to November, and they exhibit highly variable interannual recruitment success.

Age and size at maturity: These factors may vary with geographic location. In Oregon, 50% maturity is achieved at approximately 5 years old for males (29.6 cm TL) and 8 years old for females (36.5 cm TL). In California, 50% maturity in females may be reached at 4 years old. In BC, 50% maturity is reached at approximately 8 years for males (32.19 cm TL) and 9 years for females (35 cm TL). The size at 50% maturity decreased with increasing latitude from Oregon to Alaska. Along the US west coast south of Canada, the opposite trend (size at 50% maturity increased with increasing latitude) seems to occur as fish caught in California are generally smaller than fish caught at the equivalent age in Oregon and Washington. In BC, the maximum recorded age for the species is 48 years.

2.2.1.10 *Sebastes dalli*

Reproduction: Male and female calico rockfish become sexually mature at age 7 and 9, respectively. Calico rockfish are single brooders and release their pelagic larvae from January through May with a peak in February in the Southern California. An 11.6-cm female spawned 3878 larvae and a 15.5-cm female spawned 18,000 larvae. The larval stage lasts from less than 4 weeks to 2 months (McCain et al., 2005).

Growth and development: For males, length at first maturity is 7 cm. About 50% of them mature at 9 cm, and all mature at 14 cm. For females, length at first maturity is 9 cm and all mature at 10 cm. The maximum length for calico rockfish of Southern California is 20 cm. They have been aged to 12 years, but some may live longer (McCain et al., 2005).

2.2.1.11 *Sebastes elongatus*

Reproduction: Greenstriped rockfish are multiple brooders, that is, they spawn two or more times per season. The peak spawning month for this species off the southern California is April (ranging from January to July). Off central and northern California, May is the peak spawning month (ranging from May to July). This fish can spawn up to 250,000 eggs per year. Young fish are probably also born in late spring and early summer off Oregon, Washington, and British Columbia (McCain et al., 2005).

Growth and development: Newly released greenstriped rockfish larvae are about 5 mm in length. Adults can grow to 38 cm. Males live to a maximum of 37 years, and females to 28 years. Off California, males grow faster and females grow slower than in other areas of their range. There is no size difference between the sexes once they have grown fully. Males reach the size of 50% maturity at 18 cm TL and 100% maturity at 26 cm TL; and females reach the size of 50% maturity at 19 cm TL and 100% maturity at 25 cm TL (McCain et al., 2005).

2.2.1.12 *Sebastes goodei*

Reproduction: Chilipeppers are viviparous with planktonic larvae. In California, fertilization of eggs begins in October. Spawning occurs from September to April, and peak spawning is December to January. This species may spawn multiple broods in a single season. A 30-cm female may give birth to as many as 29,000 larvae and a 52–56-cm female may give birth to as many as 538,000 young (McCain et al., 2005).

Growth and development: In this species, eggs develop internally for 40–50 days. Larvae are extruded at 4.7–5.8 mm in length. Larvae are considered to be juveniles typically around 22–26 mm in length. Females of the species are significantly larger, reaching lengths of up to 59 cm in California. Males are usually smaller than 40 cm. Males mature at 2–6 years of age and 50% of them mature at 3–4 years. Females mature at 2–5 years with 50% maturity at 3–4 years. Females may attain an age of about 27 years, whereas the maximum age for males is about 12 years (McCain et al., 2005).

2.2.1.13 Sebastes helvomaculatus

Reproduction: It is oviparous with pelagic eggs. It showed an annual cycle with group synchronous development producing only one brood per year. The GSI peaked in November for males and in May for females. Parturition occurs during May and June in northern and central California, and primarily in June from Oregon to British Columbia. The length at 50% maturity was 22.8 cm (fork length, FL) for males and 20.8 cm (FL) for females. The rate of natural mortality (M) was estimated to be 0.04 using the GSI method and 0.07 using the maximum age method (McCain et al., 2005; Shaw, 2008).

Growth and development: Young rosethorn rockfish are pelagic until about 4.0–6.0 cm (SD). The largest pelagic juvenile taken off Oregon was 4.2 cm and the smallest benthic juvenile was 13.6 cm. Small larvae (<1.0 cm) are taken only in July and August. Pelagic juveniles are captured in August, September, and November. Off California, male rosethorn rockfish first mature at age 7 and all mature by age 10. For females, the age of first maturity is 5 years; 50% of them mature at age 8; and all mature at age 10. This species can grow to a maximum length of 33 cm, and it can reach an age of 87 years (McCain et al., 2005).

2.2.1.14 Sebastes levis

Reproduction: Like all members of the genus *Sebastes*, the cowcod is an internal fertilizer and gives birth to live young (viviparous, with planktonic larvae). Large female cowcod may produce up to three broods per season. Spawning peaks in January in the southern California. A 46-cm female may produce up to 181,000 young per brood, and an 80 cm female may give birth to nearly 2 million young. The larvae spend about 100 days as plankton (McCain et al., 2005).

Growth and development: This species grows to 94 cm. The length at 50% maturity for both sexes is 43–44 cm in the southern California. Extruded larvae are about 5.0 mm (McCain et al., 2005).

2.2.1.15 Sebastes maliger

Reproduction: Quillback Rockfish, like all rockfishes, is matrotrophically viviparous, supplying nutrients to the developing embryos. Mating probably occurs in March in Puget Sound. Spawning may take place at anytime from March to July. The juvenile rockfish live in eelgrass, eating zooplankton, shrimp, and krill. They move to more rocky habitats as they mature (McCain et al., 2005). In BC, mating probably occurs in December. Females may store the sperm for weeks prior to fertilization, and parturition occurs in April and May. A prolonged pelagic larval phase may last for up to 2 months, after which settlement occurs to benthic habitats. Juveniles usually occur in shallower waters than their conspecific adults.

Growth and development: The size of quillback rockfish varies with location. In Alaskan and Californian coasts, the overall length of this species has been reported to be 30.5 cm and 18.5 cm respectively at the end of 12 years. The overall length at which 50% of the fish mature has been reported to be 30.5 cm off the coast of Alaska and 23.5 cm off the California coast. The maximum recorded size for a male was 60 cm. Off California, this species lives up to 90 and 95 years in BC (McCain et al., 2005). In BC, these fishes reach 50% maturity at about 11 years of age. On average, females tend to be larger and older than the males and can reach a maximum length of 50 cm in BC. Natural mortality rate has been estimated at $M = 0.02$ for this species in BC, although for other rockfishes, estimates of 0.05 are considered appropriate. Total mortality rate estimates range from 0.05 to 0.09 based on simple catch curve analysis, and from 0.06 to 0.12 using a more complex model. Generation time is estimated at 30–60 years, depending on estimate of natural mortality used. The mean age of reproductive females in recent years has been 22.8 years.

2.2.1.16 Sebastes melanops

Reproduction: Black rockfish, like most members of the genus *Sebastes*, is viviparous (internal fertilization), with the female giving birth to very large numbers of planktonic larvae during annual spawning. Parturition occurs from February to April off British Columbia, January to March off Oregon, and January to May off California. Spawning areas are unknown, but spawning may occur in offshore waters because gravid females have been caught well offshore (McCain et al., 2005).

Growth and development: Age of 50% maturity for black rockfish off California is estimated at 7 years, and it can live to about 50 years of age. Off California, males may sexually mature at 3 years (250 mm), and all mature by 10 years (430 mm). Females off California may mature at 5 years (300 cm), and all mature by 11 years (480 mm). Off Oregon, males may mature in 5 years, and females in 6 years. The maximum length attained by the black rockfish is 65 cm. After age 7, females are larger than males of the same age (McCain et al., 2005). Specimens collected off Alaska have been aged to 49 years old. In addition, they benefit from a low natural mortality rate, at only about 7% of the total population.

2.2.1.17 Sebastes melanops

Moreno (1990) reported on the size and pigmentation of 1–3-day old laboratory-produced larvae. The larval sizes ranged from 3.4 mm at birth to 4.1 mm at day 3. All newborn larvae of this species showed pigmentation on the ventral and dorsal surfaces of the gut, the posterior surface of the hindgut, the finfold between the gut and anus, and the posterioventral surface of the notochord. Only a small proportion of these larvae had pigment on the anus, and the lower lobe of the caudal fin. By day 3, pigmentation at the same locality persisted, and the frequency of occurrence of melanophores on the anus and the anteriolateral section of the trunk increased (Moreno, 1990).

2.2.1.18 *Sebastes melanostomus*

The larvae of *Sebastes melanostomus* are extruded at a length of about 4.5 mm. Newborn larvae showed a single row of melanophores on the ventral midline of the tail, a patch above the brain, a heavy blotch at the tip of the lower jaw, a streak along the posterior region of the upper jaw, a pattern of minute melanophores on the pectoral fins, and a group along the dorsal and lateral surfaces of the gut. A large blotch appeared on the opercle at about the completion of notochord flexion, along with the upper jaw streak. Older larvae and pelagic juveniles developed a striking banding which was retained even in adults. Larvae and pelagic juveniles developed prominent head spination. The beginning of the pelagic juvenile stage was marked by the appearance of anterior lateral-line scales especially in 16.0-mm sized individuals. The largest pelagic juvenile of this species measured 46 mm and the smallest demersal juvenile measured 36 mm. The larvae and pelagic juveniles of this species are comparatively deep-bodied, big-headed, and have eyes of moderate size.

Fin development: The pectoral fins of this species are of moderate length. They increase in relative length during development. The pectoral fin base is relatively deep. The pelvic fin is also of moderate length and it further enlarges during development. The full complement of 13–14 spinous rays is achieved in all larvae with a size of 8.2 mm.

Pigmentation: The dorsal and lateral surfaces of the gut are covered with melanophores. A ventral midline row on the tail shows fewer melanophores. The pectoral fin is partially covered with a patch of fine melanophores. One or two spots were observed on the nape. A total of 8 melanophores were seen on the dorsal aspect of the head. With further development, the number of melanophores in the ventral midline is reduced, and they are absent in the larvae longer than 14.0 mm. The most striking pigmentation is the pattern of melanistic bars that develops on the body. The fins are also characteristically pigmented. Characteristic blotches develop on the dorsal and fins of pelagic juveniles. At the beginning of the pelagic juvenile stage, a large blotch develops. At about 30.0 mrn, a blotch appears on the anal fin (Moser and Ahlstrom, 1978).

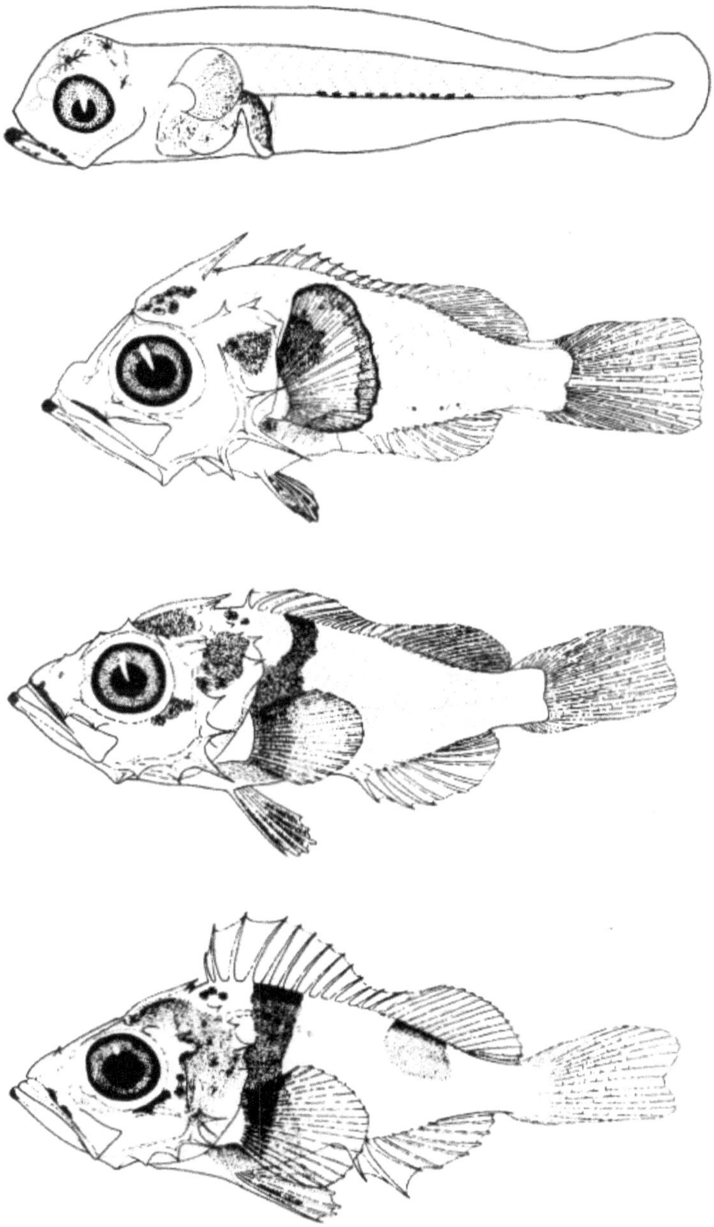

4.4-mm sized larva, 8.2-mm sized larva, 12.3-mm sized larva, 17.4-mm sized larva. *Image courtesy: H. Geoffrey Moser and Elberth Ahlstrom.*

2.2.1.19 *Sebastes miniatus*

Reproduction: Vermilion rockfish appear to mature and spawn for the first time when they are 35 cm long (5–6 years old). As with all other rockfish, fertilization is internal, and they give birth to living young. A vermilion rockfish that is 31 cm long is estimated to contain 63,000 eggs; 53 cm long fish is expected to contain 1600,000 eggs; and 76-cm female might hold as many as 500,000 eggs. The principal reproductive period lasts from December through March. The young are mobile with the free-swimming young feeding primarily upon shrimp-like organisms.

Growth and development: Young-of-the-year of this species appears in inshore water beginning in February. Vermilion rockfish can grow to 76 cm and 6.8 kg. The oldest individual aged was 25 years old (McCain et al., 2005).

2.2.1.20 *Sebastes mystinus*

The laboratory-produced larvae of this species ranged in size from 3.2 mm at birth to 5.5 mm at day 31. The mean SL at birth was 3.8 mm. The later stages have characterized pigmented pectoral fins which extended to the back of the anus at 21st day when the larvae reached a size of a 5.0 mm. At birth, all larvae had pigmentation on the dorsal surface of the gut and anus, and the short ventral midline series was composed of relatively few melanophores. Further, at a mean length of 5 mm (day 21), the majority of larvae had also developed pigmentation on the lower jaw, forebrain, nape, otic capsule, and cleithral symphysis. Melanophores were also beginning to appear posteriorly on the dorsal surface of the trunk and on the margin and blade of the pectoral fins. Head spines were first observed on larvae with a 4.7 mm mean length at 31 days, and the first and second posterior preoperculars, and pterotic spines were also present (Moreno, 1990).

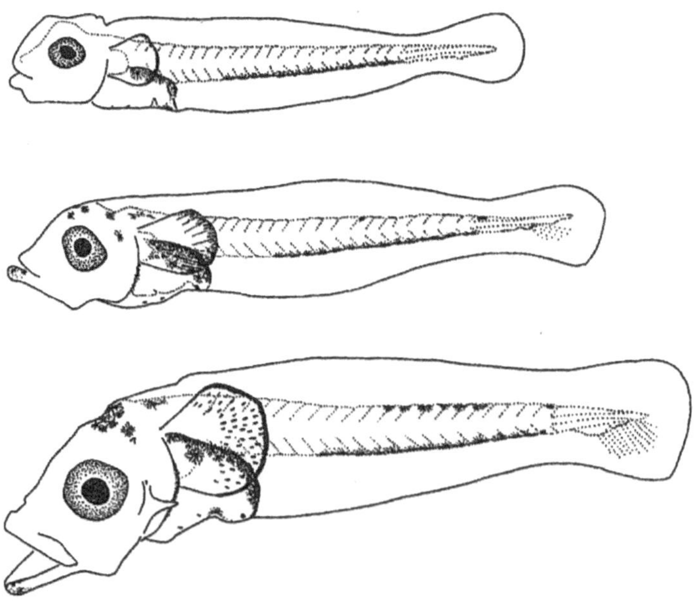

Top: 1-day old larva (4.2 mm), middle: 21-day old (4.8 mm), bottom: 31-day old (5.6 mm). *Image courtesy: Guillermo Moreno.*

2.2.1.21 Sebastes nebulosus

Reproduction: Spawning occurs from January to July throughout most of its range, with a January peak. Parturition occurs in April and May off central California (McCain et al., 2005).

Growth and development: Male and female China rockfish mature at the same size. Half of them mature at 28 cm and all mature at 30 cm. Off central California, the smallest sexually mature female was 26 cm, and the smallest sexually mature male was 34 cm. China rockfish grow up to 45 cm and reach an age of 78 years (McCain et al., 2005).

2.2.1.22 Sebastes paucispinis

Reproduction: Bocaccio are viviparous and has a protracted spawning season which lasts almost year-round (>10 months). Parturition occurs during January to April off British Columbia and Washington, November to March off northern and central California, and October to March off southern California. In California, bocaccio may become pregnant in October, give birth in November, and prepare immediately for a second brood to be born in March. Two or more broods may be born in a year in California. The spawning season is not well-known in northern waters. Although age-at-size calculations were not given, a 38.1 cm female may give birth to 20,000 young, while a 77.5 cm female may give birth to 2.3 million young. It is also estimated that a female may produce 339 eggs/g body weight (McCain et al., 2005).

Growth and development: Mature eggs of this species measure about 0.55 mm in diameter. Eggs develop for 40–50 days in the ovary, hatch, and yolkless larvae (4–6 mm) are released about 1 week later. Larvae are pelagic for up to 150 days. Metamorphosis to a semi-demersal juvenile stage occurs at about 30 mm TL. Males mature at 3–7 years with 50% mature in 4–5 years. Females mature at 3–8 years with 50% mature in 4–6 years. This species has been suspected to live as long as 50 years (McCain et al., 2005).

2.2.1.23 Sebastes pinniger

Reproduction: In this species, fertilization is internal (ovoviviparous). Off California, canary rockfish spawn from November to March and from January to March off Oregon, Washington, and British Columbia. A wide range in larval size suggests that this species may have protracted and variable spawning. The age of 50% maturity of this species is 9 years, and nearly all mature by age 13. Maximum age of this species has been estimated as 60–84 years (McCain et al., 2005).

Growth and development: The mean length of newly extruded larvae of his species is 3.7 mm (SL). The transformation to pelagic juvenile occurs at sizes greater than 12.5 mm (SL), and transformation to benthic juveniles occurs after 59 mm, during June to August. The growth of this species does not vary with latitude. The maximum length canary rockfish has been estimated as 76 cm (McCain et al., 2005).

2.2.1.24 *Sebastes rastrelliger*

The size of the laboratory-produced larvae of this species ranged from 4.3 mm at birth to 7.7 mm at day 52. The mean length at birth was 4.6 mm and at day 56, it was 6.6 mm. The newborn larvae showed completely pigmented dorsal and ventral midline series and pigmentation on the forebrain and nape. At birth, melanophores were sometimes present on the jaw, maxilla, and posterior half of the jaw. These melanophores were present on all larvae by the time they reached a size of 6.0–6.4 mm. Distinctive posteriolateral trunk pigment was present on most 5.0–5.4 mm larvae and present in all 6.5–6.9 mm larvae. This pigmentation was also observed on similar-sized *S. carnatus* larvae. Spine formation was evident on larvae of 6.3 mm mean SL at 29 days (Moreno, 1990).

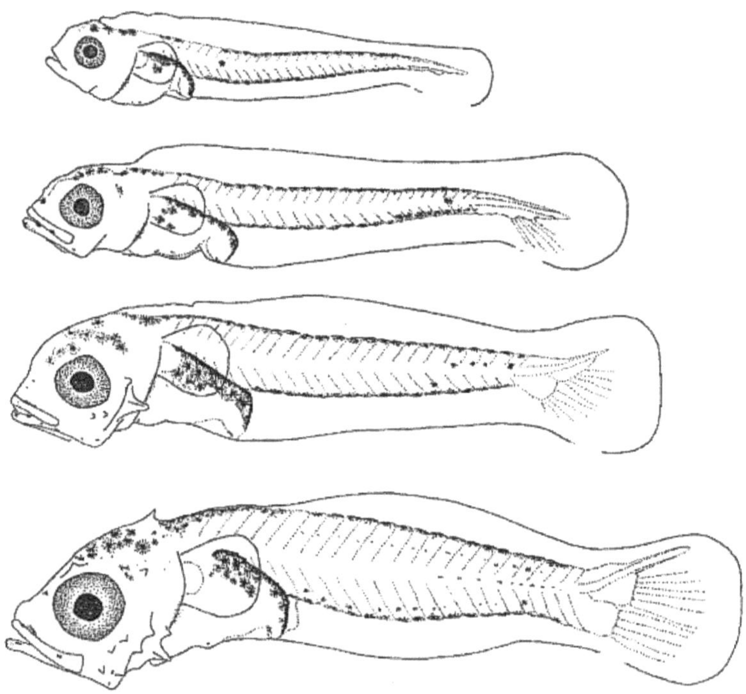

Top: 1-day old larva (4.6 mm), middle: 20-day old (5.8 mm), bottom (6.4 mm), 34-day old larva (7.2 mm). *Image courtesy: Guillermo Moreno.*

2.2.1.25 *Sebastes rosaceus*

Reproduction: It is a livebearer (viviparous). Spawning occurs in southern California from January to September with a peak in May; and at farther north, it takes place from April to July, peaking in June. Female rosy rockfish spawn from 13,000 (15 cm size) to 95,000 eggs (23 cm size). This species is a multiple brooder (McCain et al., 2005).

Growth and development: Off southern California, 50% of rosy rockfish mature by 15 cm and all mature by 20 cm. Off central and northern California, 50% of rosy rockfish mature at 20 cm and all mature by 25 cm. This species has been reported to reach a size of 36 cm, but individuals over 25 cm are rare. It is known to live for 14 years, and it is likely to live longer. Estimated age of first maturity is 4 years, and age of 100% maturity is 8 years (McCain et al., 2005).

2.2.1.26 Sebastes saxicola

Reproduction: It is an ovoviviparous species. Off California, the estimated age of first maturity for males is 3 years and all mature by age 4; females first mature at age 2 and all mature by age 3. Females produce about 15,000 young at 18 cm and 200,000 at 32 cm. Young about 4.3 mm in length are released mainly in February in British Columbia and January and February off Oregon. The larval release period is much longer in northern and central California, i.e., from November through March (McCain et al., 2005).

Growth and development: Stripetail rockfish can grow up to 41 cm and live to at least 38 years. Larvae are extruded at 3.3–5.2 mm notochord length (NL). Benthic juveniles have been estimated to grow 0.32 mm/day (McCain et al., 2005).

Laidig et al. (1996) reported on the pigmentation in the larvae of this species collected from off central California.

Larvae were extruded at a size of 3.3–5.2 mm NL. The pectoral, and pelvic fins formed at 7.5 mm NL and a full adult complement was present approximately 9.0 mm SL. Lateral line pores first appeared around 17 mm SL and a complete series was seen by approximately 38 mm SL. Changes in body shape were related to notochord flexion. Head spines appeared at 9 mm SL. At approximately 12 mm SL, the nape and dorsal midline pigments merged to form a solid line along the dorsal surface. This dorsal body pigment began to increase in intensity as the larvae grew. A distinct pattern of five body bars (1–5 from anterior to posterior) was apparent in juveniles with sizes greater than 25 mm SL. Pigment on the dorsal fin appeared at 17 mm SL with a few melanophores on the fin. Head pigmentation appeared at 5 mm NL with the appearance four to five melanophores at top of cranium. The number and size of melanophores increased with increased fish size and by 7 mm NL, the head had become highly pigmented. At 7 mm NL, pigments began to develop on the anterior portion of the head between the eyes. As fish size increased, more pigment was added to this area. Pigments started appearing on the upper jaw at about 8 mm SL. By 10 mm SL, pigmentation on the upper jaw and the anterior section of the head was well developed. Pigment started forming on the tip of the lower jaw at about 5 mm NL. At 30 mm SL, the upper and lower jaw pigments merged to encircle the mouth. Pigment on the operculum formed at 8 mm SL. Pigments along the ventral and posterior regions of the eye orbit started appearing at 12 mm SL. The dorsal cheek bar was formed first at 14 mm SL. By 36 mm SL, this bar extended from the eye orbit across the preopercle to the operculum ending just below the operculum pigment. The ventral cheek bar appeared at 28 mm SL at the base of the eye orbit and it extended ventrally.

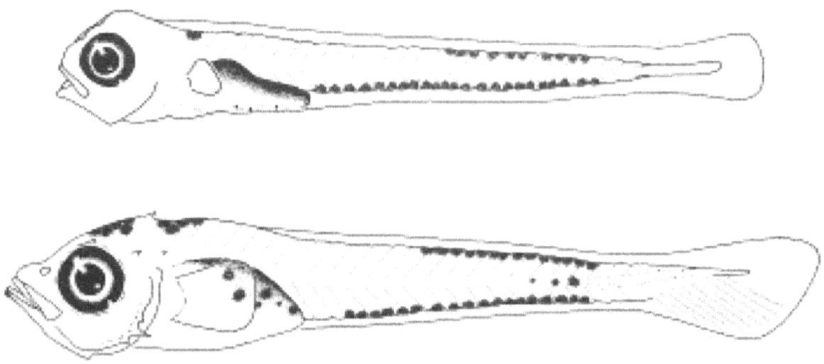

Larvae showing pigmentation, top: 4.7 mm NL larva, bottom: 6 mm NL larva. *Image courtesy: Thomas E. Laidig Keith M. Sakuma and Mary M. Nishimoto.*

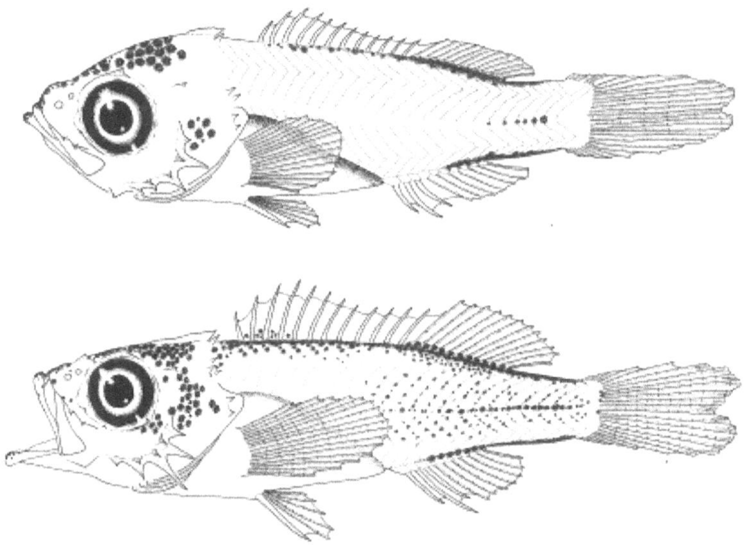

Larvae showing pigmentation, top: 9.9 mm SL larva, bottom: 17.8 mm SL larva. *Image courtesy: Thomas E. Laidig Keith M. Sakuma and Mary M. Nishimoto.*

2.3 Biology of the Lionfishes (Turkeyfishes) (Subfamily Pteroinae)

Though the lionfishes *Pterois miles* and *P. volitans* are popular in the marine ornamental aquarium trade, little is known regarding the biology and ecology of these fishes. Information on their dietary habits, predators, and seasonality of reproduction is scarce. Most of what has been published on lionfish relates largely to lionfish envenomations, which commonly occur during aquarium husbandry or as a result of poor handling by home aquarists.

Characteristics of the subfamily Pteroinae: The scorpaenid Pteroinae is composed of the following five genera, viz., *Brachypterois, Dendrochirus, Ebosia, Parapterois,* and *Pterois.* These lionfishes particularly *P. volitans* (red lionfish) and *P. miles* (devil firefish) are commonly called as zebrafish, firefish, turkeyfish, red lionfish, butterfly cod, ornate butterfly-cod, peacock lionfish, red firefish, scorpion volitans, or devil firefish. These species have distinctive brown or maroon, and white stripes or bands covering the head and body. They have fleshy tentacles above their eyes and below the mouth; fan-like pectoral fins; long, separated dorsal spines; 13 dorsal spines; 10–11 dorsal soft rays; 3 anal spines; and 6–7 anal soft rays. The bodies of these fishes are covered with cycloid scales of lionfish. In this type of scale, the anterior part of each scale is usually overlapped by the posterior portion of the scale in front of it, giving the fish greater flexibility than fishes with other types of scales. An adult lionfish can grow as large as 45 cm, while juveniles may be as small as 2.5 cm or less. Lionfish have cycloid scales (fish scales that are oval or elliptical in shape with a smooth edge). In contrast to their close relatives, the lionfishes are not reclusive. Relying on their conspicuous warning coloring, they swim slowly and majestically through the reef, with their fins spread wide. The long, thin fin rays surround the whole body like a shield, and their appearance alone gives the impression of impenetrability. The fin rays located in the dorsal, anal, and pelvic fins of these fishes are venomous.

Native range: The native range of the lionfish includes the South Pacific and Indian Oceans (i.e., the Indo-Pacific region). The range of the lionfish covers a very large area from western Australia and Malaysia east to French Polynesia and the United Kingdom's Pitcairn Islands, north to southern Japan and southern Korea and south to Lord Howe Island off the east coast of Australia and the Kermadec Islands of New Zealand. In between, the species is found throughout Micronesia.

Nonnative range: Lionfish have also been reported along the southeastern United States coast from Florida to North Carolina. Juvenile lionfish have been collected in waters off Long Island, New York, and Bermuda. Lionfish are a popular marine ornamental fish and were possibly intentionally released into the Atlantic. The first lionfish was reported in South Florida waters in 1985 with many additional sightings occurring until they were documented as established in the early 2000s (Anon. https://oceanservice.noaa.gov/facts/lionfish-facts.html).

Habitat: Lionfish are found in mostly all marine habitat types such as hard bottom, mangrove, seagrass, coral, and artificial reefs (like shipwrecks at water depths from 0 to 100 m.

Behavior: Lionfish are slow-moving and conspicuous. They often rely on their unusual coloration and fins to discourage would-be predators from eating them. These fishes are now one of the top predators in many coral reef environments of the Atlantic. These fishes were thought to be nocturnal hunters, but they have been found with full stomachs during the day in the Atlantic. They move about by slowly undulating the soft rays of the dorsal and anal fins. During the day, they sometimes retreat to ledges and crevices among the rocks and corals, although in the Atlantic, lionfish are often seen moving about during the day, both alone and in small groups.

Food and feeding: The lionfish are largely piscivorous but also feed on a number of crustaceans. In the Red Sea, lionfish (*P. miles*) have been reported to feed on assorted taxa of benthic fishes including damselfish and cardinal fish. However, in the Pacific Ocean, *Pterois lunulata* were observed to feed primarily on invertebrates including penaeid and mysid shrimps. They have been reported to consume over 50 species of fish including some economically and ecologically important species. Further, they are active hunters who ambush their prey by using their outstretched, fan-like pectoral fins to slowly pursue and "corner" them. The daily consumption rates in the laboratory for six size classes of lionfish ranging from 30 to 300 g showed that lionfish consumed approximately 2.5%–6.0% of their body weight per day at 25–26°C. Preliminary observations suggest that lionfish in their invaded range can consume piscine prey (Morris et al., 2008).

Parasites: The lionfishes are normally infected by ectoparasites such as monogeneans and copepods. A new endoparasite, myxosporean species, *Sphaeromyxa zaharoni* (Sphaeromyxidae) has been recorded from the lionfish gall bladder from the Red Sea (Morris et al., 2008).

Damage to ecosystem: The lionfish is a flourishing invasive species in U.S. Southeast and Caribbean coastal waters. This invasive species has the ability to harm reef ecosystems because it is a top predator that competes for food and space. These lionfishes have also the potential to kill off helpful species such as algae-eating parrotfish, allowing seaweed to overtake the reefs. In the United States, the lionfish population is continuing to grow and increase its range. This is largely because lionfish have no known predators and reproduce all year long; a mature female has been reported to release about 2 million eggs a year (Anon. https://oceanservice.noaa.gov/facts/lionfish-facts.html).

2.3.1 Internal Anatomy of Lionfish

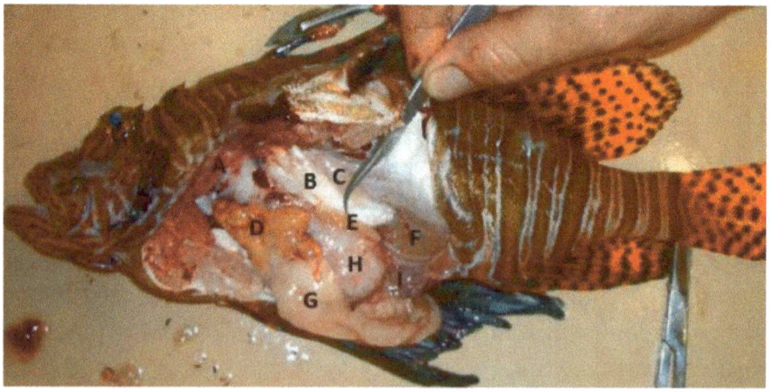

(A) Gill rakers and filaments; (B) swim bladder; (C) swim bladder muscle; (D) liver; (E) gonads; (F) urinary bladder; (G) interstitial fatty deposit; (H) stomach; (I) intestine. Lionfish internal anatomy. *Photo credit: James A. Morris, Jr. and C. Calloway.*

Digestive system in red lionfish: The Red Lionfish has large fins which help it corner the prey. Then the fish swallows its prey whole rather fast owing to its elongated esophagus. Fish that swallow their prey whole are usually known to be more aggressive predator because they try to take on larger fish. Since the Red Lionfish swallows its prey whole, they have their teeth in their pharynx. These teeth constitute a few rows and are used to grind and smash the fish that was eaten. This makes it much easier for the fish to swallow. The food then continues to move down the passage ways of the digestive system where it encounters the first part of the stomach. Here the muscular gizzard grinds down the food even more and injects enzymes into it to make it easier for later digestion. The rest of the stomach is more of just a sac full of acidic qualities which break down the eaten fish even more. The stomach of the fish is serving as a holding cell where the contents are held until they are able to pass on to the intestines to be absorbed.

2.3.2 Reproduction in Lionfish

Courtship and formation of embryos: The Pteroines, including *P. miles* and *P. volitans* are gonochoristic; and males and females exhibit minor sexual dimorphism only during reproduction. Lionfish courtship, which includes circling, side winding, following, and leading, begins shortly before dark and extends well into night time hours. Following the courtship phase, the female releases two buoyant egg masses that are fertilized by the male and ascend to the surface. The eggs and later embryos are bound in adhesive mucus that disintegrates within a few days, after which the embryos and/or larvae become free floating.

Early life history and dispersal: The traits of Pteroine larvae include: large head, relatively long and triangular snout, long and serrated head spines, robust pelvic spine, and pigment confined to the pectoral fins and postanal ventral and dorsal midlines. The size of *P. miles* or *P. volitans* larvae at hatching is about 1.5 mm. The specific planktonic larval duration of lionfish may be between 25 and 40 days (Morris et al. 2008).

2.3.2.1 Reproduction in Pterois volitans

Size at maturity in this species was found to be 19 cm (TL) and fecundity increased rapidly with size. Although this species is usually found alone in the nonbreeding season, during courtship, male lionfish will aggregate with multiple females to form groups of three to eight fishes, performing a suite of complex courtship and mating behaviors. This species is a broadcast spawner, and it is capable of reproducing year-round. Female lionfish releases two mucus-filled egg clusters, each containing between 2000 and 15,000 eggs, which are fertilized externally by the male. The adhesive mucus that binds the clusters together dissolves after several days, releasing the eggs into the water and allowing them to develop as planktonic larvae. Dispersal of the common lionfish occurs during this pelagic larval phase, during which individual planktonic larvae can travel great distances in the water column. The pelagic larval duration of this species has been estimated to be 20–35 days which is moderate to low compared to other shallow marine fishes. Juveniles and adults appear to be generalists in both habitat and diet.

2.3.2.2 Developmental Stages

These lionfishes lay 15,000–30,000 eggs per spawning. The spawning is year-round, and in every 4 days, the female has been reported to release the eggs in two mucous-encapsulated egg masses that are fertilized by the male. Approximately 36 hours later, the eggs hatch into larvae which are dispersed by ocean currents. The juveniles spend most of their time in one small area but can live in a wide range of habitats. Although lionfishes can live up to 15 years in their native range, little is known about the detailed life history and life span of these fishes in Atlantic invaders.

Developmental stages of lionfish: (1) Eggs, (2) 20–40-day old larva (2 cm approx.), (3) 10-month old juvenile (20–10 cm sized); (4) 1-year old adult (10–20 cm sized). *Author.*

Imamura and Yabe (1996) described five larvae of *P. volitans,* collected from the surface and midwater, off northwestern Australia, in the eastern Indian Ocean. Their description is given below.

Flexion larval stage (3.8 mm NL): This flexion stage is characterized by having pectoral fin. In this stage, the head and body are compressed. Body is elongated. Snout is longer than eye diameter. Exposed bony surface on head is with following spines: 1 weak postocular spine without serration; one very long parietal spine with serration; one sharp pterotic spine; second anterior preopercular spine; and second to fourth posterior preopercular spines without serration. Suborbital stay is not visible. Large nostril is present in front of eye. Pectoral fin rays are complete. Pelvic fin is weakly developed, but rays are not visible. Several weak dorsal, anal and caudal fin rays are visible (Imamura and Yabe, 1996).

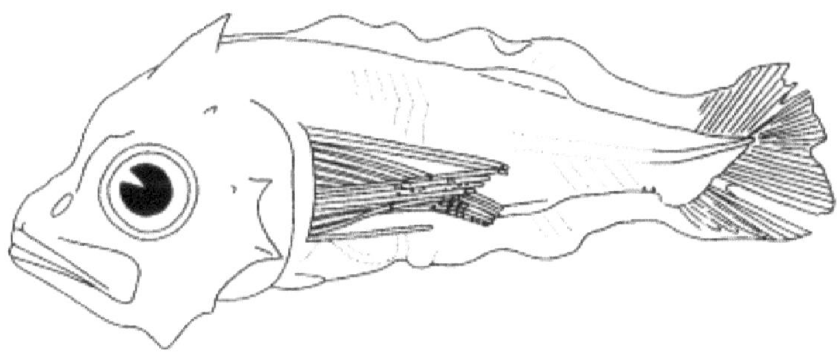

Flexion larval stage (size 3.8 mm NL).

Early postflexion larval stage (4.5 mm SL): This flexion stage is characterized by having pectoral fin. Ascending process of upper jaw long and stout. Posterior end of lower jaw strongly pointed. Postocular spine strongly developed. Third posterior opercular spine with serration. Blunt upper opercular spine present. Pelvic fin rays weakly developed.

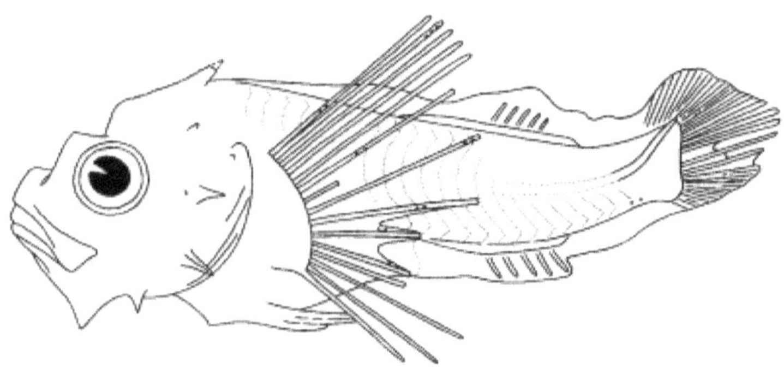

Early postflexion larval stage (size 4.5 mm SL).

Late postflexion larval stage (9.5–11.0 mm SL): In late postflexion larval stage, dorsal fin rays XIII, 11–12 (last ray double); anal fin rays III, 7 (last ray double); pectoral fin rays 13–15; vertebrae 24; long pectoral fin; dorsal and anal fins not continuous with caudal fin; suborbital stay with broad posterior margin; and with melanophores on bases of posterior half of dorsal fin and posterior anal fin rays, and anterior middle portion of caudal peduncle. Following spines on head at 9.5 mm SL: First lower infraorbital spine, fourth upper infraorbital spine, one nuchal spine, tympanic spine, third and fourth anterior preopercular spines, and first posterior preopercular spine (lower posttemporal present on right side, absent on left side). A sensory canal between bases of parietal and nuchal spines present at 9.5 mm SL. Second lower infraorbital spine and two lower opercular spines present at 11.0 mm SL. Postocular spines becoming shorter with growth. Suborbital stay developed, posterior end broad, and not reaching preopercle at 9.5 mm SL. Two nostrils present in front of eye; anterior nostril without a flap on posterior margin at 9.5 mm SL. Body scales not yet developed. Dorsal, anal, and pelvic fin rays complete at 9.5 mm SL. Dorsal, anal, pectoral, and caudal fin rays with segments at 10.5 mm SL. Pectoral fin long, at least reaching middle of anal fin at 9.5 mm SL (most fin rays broken). No fin rays branched yet. Dorsal fin originating behind posterior margin of opercle (Imamura and Yabe, 1996).

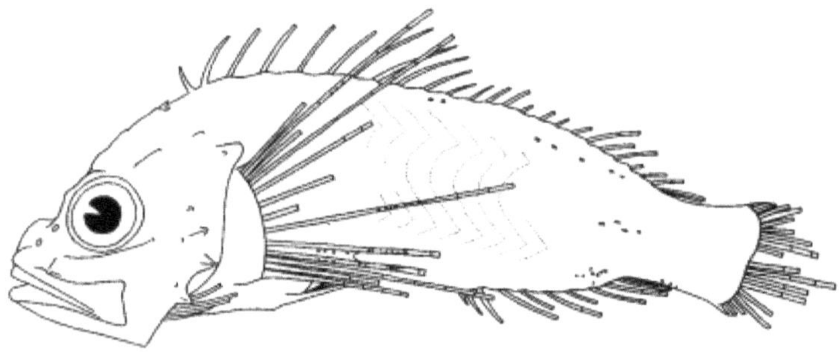

Late postflexion larval stage (size 10.7 mm SL). *Image credit: Imamura and Yabe.*

Reproduction of *Pterois volitans* and *P. miles* in western Atlantic: The females of this region matured at approximately 175 mm TL or 1 year of age and released approximately 25,000 eggs per spawning event. Based on the presence of hydrated oocytes, mature females appeared capable of spawning every 3–4 days throughout the year, although the proportion of females with ovaries in spawning condition was higher in summer [June–August (Gardner et al., 2015)].

The available information on the biology and ecology of the individual species of this subfamily are given elsewhere.

3

Biology and Ecology of Venomous Marine Scorpionfishes (Family Scorpaenidae)

3.1 Scorpionfishes (Subfamily Scorpaeninae)

Hoplosebastes armatus (Schmidt, 1929) (=*Hoplosebastes pristigenys*)

Photo credit: Smithsonian Institution, National Museum of Natural History.

Order Scorpaeniformes	Suborder Scorpaenoidei
Family Scorpaenidae	Subfamily Scorpaeninae

Common name(s): Flower scorpionfish
Distribution: Northwest Pacific: Japan and the East China Sea.
Ecology: This demersal species is found on the continental shelf edge region.

Biology

Identification: This species is the only known member of its genus. This species is diagnosed by its dorsal fin which is with 13 spines and 9 soft rays; anal fin is with 2 spines and 6.5 or 7.5 soft rays (occasionally with 3 spines and 5.5 rays); rudimentary first anal-fin spine which is about 20% length of second spine; pectoral fin is with 17–18 soft rays; and 24–26 lateral line scales. Body is usually weakly, rather than strongly, compressed with is red color with six dark reddish bands. Eyes are with 3–4 radial bands. This species grows to a length of 13 cm (TL).

Biology and Ecology of Venomous Marine Scorpionfishes. DOI: https://doi.org/10.1016/B978-0-12-815475-5.00003-0

Fisheries/aquarium/food/pharmaceutical values: It is not a commercial species but edible.

Idiastion hageyi (McCosker, 2008)

Common name(s): Not designated

Distribution: Pacific Ocean: Galápagos Islands.

Ecology: It is a reef-associated deep water species (benthopelagic, bathydemersal) inhabiting depths from about 500 m. It usually occupies rock, steep boulder, and sediment-laden bottoms.

Biology

Identification: This species is diagnosed by its dorsal spines (total): 12; dorsal soft rays (total): 9; anal spines: 3; anal soft rays: 6; and vertebrae: 25. Body of this species is deep, compressed posteriorly, and humpbacked. Orbit diameter is large (15% SL). Head spines are strong and well developed. There are no distinct small papillae on head. However, a distinct row of papillae nearly surrounds outer edge of pupil. A small swim bladder is present. Lateral line is complete with tubes on scales. Body coloration is reddish-orange, with irregular white blotches on head, body, and pectoral fin bases. It grows to a maximum length of 9.5 cm (SL) and 11.8 cm (TL).

Food and feeding: It is a carnivore feeding on crustaceans (shrimps and crabs) and molluscs (octopus, squid and cuttlefish).

Fisheries/aquarium/food/pharmaceutical values: Not reported.

Idiastion kyphos (Eschmeyer, 1965)

Common name(s): Sharpcheek scorpionfish

Distribution: Western Atlantic: Caribbean Sea.

Ecology: It is a small, deep water (bathydemersal) scorpionfish occupying rocky or soft bottoms around the continental slope at a depth range, 439–622 m. It is often seen hiding in coral and rocks. It is a master of camouflage and is hard to observe in the wild.

Biology

Description: This species is diagnosed by its dorsal fin which is with 12 spines and 9.5 soft rays (last split to base and counted as 1.5); and pectoral fin has 17 or 18 rays and there are 25 vertebrae. The condition of the branching of pectoral rays is variable in this fish. Villiform teeth are present on premaxillae, dentary, vomer, palatine, and pharyngeal bones. Swim bladder is present. Interorbit is with scales. Head spines are strongly developed, often with multiple points. There is no occipital pit. Peritonium is unpigmented. All scales are ctenoid (rough to touch), except on breast. Small slit is seen behind fourth gill arch. Coloration of body is mostly red, with some dark markings on dorsum. It grows to a maximum size of 15.3 cm (SL).

Food and feeding: It typically feeds on benthic invertebrates and small fishes.

Fisheries/aquarium/food/pharmaceutical values: Very little is known about the biology and fisheries of this species because it was only recently observed on deep reefs of the Gulf of Mexico.

Idiastion pacificum (Ishida and Amaoka, 1992)

Common name(s): Flame humpback scorpionfish

Distribution: North Central Pacific: Kyushu-Palau Ridge and Emperor Seamounts.

Ecology: It is a bathydemersal species occupying a depth range of 350–430 m. Its main habitat is in the ridges.

Biology

Identification: This species is characterized by having 25 rather than 24 vertebrae, branched pectoral rays, humpbacked body outline, no occipital pit, a small slit behind the fourth gill arch, a swim bladder, teeth on the palatines, and well-developed spines on head. Body length is oval and its side is flat. There is a large head without any flap. Eye and mouth are large. Dorsal fin is continuous. Tail fin is circular and truncated. Body is light red. It grows to a maximum length of 12.8 cm (SL) (Ishida and Amaoka, 1992).

Fisheries/aquarium/food/pharmaceutical values: This fish has no economic value.

Iracundus signifer (Jordan and Evermann, 1903)

Photo credit: John E. Randall.

Common name(s): Decoy scorpionfish

Distribution: It is native to the Indian and Pacific oceans; Western Indian Ocean: South Africa, Mauritius, Reunion; Pacific Ocean: Ryukyu Islands, Taiwan, and Hawaiian Islands.

Ecology: It is found associated with the sand and rubble under ledges of seaward reefs. It is a nonmigratory species having a depth range 20–110 m.

Biology

Identification: This species is diagnosed by its 12 dorsal spines, 8–9 dorsal soft rays. 3 anal spines; and 5 anal soft rays. Anterior part of dorsal fin resembles a small fish and this fin also has a spot between spines 1 and 2 or 1 and 3. Lachrymal bone is with 2 spines over maxillary, first spine points forward, and second spine is broad, pointing out and to rear; suborbital ridge is without spines except one at rear before opercle; preopercular spines are short, usually only 3 developed; there is no supplemental preopercular spine at base of first spine; presence of a dark spot on the spinous dorsal fin between spines 1 or 2 and 3; and vertical scale rows of 65–75. This species grows to a length of 13 cm (TL).

Major threats: Due to its affinity with coral reefs, this species may be experiencing localized population declines due to habitat loss in parts of its range.

Fisheries/aquarium/food/pharmaceutical values: Due to its uncommon nature and habitat preferences, this species is not exploited.

Neomerinthe amplisquamiceps (Fowler, 1938)

Photo credit: Sainsbury, Keith.

Common name(s): Orange scorpionfish

Distribution: Indo-West Pacific: Malaysia, the Philippines, Indonesia, the Arafura Sea, Papua New Guinea, and north-western Australia.

Ecology: This demersal species is occupying gulf, deep waters, and hard bottoms of the continental shelf at a depth of about 40 cm.

Biology

Identification: Body of this species is elongated and somewhat compressed. Head is with many spines and ridges. There is a spiny hard ridge below eye (suborbital stay) to preopercle. Body coloration is red to orange with brownish mottling. Body is also with brown, green or yellowish spots. Most fins are with spots. 3–4 large dusky blotches are seen on back and sides. A dark pigment is also seen on inside of pectoral fin. It grows to a maximum length of 16.2 cm (SL).

Fisheries/aquarium/food/pharmaceutical values: Not reported.

Neomerinthe bathyperimensis (Zajonz and Klausewitz, 2002)

Image courtesy: Motomura, H.

Common name(s): Not designated as it is a new species

Distribution: Southern Red Sea; north-western Indian Ocean.

Ecology: It inhabits deep-waters.

Biology

Identification: The type specimen is aberrant in that it lacks the anal-fin spines and their supporting pterygiophores (Zajonz and Klausewitz, 2002).

Fisheries/aquarium/food/pharmaceutical values: Not reported.

Neomerinthe bauchotae (Poss and Duhamel, 1991)

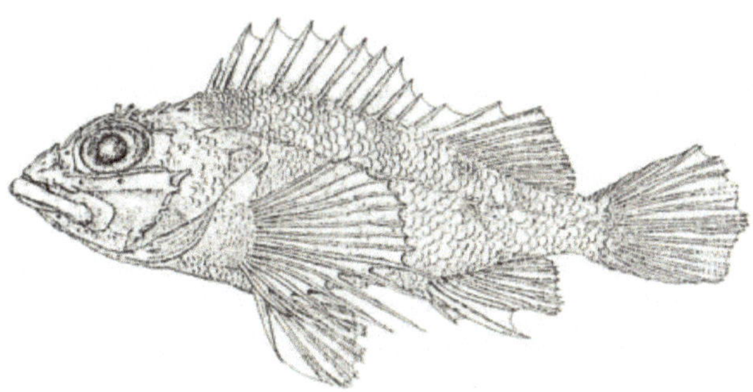

Image credit: Stuart G. Poss and Guy Duhamel.

Common name(s): Not designated as it is a new species

Distribution: Southern Indian Ocean: Saint Paul and Amsterdam Islands.

Ecology: It is bathydemersal species occupying a depth range of 50–510 m.

Biology

Identification: It is distinguished from other species of this genus by its relatively low pectoral fin-ray count (usually 16), lack of a slit behind the last gill arch, the lack of a swim bladder, and in having a second preopercular spine. Head is moderately large. Maxilla extends to halfway between middle and posterior margin of pupil without scales. Operculum is covered with scales except posteriorly. Body scales are of moderate size. Dorsal fin originates dorsal to about middle of opercle. The third and fourth spines are the longest. Caudal fin is slightly rounded. Head and body are reddish with distinctly darker markings primarily over the dorsum. Body is buff white or tan ventrally. A distinct red blotch is seen on the pectoral fin. Anal fin is with a few diffuse spots. Caudal fin is with elongate markings. This species grow to a maximum size of 10.1 cm (TL) (Poss and Duhamel, 1991).

Fisheries/aquarium/food/pharmaceutical values: Not reported.

Neomerinthe beanorum (Evermann and Marsh, 1900)

Common name(s): Spotwing scorpionfish, Western Atlantic fish

Distribution: Western Central Atlantic: Antilles, including Cuba.

Ecology: It is a demersal species having a depth range, 92–375 m. It lives on compact sand bottoms and deep reefs especially in clear water.

Biology: As it is being infrequently taken, little else is known of its biology.

Identification: Body of this species is deep (deepest at gill opening) with a huge head. Eye diameter is 0.7–1.3 times in snout length. Dorsal profile is convex. Ventral profile is flattened. Eye and mouth are large. Head is with multiple strong ridges and short spines. A strong cheek ridge is present (with 3–4 spines). Preopercle is with a few strong spines. Body is covered in ctenoid scales. There is a spiny dorsal fin with longest middle spines. Soft dorsal fin is rounded. Pectoral fin is fanlike, with the middle rays being the longest. Upper 3–9 pectoral rays are branched at tip (otherwise unbranched). Caudal fin is truncate. Body coloration is dark red above, grading to white on the belly. Black blotches and white mottling are seen on the body. Dorsal fin is red with darker and lighter spots. Pectoral, anal, pelvic, and caudal fin are blood red with no markings. Its maximum length is 15.5 cm (SL).

Fisheries/aquarium/food/pharmaceutical values: This species rarely occurs as by-catch in Gulf of Mexico.

Neomerinthe erostris (Alcock, 1896) (=*Scorpaena gibbifrons; Neomerinthe rotunda; Neomerinthe bathyperimensis*)

Image credit: Motomura, H., Causse R, Béarez P, Mishra SS.

Common name(s): Round scorpionfish

Distribution: Indo-West Pacific Ocean: Madagascar, Réunion, Yemen, Sri Lanka, Taiwan, the Philippines, Indonesia, the Solomon Islands, New Caledonia, Vanuatu, and the Wallis and Futuna Islands.

Ecology: It is a benthopelagic (bottom dweller, incl. seamounts) species having a depth range, 52–505 m.

Biology

Identification: It is a scorpionfish with a distinct longitudinal ridge on the lateral surface of the maxilla and a strongly rounded dorsal profile of the head. Body of this species is moderately compressed anteriorly and is progressively more compressed posteriorly. Nape and anterior body are arched. Body depth is relatively deep but slightly less than head length. Tentacles are seen on posterior ends of bases of most head spines. There is a short, broad tentacle, with several short branches distally. No tentacles are present on eye membrane, anterior margin of lower snout in anterior view, cheek, maxilla, lips, underside of lower jaw, opercle, midinterorbital space, and occiput. Similarly, there are no tentacles on lateral surface of trunk, except for several slender tentacles associated with pored lateral-line scales. Well-exposed ctenoid scales are covering posterior half of lateral surface of head, including cheek, behind eye, opercle, and an area surrounded by parietal, nuchal, pterotic, and lower posttemporal spines. Mouth is large, and slightly oblique, forming angle of about 20° to horizontal axis of body. Posterior margin of maxilla is just reaching to (or extending slightly beyond) vertical through posterior margin of orbit. Dorsal profile of snout is steep, forming an angle of about 50° to horizontal axis of body. Origin of first dorsal-fin spine is above supracleithral spine. Posterior margin of opercular membrane is extending beyond vertical through fourth dorsal-fin spine base. Origin of pelvic-fin spine is slightly anterior to base of first pectoral-fin ray. Origin of first anal-fin spine is slightly posterior to origin of last dorsal-fin spine. Body coloration is strongly variegated, mainly brownish red, suffused with irregular gray, dark red, and reddish white blotches. Spinous portion of dorsal fin is with indistinct, broad, longitudinal, gray band, with poorly defined black blotch posteriorly. Soft-rayed portion of dorsal fin, pectoral fin, and caudal fin are with indistinct, irregular, small blotches scattered over entire fin. Pelvic fin is reddish white and grayish distally. Anal finis grayish anteriorly and whitish posteriorly. It grows to a maximum size of 47 mm (SL) (Motomura et al., 2002).

Fisheries/aquarium/food/pharmaceutical values: Not reported.

Neomerinthe folgori (Postel and Roux, 1964)

Photo courtesy: Cambraia Duarte and Pedro Miguel Niny; Denis Tweddle and M. Eric Anderson.

Common name(s): Folger's scorpionfish

Distribution: Eastern Atlantic: Cape Verde and off Mauritania; northern Namibia.

Ecology: It is a bathydemersal species having a depth range, 180–310 m. This species is probably found on rocky and other hard-bottoms. There is no other information concerning its habitat/ecology.

Biology

Identification: This species has significantly shorter predorsal and snout lengths and fewer gill rakers (lowermost somewhat coalesced). Cirri on head are weakly developed and they are mostly absent. Background body coloration is yellowish or reddish with black and brownish variegations. Four variegated black bands are seen on body. Caudal fin is with narrow bands. It has a maximum length of 59.0 cm (TL).

Fisheries/aquarium/food/pharmaceutical values: Not reported.

Neomerinthe hemingwayi (Fowler, 1935)

Photo credit: Wikipedia.

Common name(s): Spinycheek scorpionfish

Distribution: Western Atlantic: New Jersey and northern Gulf of Mexico to southern Florida in USA and Cuba.

Ecology: It is a demersal species inhabiting gulf, deep waters, hard bottoms and reefs, or associated with coral rubble; depth range 45–230 m.

Biology

Identification: Body of this species is deep (deepest at gill opening) with a huge head. Eye diameter is 1.1–1.7 times in snout length. Dorsal profile is convex. Ventral profile is flattened. Eye and mouth are large. Head is with multiple strong ridges and short spines. A strong cheek ridge is present (with 3–4 spines). Preopercle is with a few strong spines. Body is covered in ctenoid scales. Spiny dorsal fin is with middle spines the longest. Soft dorsal fin is rounded. Pectoral fin is fanlike, with the middle rays being the longest. Upper 3–9 pectoral rays are branched at tip (otherwise unbranched). Caudal fin is truncate. Body coloration is dark red, grading to white on the belly. There are numerous black to dark brown blotches on the body that become yellowish laterally. The pinkish fins are covered with black, dark brown, or yellowish spots. Maximum length of this species is 40.0 cm (SL). It is an oviparous species.

Fisheries/aquarium/food/pharmaceutical values: Not reported.

Neomerinthe megalepis (Fowler, 1938)

Photo courtesy: Gloerfelt-Tarp, Thomas.

Common name(s): Bigscale Scorpionfish

Distribution: Western Pacific: Indonesia.

Ecology: It is a deep-water offshore benthic fish (demersal) species, commonly found on sandy bottoms. Its habitat depth range is 215–505 m.

Biology

Identification: Body of this species is oval and side is flat. It has a big head. Mouth is large and tip of lower jaw is protruding slightly beyond tip of upper jaw. Posterior tip of maxilla is reaching vertical to the posterior margin of eye. Villiform teeth are present in bands on both jaws. Snout is short and convex. Scales on body are moderately large size. Dorsal fin originates vertical to just above the last opercular spine tip. Fourth dorsal spine is the longest. Anal fin originates just below the third dorsal soft ray and its second spine is the longest and strongest. Pectoral fin tip is rounded. Few pectoral fin rays are branched. Ventral fin origin is vertically more or less just below the dorsal fin origin and ventral fin tip is reaching to anus. Caudal fin is rounded. Body is with variegated reddish brown in color, more deep on sides and pale in ventral part. Anal fin is pale red with median dark band. Pectoral fin is pale red with numerous dark spots. Ventral fin is whitish and pale red in tip. Caudal fin is reddish white with dark spots forming indistinct bands. It reaches a maximum standard length (SL) of 9.9 cm.

Fisheries/aquarium/food/pharmaceutical values: Small fish, in addition to academic research, have no economic value.

Neomerinthe naevosa (Motomura et al., 2011c)

Photo courtesy: H. Motomura.

Common name(s): Blotched Polynesian scorpionfish

Distribution: Eastern Central Pacific: French Polynesia.

Ecology: It is a pelagic-neritic species, found at depths from 80 to 305 m on sandy bottoms.

Biology

Identification: Body of this species is moderately compressed anteriorly, and progressively more compressed posteriorly. A relatively small eye is seen. Nape and anterior body are moderately arched. Body depth is relatively shallow, less than head length. Pectoral-fin axil is without skin flaps. Well-exposed ctenoid scales are covering the surface of body. Cycloid scales are covering pectoral-fin base and anteroventral body. Exposed ctenoid scales are present on cheek, behind orbit, opercle, and occiput. Lateral line is complete, extending onto base of caudal fin. Lateral line is sloping slightly downward above posterior tip of opercle. Gill rakers are relatively long with spinous tips. Swim bladder is absent. Mouth is large, oblique, forming an angle of about 20° to horizontal axis of head and body. Posterior margin of maxilla is just short of a vertical through posterior margin of orbit. Lower jaw is with an indistinct symphysial knob. Villiform teeth are present on upper and lower jaw. In f preserved specimens, the head and body are strongly mottled dorsally and whitish ventrally. Inside mouth is white. An elongate black blotch is seen above first and second pored lateralline scales. Spinous portion of dorsal fin is with poorly defined black blotches. Soft-rayed portion of dorsal fin is whitish with a grayish band basally and a poorly defined blotch at posterior tip of fin. Pectoral, anal and caudal fins are with scattered melanophores. Pectoral-fin axil is without black markings. Pelvic fin is whitish. This species reaches a maximum length of 7.6 cm (SL) (Motomura et al., 2011a,b).

Fisheries/aquarium/food/pharmaceutical values: Not reported.

Neoscorpaena nielseni (Smith, 1964)

Image not available

Common name(s): Not designated

Distribution: Western Indian Ocean: South Africa, Madagascar, the Seychelles and Réunion.

Ecology: It is a demersal species; depth range 40–507 m.

Biology

Identification: The taxonomic characters of this species are poorly understood. This species is characterized by the presence of 4–8 suborbital spines and 5 preopercular spines. Preocular, supraocular, postocular, and tympanic spines are well developed. Interorbital ridges join posteriorly to tympanic spine bases. Nuchal spine is large, well developed and is originating from posterior end of tympanic spine. Teeth are seen on vormer and palatines. Pectoral fin is bilobed. There is usually a large black blotch on dorsal fin basally between fifth–seventh and tenth spines. Irregular yellowish blotches are found scattered dorsally on the head and body, and on the dorsal-fin membranes and upper half of the caudal fin. This species grows to a maximum length of 21.6 cm (TL).

Fisheries/aquarium/food/pharmaceutical values: Not reported.

Neomerinthe pallidimacula (Fowler, 1938)

Photo credit: Hiroyuki Motomura.

Common name(s): Not known

Distribution: Western Central Pacific: Philippines.

Ecology: It is a demersal species inhabiting to a depth of 27 m.

Biology

Identification: Body of this species is moderately compressed anteriorly and is progressively more compressed posteriorly. Nape and anterior body are moderately arched. Body depth is relatively deep but less than head length. Pectoral-fin axil is without skin flaps. Mouth is large, oblique, forming an angle of about 20° to horizontal axis of head and body. Posterior margin of maxilla is just reaching to vertical through posterior margin of orbit. An indistinct longitudinal median ridge is seen on lateral surface of maxilla. Lower jaw is with well-developed symphysial knob. Villiform teeth are seen on upper and lower jaws. Tooth band is narrowing posteriorly. Villiform teeth are present on vomer and palatines. Underside of lower jaw is without ridges. Dorsal profile of snout is steep, forming an angle of about 70° to horizontal axis of head and body. Nasal spine is simple, directed dorsally. Body coloration is brown (brownish white) with three slightly curved brownish white (white) stripes on trunk. No black blotches are seen on body or fins.

Fisheries/aquarium/food/pharmaceutical values: Not reported.

Neomerinthe procurva (Chen, 1981)

Photo credit: Gloerfelt-Tarp, Thomas.

Common name(s): Curved spine scorpionfish
Distribution: Western Pacific: Indonesia, Taiwan.
Ecology: It is a demersal species found in the continental shelf; depth range 20–920 m.

Biology

Identification: It has oblong and moderately compressed body. Posterior lacrimal spine is well developed. Lateral lacrimal spine is present and there is no median ridge on lateral surface of maxilla. The third dorsal spine is the longest and lower jaw is not protruding. Its body color is variable, blackish to reddish and variously blotched dorsally. It grows to a maximum length of 7.1 cm (SL).

Fisheries/aquarium/food/pharmaceutical values: It is usually taken by trawl in its range.

Neomerinthe rufescens (Gilbert, 1905)

Image not available

Common name(s): Not designated
Distribution: Eastern Central Pacific: known only from the Hawaiian Islands.
Ecology: It is an offshore, demersal species found at relatively deep depths on seamounts and knolls; depth range, 108–302 m.

Biology

Identification: In this species, scales are present in about 50 vertical scale rows. Lachrymal bone is with two spines over maxillary, both point back and down. Suborbital ridge is with 2–4, usually three, spines; one on lateral face of lachrymal bone is usually absent. There is no occipital pit. It can reach a maximum length of 10.0 cm

Fisheries/aquarium/food/pharmaceutical values: Not reported.

Parascorpaena aurita (Rüppell, 1838) (=*Sebastapistes bynaensis*)

Photo credit: Richard Winterbottom.

Common name(s): Golden scorpionfish

Distribution: Indo-Pacific: Red Sea, Mozambique, and Papua New Guinea; East Asia, and Yakushima Island.

Ecology: This reef-associated species is also found in brackish waters. It is found in shallow nearshore waters, coastal lagoons, large intertidal pools or estuaries; also in silty reefs or amongst large coral rubble; depth range, 1–30 m.

Biology

Identification: Body of this species is oblong and moderately compressed. Posterior lacrimal spine is directed anteriorly in adults; ventrally or anteroventraly in young. Suborbital ridge is with two spines and both the spines are located behind orbit. Interorbital ridges are well developed from middle of orbit posteriorly, joining to form a broad loop at rear of interorbit and enclosing a depression. Tympanic spines are usually closer together than postocular spines. It is a mostly grayish to brownish scorpionfish (sometimes reddish) with indistinct darker blotches. An indistinct dark blotch is seen on the upper side in front of the first dorsal fin. Snout is yellowish and many radiating lines are seen on the iris. It grows to a maximum length of 15.0 cm (TL).

Major threats: Due to its affinity with coral reefs, this species may be experiencing population declines due to habitat loss in parts of its range.

Fisheries/aquarium/food/pharmaceutical values: Not reported.

Parascorpaena bandanensis (Bleeker, 1851)

Image not available

Common name(s): Banda scorpionfish

Distribution: East Indies: Singapore and Papua New Guinea; Philippines to Indonesia.

Ecology: It is a demersal species occupying mainly estuarine, shallow rubble habitats; also found in coral reefs; depth range 1–15 m.

Biology

Identification: This species is diagnosed by its dorsal fin which is with 12 spines and 9(10) soft rays; anal fin is with 3 spines and 5 rays; and pectoral fin is with 15 rays. Mouth is somewhat oblique. Villliform teeth are present in narrow bands on palatines. Head is naked with exception of deciduous scales on upper part of opercle. Body is totally covered with ctenoid scales. Three to four rows of scales are present before dorsal. Origin of dorsal is above middle of upper border of opercle. First dorsal spine is somewhat less than eye. Soft dorsal is rounded. Origin of anal fin is below last dorsal spine or first dorsal ray. Pectoral fins are pointed. Caudal is slightly rounded. Color of alcohol-preserved specimens is yellowish-brown, marbled with dark brown; lower jaw is sometimes distinctly streaked white and brown. Fins are marbled or indistinctly banded with brown. It grows to a maximum length of 10.0 cm (TL).

Fisheries/aquarium/food/pharmaceutical values: Not reported.

Parascorpaena maculipinnis (Smith, 1957)

Image not available

Common name(s): Not reported
Distribution: Indo-West Pacific.
Ecology: This benthic species is found in inshore reefs.

Biology

Identification: This species is diagnosed by its 12 dorsal spines and 9 dorsal soft rays; and 3 anal spines and 5 anal soft rays. A black ocellus is present between 8th and 11th dorsal spines. It grows to a maximum length of 5.0 cm (SL). No other information relating to its biology is available.

Fisheries/aquarium/food/pharmaceutical values: Not reported.

Parascorpaena mcadamsi (Fowler, 1938)

Photo credit: John E. Randall.

Common name(s): McAdam's scorpionfish, Adam's scorpionfish, ocellated scorpionfish

Distribution: Indo-Pacific: East Africa to the Ducie Islands, north to Ryukyu Islands, south to Rapa; Belau [=Palau] and Marshall Islands in Micronesia.

Ecology: It is a reef-associated species found in shallow, near shore waters in lagoons to outer reef slopes; depth range 1–40 m. This species is quite secretive, as it hides in coral and rocky reefs during the daytime, and only ventures out at night.

Biology

Identification: This species is diagnosed by its 12 dorsal spines and 8–9 dorsal soft rays; and 3 anal spines and 5 soft rays. It has quadrangular, naked depression behind the eyes. Large males of this species are with a large black spot at rear of spinous dorsal fin between spines 8 and 10. Females and juveniles are with no spot or a faint one. It grows to a maximum length of 8.0 cm (TL). Other biological aspects of this species are not known.

Major threats: Due to its affinity with coral reefs, this species may be experiencing population declines due to habitat loss in parts of its range.

Fisheries/aquarium/food/pharmaceutical values: It has minor commercial value, but it is occasionally seen in subsistence fisheries.

Parascorpaena mossambica (Peters, 1855)

Photo credit: John E. Randall.

Common name(s): Mozambique scorpionfish; Golden Scorpionfish

Distribution: It is native to the Indian Ocean and the Pacific Ocean to Micronesia.

Ecology: It is a reef-associated species inhabiting areas with mixed sand and rubble in reef flats, shallow lagoons, and channels; depth range 0–20 m. It hides during the day and ventures out in the open at night.

Biology

Identification: In this species, the posterior lacrimal spine is directed anteriorly, and its suborbital ridge is with three spines. Occipital pit is shallow. A large supraorbital tentacle is present, and it is longer than eye diameter and is curved forward, with fringes and appendages. Head and body are yellowish brown; mottled with creamy white; and also with dark brown spots. Supraorbital tentacle is light brown, and its appendages are creamy white. Suborbital region is with an angular dark brown streak covered with blackish spots. Fins are mottled with creamy white and are bearing oblique brown streaks. Membranes of spinous part of dorsal fin are with distal orange streaks. It can reach a maximum length of 12 cm (SL).

Major threats: Due to its affinity with coral reefs, this species may be experiencing population declines due to habitat loss in parts of its range.

Fisheries/aquarium/food/pharmaceutical values: It occasionally makes its way into the aquarium trade. Although this species makes its way into local markets and is consumed as food, no significant fisheries exist for it, due to its small size.

Parascorpaena picta (Cuvier, 1829)

Photo credit: John E. Randall.

Common name(s): Northern scorpionfish, painted Scorpionfish

Distribution: It is native to the Indian Ocean and the western Pacific Ocean; Indo-West Pacific: restricted to Southeast Asia and Australia.

Ecology: It is a reef-associated, solitary, benthic species, and it occurs along rocky bottoms and in protected bays on silty substrates; depth range 1–15 m.

Biology

Identification: Body of this species is robust with large eyes that bulge out of the head. Wide mouth is on a large head. There are skin flaps and tentacles on the head and body. Dorsal fin which has 12 spines starts well behind the eyes. Scales are cycloid. It is solitary carnivore on reefs. This species grows to a total length (TL) of 17 cm.

Fisheries/aquarium/food/pharmaceutical values: Not reported.

Phenacoscorpius adenensis (Norman, 1939)

Source: Australian National Fish Collection, CSIRO.

Common name(s): Toothed no-line scorpionfish

Distribution: Indo-West Pacific.

Ecology: It is a bottom dweller (bathydemersal) on seamounts and knolls; depth range 139–736 m.

Biology

Identification: This rare species is identified by its rounded black blotch, on the spinous part of the dorsal fin. The chief characteristics include presence of palatine teeth; distinct nuchal and parietal spines; suborbital spines 3–5. The nape and anterior part of body are moderately arched throughout life. Black spots are seen on posterior half of caudal peduncle. Body of preserved specimens is with 4 distinct saddles. It grows to a maximum size of 10 cm (TL).

Fisheries/aquarium/food/pharmaceutical values: Not reported.

Phenacoscorpius eschmeyeri (Parin and Mandrytsa, 1992)

Photo courtesy: Hiroyuki Motomura, Naoko Kanehira and Hisashi Imamura.

Common name(s): Not designated

Distribution: Southeast Pacific: Sala y Gomez Ridge.

Ecology: It is a deep-water (demersal) species occurring up to a depth of 630 m.

Biology

Identification: Body of this species is moderately deep, but body depth is less than head length. Uppermost ray and lower 9 rays of pectoral fin are unbranched, and the remaining rays are branched. Ninth or tenth ray is the longest. Second soft rays are the longest among pelvic and anal fin rays. There are no distinct papillae, tentacles, or cirri on head and body. Similarly, there are no supraocular tentacles. Pectoral fin axil is without skin flap. Mouth is large and is slightly oblique, forming angle of about 20° to longitudinal axis of head and body. Posterior margin of maxilla is just reaching to a vertical drawn through posterior margin of orbit. Dorsal profile of snout is steep, forming an angle of about 50 degrees to longitudinal axis of head and body. Nasal spine is simple, somewhat conical and is directed upward. Anterior nostril is with low membranous tube and no tentacle. Origin of first dorsal-fin spine is above first pored lateralline scale. Posterior margin of opercular membrane is extending beyond vertical drawn through origin of second dorsal-fin spine. Posterior tip of pectoral fin is reaching vertical drawn through middle of soft-rayed portion of dorsal fin. Origin of pelvic fin spine is anterior to origin of pectoral fin. Origin of first anal-fin spine is slightly posterior to origin of last dorsal-fin spine. Body and fins are pale reddish and are mottled with whitish blotches, without black stripes, bands, or blotches. Dark areas are faintly seen through opercle and abdomen. It grows to a maximum length of 13.3 cm (SL) (Motomura et al., 2012).

Fisheries/aquarium/food/pharmaceutical values: Not reported.

Phenacoscorpius longilineatus (Motomura et al., 2012)

Photo courtesy: Hiroyuki Motomura.

Common name(s): Southern no-line scorpionfish

Distribution: Southwestern Pacific: New Caledonia and New Zealand; Indian Ocean.

Ecology: This is a bathydemersal species found at depths of 345–1089 m.

Biology

Identification: This species is distinguished by the following set of characters: pectoral-fin rays are 16–18; middle rays are branched in young and adults; pored lateral-line scales are 8–18; 50–52 scale rows in longitudinal series; gill rakers are 18–21; presence of palatine teeth; absence of second preopercular spine; distinct nuchal and parietal spines; nape and anterior part of body are strongly arched in adults of over c. 80 mm SL; relatively short caudal fin length; and 1–5 black spots on posterior half of caudal peduncle. Body is usually uniformly whitish without distinct dark saddles in preserved specimens

Fisheries/aquarium/food/pharmaceutical values: Not reported.

Phenacoscorpius longirostris (Motomura and Last, 2009)

Photo courtesy: Australian National Fish Collection, CSIRO.

Common name(s): Longsnout no-line scorpionfish

Distribution: Southwest Pacific: south of Northfolk Island, Norfolk Island Ridge, Tasman Sea.

Ecology: It is a bottom dweller (demersal) at depths of 300–322 m.

Biology

Identification: Body of this species is strongly compressed anteriorly and is progressively more compressed posteriorly. Body width at pectoral fin bases is less than upper-jaw length. Nape and anterior body are not strongly arched. Body depth is relatively shallow and is less than head length. Very few small papillae are seen on head. Some small papillae are found scattered on interorbital space and occipital pit. There is no skin flap on outer margin of anterior and posterior nostrils. A short, slender, unbranched tentacle is seen on posterior end of preocular spine base. Tentacle on posterior end of supraocular spine base is unbranched, short, and slender. Several small, short, unbranched tentacles are found on upper and anterior margin of eye membrane. Mouth is large and is slightly oblique, forming an angle of about 20° to horizontal axis of head and body. Posterior margin of maxilla is just reaching a vertical through posterior margin of pupil. Upper edge of posterior maxilla is slightly swollen laterally. Anterior part of maxilla is slightly convex. Lower jaw is with a symphysial knob. Dorsal profile of snout is elevated slightly, forming an angle of about 20° to horizontal axis of head and body. Nasal spine is simple, sharp and is directed dorsally. Origin of first dorsal-fin spine is above supracleithral spine. Posterior margin of opercular membrane is reaching a vertical through third (second) dorsal-fin spine base. Posterior tip of pectoral fin is extending beyond a vertical through third anal-fin spine base. Origin of pelvic-fin spine is slightly anterior to origin of pectoral fin. Origin of first anal-fin spine is slightly posterior to origin of last dorsal-fin spine. Body coloration is pinkish (reddish), with four, indistinct orange or yellowish saddles; first saddle on nape, second extending from fifth to eleventh dorsal-fin spine bases to middle of body, third extending from soft-rayed portion of dorsal-fin base to anal fin, and fourth on caudal peduncle. Spinous portion of dorsal fin is pinkish, without distinct marking. Soft-rayed portion of dorsal fin, pectoral fin and caudal fin are with indistinct reddish spots. Pelvic fin is whitish. Anal fin is whitish with an indistinct reddish band basally. Small, but distinct black spots are found scattered on posterior portion of caudal peduncle. It can reach a maximum length of 6.3 cm (SL) (Motomura and Last, 2009).

Fisheries/aquarium/food/pharmaceutical values: Not reported as it is a new species.

Phenacoscorpius mccoskeri (Wibowo and Motomura, 2017)

A: Lateral view; B: Dorsal view. *Photo Courtesy: Kunto Wibowo and Hiroyuki Motomura.*

Common name(s): McCosker's no-line scorpionfish

Distribution: Off Darwin Island, Galápagos Islands, Ecuador, and eastern Pacific.

Ecology: These fishes are found in very steep rocky wall areas with white sponges and stony corals, and at 515 m depth.

Biology

Identification: This species is diagnosed by its dorsal fin which has 12 spines, of which third spine is the longest but is shorter than pelvic spine length; 19–20 pectoral fin rays in which second to ninth rays are branched; 8 soft dorsal, branched fin rays; anal fin is with 3 spines in which second spine is the longest; anal fin is with 5 soft, branched rays; pelvic fin is with 1 spine and 5 soft rays, in which last ray is unbranched and other rays are branched; posterior margin of caudal fin is slightly rounded. Body is moderately compressed anteriorly and is progressively more compressed posteriorly. Nape and anterior body is strongly arched. Body depth is relatively shallow and is less than head length. There are 2–3 slender unbranched tentacles, of which first is on anterior of posterior nostril and is shorter than second; second is on upper posterior end of preocular spine; and third is on upper posterior end of supraocular spine. No other distinct tentacles are seen on head and body. Exposed ctenoid scales are covering opercle and area enclosed by posterior margin of orbit, upper preopercular margin and pterotic spine base. Exposed cycloid scales are also covering cheek. Remainder of head is however scaleless. Well-exposed ctenoid scales are on body. Scales are becoming cycloid on abdomen, pectoral fin base and between pelvic fin bases. Body scales are extending onto pectoral and caudal fin bases. 5–19 pored lateral-line scales; 46–48 scale rows in longitudinal series; presence of palatine teeth; presence of small slit behind fourth gill arch; and absence of preopercular spine. Caudal peduncle is deep and its depth is 10.0%–11.0% of SL. Head is large, long, and its values of width and length are 17.7%–19.5% of SL and 46.1%–51.9% of SL respectively. Interorbital space is broad. Mouth is large and is slightly oblique. Vomer and palatine are with villiform teeth.

Dorsal profile of snout is moderately elevated forming an angle of about 30–40° to horizontal axis of head and body. Origin of first dorsal fin spine is above supracleithral spine. Origin of pelvic fin is below origin of pectoral fin. Body coloration is bright red, with white ventrally. There are poorly defined white blotches on lateral surface of body and head. Spinous portion of dorsal fin is red. Soft rayed portions of dorsal and anal fins and caudal fin is red with white distally. Pectoral fin is red with white basally and semitransparent distally (Wibowo and Motomura, 2017).

Fisheries/aquarium/food/pharmaceutical values: Not reported as it is a new species.

Phenacoscorpius megalops (Fowler, 1938) (=*Scorpaenopsis stigma*)

Photo credit: John E. Randall.

Highly dangerous and poisonous with venomous spines along its back if trodden on etc.

Common name(s): Noline scorpionfish

Distribution: Pacific: Taiwan, Vietnam, Philippines, Indonesia, and Hawaii.

Ecology: It is a deep-water (bathydemersal) species found over coralline rubble bottoms at depths of 68–622 m.

Biology

Identification: This species is distinguished from other Hawaiian scorpionfishes by having only the anterior 4 or 5 tubed lateral line scales and presence of large melanophores on the caudal peduncle. Other characters of this species include dorsal fin with 12 spines and 9 soft rays; anal fin with 3 spines and 5 soft rays; 7 predorsal scale rows; a small slit behind fourth gill arch; posterior margin of maxilla extending slightly beyond a vertical through posterior margin of pupil; no distinct longitudinal ridge on lateral surface of maxilla; villiform teeth on vomer; no palatine teeth; nasal spine directed dorsally; well-developed interorbital ridges; interorbital, coronal and pretympanic spines absent; occipital pit absent; space between ventral margin of eye and suborbital ridge very narrow; preopercle with 4 spines; upper and lower opercular spines simple, lower spine with a median ridge; posterior tip of pectoral fin extending beyond a vertical through origin of anal fin; posterior tip of depressed pelvic fin not reaching anus; origin of first anal-fin spine level with slightly posterior to base of last dorsal-fin spine; distinct melanophores scattered on caudal-fin base and posterior portion of caudal peduncle; a large brownish blotch on dorsal-fin membranes between fifth and ninth spines. It grows to a maximum length of 13.5 cm (TL). It is highly dangerous and poisonous with venomous spines along its back (Motomura, 2008).

Food: It feeds on small fishes and crustaceans.

Fisheries/aquarium/food/pharmaceutical values: *Phenacoscorpius megalops* is not utilized. It is sometimes caught as bycatch of deep sea trawling.

Phenacoscorpius nebris (Eschmeyer, 1965)

Common name(s): Short-tube scorpionfish

Distribution: Western Atlantic: northern Gulf of Mexico in USA to Venezuela.

Ecology: It is a bottom dwelling, deep-water (bathydemersal) species; depth range 350–480 m.

Biology

Identification: Body of this species is very deep (deepest at gill opening) with a huge head. Dorsal profile is convex. Ventral profile is flattened. Eye and mouth are large. Head is with multiple strong ridges and short spines. A strong cheek ridge with 5–6 spines is present. Preopercle is with a few strong spines. Lacrimal bone is with two downward-pointing spines. Body is covered in ctenoid scales. Lateral line has only 3–5 scales. Spiny dorsal fin is with the longest middle spines. Soft dorsal fin is rounded. Pectoral fin is small and fan-like, with the upper rays being the longest. Upper pectoral rays are branched at tip and lower rays are unbranched. Caudal fin is truncated. Body coloration is red or pink with three irregular dark saddle markings on the dorsal surface of the body behind the gill cover. Dark markings are seen on head. Soft dorsal, caudal, and pectoral fin with dark red spots. A dark blotch is present on the rear part of the spiny dorsal fin. A diffuse bar is seen on the caudal peduncle formed by a concentration of melanophores. It grows to a maximum length of 10.0 cm (TL) (William, 1965).

Fisheries/aquarium/food/pharmaceutical values: Not reported.

Pogonoscorpius sechellensis (Regan, 1908)

Image not available

Common name(s): Not designated

Distribution: Western Indian Ocean: Seychelles.

Ecology: It is a demersal species occurring at depths down to 68 m.

Biology

Identification: It reaches a maximum length of 5.1 cm (SL). Other aspects of its biology are not known.

Fisheries/aquarium/food/pharmaceutical values: Not reported.

Pontinus accraensis (Norman, 1935)

Photo courtesy: Cambraia Duarte, Pedro Miguel Niny.

Common name(s): Ghanean rockfish

Distribution: Eastern Atlantic: Mauritania to Cabinda, Angola.

Ecology: It is a demersal species occurring on mud or sand bottom; depth range 54–450 m.

Biology

Identification: Body coloration is red, and the upper parts of the sides are spotted with darker marks. The spots tend to form irregular rows along the body. There is another row of spots along the lateral line. The soft dorsal fin and the caudal are ornamented with small dark pots. The remaining fins are uniformly colored. It grows to a maximum size of 90.0 cm (TL).

Food: It feeds on crustaceans and fishes.

Fisheries/aquarium/food/pharmaceutical values: This fish is edible and is caught by hooks, bottom trawls and bottom set longlines.

Pontinus castor (Poey, 1860)

Common name(s): Longsnout scorpionfish

Distribution: Western Atlantic: Bermuda and southeastern Florida, USA to Colombia.

Ecology: It is a demersal species inhabiting hard bottom; depth range 45–400 m.

Biology

Identification: Body is very deep (deepest at gill opening) with a huge head. Snout longer (1.3–2.0 times larger) than eye diameter. Dorsal profile is convex. Ventral profile is flattened. Eye and mouth is large. Head is with multiple strong ridges and short spines. A strong cheek ridge is present. Preopercle is with a few strong spines (second spine small or missing). Body is covered in ctenoid scales. Lateral line is complete. Spiny dorsal fin is with spines of similar length. Soft dorsal fin is rounded. Pectoral fin is fan-like, with the middle rays being the longest. Pectoral rays are all unbranched. Caudal fin is truncate. Body coloration is pale pinkish to salmon with darker pink or orangish mottling or incomplete body bars. Often with widely scattered darker brown spots and blotches. Spiny dorsal fin is with orange or pink mottling and a pale margin. Soft dorsal, anal, pectoral, and caudal fin are with spots along the fin rays. Belly is pale. It feeds mainly on other animals and macrofauna including crabs, molluscs and other benthic mobile invertebrates. It grows to a maximum size of 46 cm (TL).

Fisheries/aquarium/food/pharmaceutical values: This species is sometimes taken by hook-and-line as well as in traps and bottom trawls. It may be a minor commercial species.

Pontinus clemensi (Fitch, 1955)

Photo courtesy: Discover Life.

Common name(s): Mottled scorpionfish

Distribution: Southeast Pacific: Colombia, Ecuador, the Galapagos Islands, and northern Peru.

Ecology: This is a demersal species found in sandy substrata; depth range 91–110 m (also collected from shallow waters of 14–18 m depth).

Biology

Identification: This species is diagnosed by its dorsal fin which is with 11 spines and 9–10 soft rays; anal fin is with 3 spines and 5–6 soft rays; and pectoral is with 19 soft rays. Head and body of this species are slightly compressed. Body depth is 2.9–3.2 times or more in SL. Eyes are moderately large and, their length is less than or equal to snout length. Interorbital space is wide, generally half the eye diameter. Tip of maxilla is not reaching posterior rim of the eye. Third dorsal spine is prolonged and its tip is curved slightly forward. Body is covered with small ctenoid scales. Body is uniformly reddish with small, irregular, brown or olive green spots on the dorsal region. It feeds on mobile benthic crustaceans, octopuses, squid, cuttlefish, and bony fishes. It grows to a maximum size of 55 cm (TL) with a maximum weight of 2.6 kg and age of 8 years.

Fisheries/aquarium/food/pharmaceutical values: This species does not have commercial interest and it is captured mainly using hooks and lines.

Pontinus corallinus (Miranda Ribeiro, 1903)

Common name(s): Not designated

Distribution: Southwest Atlantic: Brazil.

Ecology: It is a demersal species inhabiting depths up to 100 m.

Biology

Identification: Not reported

Fisheries/aquarium/food/pharmaceutical values: Not reported.

Pontinus furcirhinus (Garman, 1899)

Common name(s): Red scorpionfish

Distribution: Eastern Pacific Ocean: from northern Peru to Baja California, Mexico.

Ecology: This is a deep-water (bathydemersal) species dwelling in soft and sand bottom at depths from 50 to 390 m. It is often found in large aggregations.

Biology

Identification: Body of this species is slightly compressed. Body depth is 2.9 times in SL. Head is enlarged, compressed, bulbous, and very bony with numerous spines. Eyes are large and its diameter is greater than snout length. Interorbital space is narrow, usually less than half eye diameter. Tip of maxilla is reaching posterior rim of the eye. Uppermost gill cover spine is the longest. Dorsal fin has 12 spines, the third being much longer than the others. Pectoral fin has 17–19 rays. Anal fin has 3 spines, the second being the longest and stoutest. Caudal fin is square with a ragged margin. There are 15–20 gill rakers. Lateral line is complete and prominent. Body is covered with medium sized ctenoid scales. Lateral scales are fewer (60). There are various shades of red with white mottling and dark brown and olive spotting on their upper body. Their caudal fin and the posterior portion of their dorsal fin are heavily spotted with small dark rectangular spots. It reaches a maximum length of 32.0 cm (TL).

Food: It feeds on mobile benthic crustaceans, octopuses, squid, cuttlefish, and bony fishes.

Fisheries/aquarium/food/pharmaceutical values: The Red Scorpionfish is consumed by subsistence fishermen but is generally too small and difficult to handle, thus are mostly a "catch and release."

Pontinus helena (Eschmeyer, 1965)

Photo courtesy: Keiichi Matsuura.

Common name(s): Not known

Distribution: Western Central Atlantic: Venezuela to Suriname.

Ecology: This demersal fish lives on rocky and shell soft bottoms; depth range 75–440 m.

Biology

Identification: This species is diagnosed by its dorsal fin which has 12 spines and 9.5–10.5 soft rays; anal fin is with 3 spines and 5 soft rays; and pectoral with 19–20, unbranched rays (longest rays in center). Head is very bony with many spines and is without pit behind eye. Snout is short and its length is almost equal to orbit diameter. While preopercular spine is the longest, second spine is reduced or absent. Teeth are present on roof of mouth at sides. A slit is present behind last gill arch. Scales are rough and are present on cheek, behind eye and on chest. Lateral line is complete with tubes on scales. There are 23–24 tubed scales on body and 2 such scales on tail fin. Body is reddish above and yellowish below. A few small dark specks on head and body. The maximum recorded length is 25 cm (TL).

Fisheries/aquarium/food/pharmaceutical values: Not reported.

Pontinus kuhlii (Bowdich, 1825) (=*Sebastes kuhli*) (Bowdich, 1825)

Photo courtesy: Fernando Bento, FishBase.

Common name(s): Offshore rockfish

Distribution: Eastern Atlantic: Portugal, Madeira, the Azores and Mediterranean including Canary Islands and Cape Verde; São Tomé Island.

Ecology: It is a bathydemersal species living on hard bottoms including island slopes and seamounts; depth range 100–600 m.

Biology

Identification: Head of this species is large and snout is moderate, longer than orbit diameter. Supraocular tentacle is small or absent. Dorsal fin is with 12 spines and 9 rays. Pectoral rays are usually 17. Lateral line is with tubular scales. Body is covered with ctenoid scales in 60–65 vertical rows. There are 24 vertebrae and 6–8 gill rakers on upper arch and 9–10 on lower arch. Coloration of body is reddish interspersed with yellow. Dusky blotches are seen alongside. Soft dorsal fin is usually spotted. Dark pigment is seen at bases of dorsal spines. Head is with yellow area below suborbital ridge, behind orbit and opercle flap. Females appear to remain smaller, up to 38.4 cm (TL). Males are heavier and larger. It has a maximum length of 52.0 cm (TL). Studies relating to the age and growth of this species showed that the ages of this species ranged from 6 to 18 years for males and from 6 to 14 years for females (Abellán et al., 2001).

Food and feeding: It feeds on fishes and shrimps.

Mercury accumulation: This species has been reported to concentrate mercury in its tissue. A significant sex-related difference in Hg levels was found; the rate of Hg accumulation was significantly faster in females than in males and mean Hg levels were higher in females. This species may therefore serve as an indicator in pollutant monitoring of benthic compartment of marine ecosystems (Monteiro et al., 1991).

Fisheries/aquarium/food/pharmaceutical values: This species is consumed in western Africa, although it is not specifically targeted. It is often caught in trawls and longlines. Exploitation rates of this species are probably too low to significantly affect its size structure, even in the Azores where this species is a secondary target of the commercial fishery (Catarino et al., 2013).

Pontinus leda (Eschmeyer, 1969)

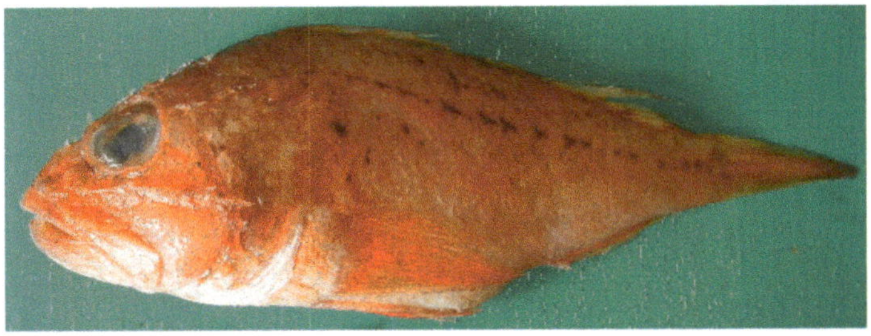

Photo credit: Wikipedia.

Common name(s): Speckled deepwater scorpionfish

Distribution: Eastern Atlantic Ocean between Guinea-Bissau and Namibia.

Ecology: It is a demersal species found over soft, sandy or muddy bottoms at 90–400 m depth.

Biology

Identification: Max length: 27.0 cm (TL). Reddish in color with black specks on dorsolateral part of head, body, dorsal fin, and caudal fin.

Body of this species is moderately compressed anteriorly and is progressively more compressed posteriorly. Nape and anterior body are not highly arched and body is relatively deep. Head is and its large, length is greater than body depth. Eyes are moderately large and are oriented somewhat dorsolaterally. There is no supraocular tentacle. A short, slender tentacle is seen behind posterior end of parietal spine base. Ctenoid scales are covering head, including cheek, opercle, area between upper and lower opercular spines, interorbital space, and occiput. Scales are absent only on spines, ridges, eye, lacrimal, lips, maxilla, and underside of mandible. Ctenoid scales are covering entire body but not extending onto rays or membranes of median fins, except caudal fin. Well-exposed cycloid scales are covering pectoral-fin base and ventral surface of body. Mouth is large and is slightly oblique, forming an angle of about 20° to horizontal axis of head and body. Posterior margin of maxilla is reaching a vertical at posterior margin of orbit. Upper jaw is with a band of short, conical teeth and a pair of tooth patches is projecting to anterior tip of upper jaw. Vomerine and palatine teeth are present. There are no tentacles on underside of lower jaw. Dorsal profile of snout is steep, forming an angle of about 40–45° to horizontal axis of head and body. Nasal spine is simple, conical and directed posterodorsally; and its length is approximately equal to anterior nostril diameter. Anterior nostril is with a tentacle. Posterior tip of pectoral fin is extending beyond a vertical through last dorsal spine base but not reaching first anal spine base. Origin of pelvic spine is slightly anterior to origin of uppermost pectoral ray. Posterior tip of depressed pelvic fin is not reaching anus. Origin of first anal spine is posterior to origin of last dorsal spine (Motomura et al., 2005a–c). Body of fish is reddish in color with black specks on dorsolateral part of head, body, dorsal fin and caudal fin. It attains a maximum length of 27.0 cm (TL).

Food and feeding: There is little information concerning its food and feeding habits.

Major threats: This species is consumed, but it is not believed to be faced with any major population reduction.

Fisheries/aquarium/food/pharmaceutical values: It is caught in trawls in local fisheries and consumed.

Pontinus longispinosus (Goode and Bean, 1896)

Common name(s): Longspine scorpionfish, Western Atlantic fish

Distribution: Western Atlantic: Georgia, USA, and northern Gulf of Mexico to Brazil.

Ecology: This coastal demersal species is found on soft and hard substrate; depth range 75–440 m.

Biology

Identification: Body of this species is deep (deepest at gill opening) with a huge head. Snout is about equal to eye diameter. Dorsal profile is convex. Ventral profile is flattened. Eye and mouth are large. Head is with multiple strong ridges and short spines. Lacrimal bone is with a pair of downward-pointed spines (point more anteriorly) and one lateral spine. A strong cheek ridge is present (with 4 spines). Preopercle is with a few strong spines (second spine small). Body is covered in ctenoid scales. Lateral line is complete. Spiny dorsal fin is fairly high, with an elongated third spine. Soft dorsal fin is truncate posteriorly. Pectoral fin is fan-like, with the middle rays being the longest. Pectoral rays are all unbranched. Caudal fin is truncate. Coloration of body is pale pink to orange with obscure darker mottling. Scattered darker spots are seen especially along lateral line. Belly is pale. Dorsal, anal, pectoral, and caudal fins are with pink or orange spots along fin rays. Its maximum size is 25 cm (TL).

Fisheries/aquarium/food/pharmaceutical values: Not reported.

Pontinus macrocephalus (Sauvage, 1882)

Photo credit: John E. Randall.

Common name(s): Large-headed scorpionfish

Distribution: Western Pacific: southern Japan to Micronesia and the Hawaiian Islands; Malaysia and from Bali Strait to Timor Sea in Indonesia.

Ecology: It is an upper slope, demersal species; depth range 174–278 m.

Biology

Identification: This species has an oval body, with large protruding eyes. Head is long with scattered spines. Mouth is large and is slightly angled. Pectoral rays are unbranched in adults There are about 45–60 vertical scale rows on body. Lachrymal bone is with 2 spines over maxillary; first points out and back and second points mostly back. Suborbital ridge is with 3 or 4 spines. Second preopercular spine is usually absent in large specimens. Coloration of body is bright orange and reddish with darker mottling and it is paler below. It reaches a maximum length of 50.0 cm (SL).

Fisheries/aquarium/food/pharmaceutical values: Not reported.

Pontinus nematophthalmus (Günther, 1860)

Common name(s): Spinythroat scorpionfish, west Atlantic fish

Distribution: Western Atlantic: Florida, USA and the Bahamas to north-eastern Brazil.

Ecology: It is a demersal species occurring on rough and hard bottoms; depth range 82–410 m.

Biology

Identification: This species is diagnosed by its dorsal fin which has 12 spines and 8.5–9.5 soft rays; anal fin is with 3 spines and 5 soft rays; 15–17 (usually 16) soft rays in pectoral fin; 1 spine and 5 rays in pelvic fin; and 42–47 scale rows above lateral line. Body of the fish is deep (deepest at gill opening) with a huge head. Snout is about equal to eye diameter. Supraorbital cirrus is long and tapered. Dorsal profile is convex. Ventral profile is flattened. Eye and mouth are large. Head is with multiple strong ridges and short spines. Lacrimal bone has a pair of downward-pointed spines and sometimes one small lateral spine. A strong cheek ridge with 3 spines is present. Preopercle is with a few small spines. Hypohyals with an extended ventral process. Body is covered in ctenoid scales. Lateral line is complete. Spiny dorsal fin is fairly high, with spines of similar length. Soft dorsal fin is truncate posteriorly. Pectoral fin is fanlike, with the middle rays being the longest. Pectoral rays are all unbranched. Caudal fin truncate. Body coloration is rosy red with several irregular darker bars. Fins are with reddish spots along fin rays. Its maximum size is 14.0 cm (TL).

Food and feeding: It consumes crabs, molluscs, and other benthic mobile invertebrates.

Fisheries/aquarium/food/pharmaceutical values: This species is only taken incidentally in traps and trawls.

Pontinus nigerimum (Eschmeyer, 1983)

Common name(s): Blacklash scorpionfish

Distribution: Western Indian Ocean: off Natal, South Africa.

Ecology: It is a demersal species found at depths up to 1200 m. Little else is known regarding its biology or ecology.

Biology

Identification: This species is diagnosed by its dorsal fin which has 12 spines and 9 rays; and anal fin is with 3 spines and 5 soft rays. It has a large tentacle above eye. Body coloration is reddish with black lower portion. No other information is available on its biology.

Fisheries/aquarium/food/pharmaceutical values: This species is not utilized.

Pontinus nigropunctatus (Günther, 1868)

Photo courtesy: Teodoro Vaske Junior.

Common name(s): Deepwater jack; St. Helena deepwater scorpionfish

Distribution: Southeast Atlantic: endemic to St. Helena.

Ecology: This demersal species occurs on hard bottoms in depths between 91 and 183 m.

Biology

Identification: This species has a maximum length 42.5 cm (TL) and a weight of 1100 g. No other information on its biology is known.

Fisheries/aquarium/food/pharmaceutical values: It is caught by hand line at night at 150 m local depth (Vaske et al., 2008). It is not often targeted by fishermen, but its soft flesh is thought to be good for eating.

Pontinus rathbuni (Goode and Bean, 1896)

Photo courtesy: Keiichi Matsuura.

Common name(s): Highfin scorpionfish

Distribution: Western Atlantic: Virginia, USA and Bahamas to Venezuela.

Ecology: This demersal species inhabits deep rocky reefs, gulf, and soft of semihard substrate at depths from 73 to 384 m.

Biology

Identification: Body of this species is elongate and somewhat compressed. It is deep (deepest at gill opening) with a huge head. Snout is about equal to eye diameter. Supraorbital cirrus is with an expanded bifurcation at the tip. Dorsal profile is convex. Ventral profile is flattened. Eye and mouth are large. Head is with multiple strong ridges and short spines. Lacrimal bone is with a pair of downward-pointed spines (point more posteriorly) and one small lateral spine. A strong cheek ridge is present (with 3 spines). Preopercle with a few strong spines (second spine is small). Body is covered in ctenoid scales. Lateral line is complete. Spiny dorsal fin is quite high, with flaps on membranes between each spine. Soft dorsal fin is rounded posteriorly. Pectoral fin is fanlike, with the middle rays being the longest. Pectoral rays are all unbranched. Caudal fin is truncate. Body coloration is pale pink with darker pink mottling or irregular body bars. Soft dorsal, anal, pectoral, and caudal fins are with pink spots along fin rays. Eyes are yellow with pink markings. Its maximum size is up to 25.0 cm (TL).

Food: It is a mobile benthic invertivore, mainly feeding on crabs, molluscs, and other benthic mobile invertebrates.

Fisheries/aquarium/food/pharmaceutical values: Not known.

Pontinus rhodochrous (Günther, 1872)

Photo courtesy: Trevor, Michael N. mtrevor@ntamar.net.

Common name(s): Not designated

Distribution: Western Pacific: Indonesia, New Caledonia, Philippines, Vanuatu and Wallis and Futuna Islands.

Ecology: This benthopelagic has a depth range, 223–369 m. It is thought to be conspecific with *Pontinus tentacularis* and possibly *Pontinus macrocephalus*.

Biology

Identification: It has a maximum length of 28.0 cm (TL). No other information is available on its biology.

Fisheries/aquarium/food/pharmaceutical values: Not reported.

Pontinus sierra (Gilbert, 1890)

Photo courtesy: Béarez, Philippe.

Common name(s): Speckled scorpionfish

Distribution: Eastern Central Pacific: Gulf of California to Panama; Peru.

Ecology: This is a demersal species that is found in rubble and soft substrata; depth range 48–351 m. Adults are demersal while juveniles are pelagic. These fishes are an exceedingly rare species that is seldom seen by humans; thus, very little is known about their behavioral patterns.

Biology

Identification: This species has spine-laden slender compressed body that tapers posteriorly. Body depth is 29%–33% of SL. Body coloration is red but shade to pink ventrally with gray-green blotches on its sides. Underside of its head and the inside of its mouth are white. Fins are pinkish-red and caudal and soft dorsal fins are heavily spotted. Head is very large, deep, and bony with numerous spines. Mouth is small with large disproportionately-sized very closely-spaced eyes. The uppermost spine on its gill cover is the longest. Anal fin has 3 spines, the second being the longest, and 5 or 6 rays. Dorsal fin has 12 spines and 8 or 9 rays. Pectoral fin has 16–19 rays. There are 8–13 gill rakers and body is covered with rough scales. It grows to a maximum length of 33 cm (TL).

Food: It feeds on mobile benthic crustaceans, octopuses, squid, cuttlefishes, and bony fishes.

Reproduction: It is oviparous, with planktonic larvae. Eggs are probably released in buoyant gelatinous mass.

Fisheries/aquarium/food/pharmaceutical values: Not known.

Pontinus strigatus (Heller and Snodgrass, 1903)

Common name(s): Stalkeye scorpionfish

Distribution: Southeast Pacific: Galapagos Islands.

Ecology: It is a bathydemersal (deep-water) species found on sandy and rocky bottoms; depth range 50–120 m.

Biology

Identification: The main identification features of this species are branched tentacle over eye; long third spine of dorsal fin; red fins and yellow dorsal fin membranes; very bony head with numerous spines; lateral line with tubes on scales; pectoral fin 17–19 rays, all unbranched; head large, with low depressed spines; large eyes ($\sim$31% of head length); short snout; bright red body and pink belly; and body, head and fins are spotted and blotched with olive brown, those on head are smaller.

Fisheries/aquarium/food/pharmaceutical values: Not reported.

Pontinus tentacularis (Fowler, 1938)

Photo courtesy: Shao, Kwang-Tsao.

Common name(s): Not known

Distribution: Indo-West Pacific: Mauritius and Reunion to the Philippines and Taiwan.

Ecology: It mainly inhabits deep-water benthic environments; depth range 170–600 m.

Biology

Identification: The main identifying features of this species are its dorsal fin which is with 12 spines and 9–10 soft rays; anal fin is with 3 spines and 5 soft rays; and pectoral fin is with 16–17 soft rays which are unbranched. A supraocular tentacle like cirrus is present. Body of the fish is long, flat and oval. Eyes are large and its mouth which is large is in end position and oblique. Upper and lower jaws are equal and upper cheekbone extends to the eyelid below. Upper and lower jaws, sacrum and sacrum have fine teeth. Distance between the eyes is not obvious and there is no frontal spine. Anterior nostrils are with branching flaps. Body scales are large and there are round scales on the chest and cheeks. First half of the lateral line is straight; second half is straight, and distal end extends to the base of the caudal fin. Dorsal fin originates slightly in front of the supracondylar spine, and the hard spines connect with the fin membranes of the fins. Base of the hard spines is longer than the base of the soft strips. Anal fin originates below the hard spiracles of the dorsal fin. The fins are shorter than those of the dorsal fins. The fins extend slightly beyond the dorsal fin base. Caudal fin is rounded. In life, body coloration is red-orange, and the ventral surface is light. There are four irregular-shaped dark brown stripes on the side of the body above the lateral line. Fins are with scattered red dots. It grows to a maximum length of 16 cm (TL).

Fisheries/aquarium/food/pharmaceutical values: This small fish, except for academic research, have no economic value. However, it is used as food in some localities of its range.

Pontinus vaughani (Barnhart and Hubbs, 1946)

Photo courtesy: John Snow.

Common name(s): Spotback scorpionfish

Distribution: Eastern Central Pacific: Baja California, Mexico and the Clipperton, Clarion and Revillagigedo Islands.

Ecology: It is predominantly deep-sea and tropical. This is demersal species is found in rubble substrata. It is a deep reefs associate. It is found in depths from 30 to 120 m.

Biology

Identification: Body of this species is wedge shaped. The dorsal contour is rather sharply elevated at the origin of the dorsal. From the tip of the snout to a point on the nape one eye's diameter in front of the dorsal origin, the contour is an almost straight line. An outstanding character of this species is the strong compression of the body. The wedge-shaped head forms in side view an angle of 65″ between its straightish dorsal contour and the closed mouth. The jaws are rather massive anteriorly. When the mouth is tightly closed the tip of the mandible projects about a millimeter beyond the upper lip. The spines of the head are moderately strong. The strolig, convergent nasal spines project into the contour. The preorbital spines, though rather large, do not project. The bluntish supraorbital spine is nearly flat. The postorbital and tympanic spines, are directed about as far outward as backward and enter the contour rather inconspicuously. The mouth is large. The upper jaw extends to the vertical that passes through the posterior rim of the orbit. The teeth are in villiform bands on the jaws and on the vomer and the palatines. The soft dorsal fin is abruptly higher than the last dorsal spines. The pectoral fin has a relatively narrow base. The pelvic fin reaches more than nine-tenths the distance to the anus. When fresh the specimen was olive-green over the head, back, sides and fins, becoming somewhat darker on the back, particularly in four blotches. The caudal becomes particularly dark outward between the spots. The lower edge of the caudal, the front edge of the anal and pelvic fins, and the thickened part of the pectoral rays are largely clear or whitish. It has a maximum length of 44.3 cm (SL) (Barnhart and Hubbs, 1946).

Fisheries/aquarium/food/pharmaceutical values: It is considered as a good food fish.

Pteroidichthys amboinensis (Bleeker, 1856) (=*Pteroidichthys godfreyi*)

Photo credit: John E. Randall.

Common name(s): Ambon scorpionfish

Distribution: It is native to the Indian and Pacific Oceans. Indo-West Pacific: from southern Japan (except Ryukyu Islands), Vietnam, Indonesia, Papua New Guinea, Fiji, Australia, Taiwan, India, and the Red Sea.

Ecology: This demersal species which is solitary and an ambush predator lives primarily in sandy, muddy and algae covered bottoms. It is shaggy and can change its color for the ideal camouflage. It can be easily overlooked because of its superb camouflage; depth range 3–43 m.

Biology

Identification: It is characterized by having deeply incised membranes of spinous portion of dorsal fin; elaborate antenna-like skin flap above eye; and snout and chin with shorter filaments. Its body depth is 2.7–2.9 of SL. Its dorsal-fin spines are flexible and bend easily under slight pressure. Snout is relatively short. Posterior margin of maxilla is reaching to or extending beyond vertical through middle of pupil. Trunk and fins are with numerous tentacles. Its body is overgrown by skin flaps, which are more prominent in males. It has a wide head with a large mouth, and large pectoral fins. Color of body is variable, matching environment. It has red, pink, yellow, brown, or green colors that relate to the algae types found in the surrounding habitats. It can reach up to 12 cm.

Behavior: It will camouflage itself, wait for some prey to come close in front of itself, and then lunge forward and inhale the prey.

Fisheries/aquarium/food/pharmaceutical values: This species is popular in the aquarium trade and fetches high prices due to is rarity.

Pteroidichthys acutus (Motomura and Kanade, 2015)

Image not available

Common name(s): Longsnout weedy scorpionfish

Distribution: Western Pacific: from the Philippines, Vanuatu, and the Wallis and Futuna Islands.

Ecology: It is a pelagic-neritic species found on sandy bottoms; depth range 73–400 m.

Biology

Identification: Body and head of this species are strongly compressed. Body width is less than length of the longest dorsal-fin spine. Body is shallow, deepest at bases of first to third dorsal-fin spines. Head is large, and its length is greater than body depth. Ventral surface of posterior lacrimal spine is with unbranched, slender tentacle. Slender tentacle is extending beyond anterior margin of posterior nostril. Mouth is oblique, forming angle of about 20° to horizontal axis of head and body. Upper edge of posterior maxilla is slightly swollen laterally, forming indistinct ridge. Central part of maxilla is nearly flat (or slightly convex) with median ridge. Lower jaw is with thickened symphyseal portion fitting into shallow median depression of upper jaw when mouth fully closed. Dorsal profile of snout is curved sinuously, initially convex, then deeply concave. Nasal spine is with 2 spinous points, directed dorsally, its tip is well exposed. Interorbital space is deep. About one-third of orbit is extending above dorsal profile of head. Lateral surface of lacrimal is with 6 low ridges radiating from center. Origin of first dorsal-fin spine is above posttemporal spine base. Posterior margin of opercular membrane is extending beyond vertical through base of third dorsal-fin spine. Posterior tip of the longest pectoral-fin ray is extending well beyond vertical through base of first anal-fin soft ray. Origin of pelvic fin is anterior to base of uppermost pectoral-fin ray. Posterior tip of depressed pelvic fin is not reaching to anus. Origin of first anal-fin spine is below origin of last dorsal-fin spine. In preserved specimens, body is brownish white. There is no distinct marking on head. Black blotch distally on membranes is found between first and fifth (or fourth) soft rays of dorsal fin. Lower halves of pectoral and anal fins are blackish. Pelvic fin is black, except for whitish base. Lower one-third of caudal fin is black. This species attains a maximum length of 4.1 cm (SL) (Motomura and Kanade, 2015).

Fisheries/aquarium/food/pharmaceutical values: Not reported.

Pteroidichthys noronhai (Fowler, 1938)

Image not available

Common name(s): Noronha's scorpionfish

Distribution: Indo-West Pacific: Japan, China, Vietnam, the Philippines, Vanuatu, and Western Australia; recorded from Saya de Malha Bank and Taiwan.

Ecology: This demersal species occurs on sandy bottoms; depth range 50–215 m.

Biology

Identification: This species has rigid dorsal-fin spines and relatively short snout with a length of 13.5%–16.9% (mean 15.3%) of SL. Posterior margin of maxilla is reaching to or extending beyond vertical through middle of pupil. Distance between tips of lateral lacrimal and first suborbital spines is shorter than or subequal to that between tips of first and second suborbital spines. Poorly developed tentacles are seen on supraocular and posterior lacrimal spines. There are a few tentacles on trunk and fins. Lateral lacrimal and suborbital spines are well developed with pointed tips. There is no distinct black blotch on anterodistal part of soft-rayed portion of dorsal fin. This species can reach a maximum length of 6.0 cm (SL).

Fisheries/aquarium/food/pharmaceutical values: Not reported.

Rhinopias aphanes (Eschmeyer, 1973)

Photo courtesy: Nick Hobgood and Wikipedia.

Common name(s): Weedy scorpionfish, lacy scorpionfish, Merlet's scorpionfish

Distribution: Western Pacific: Papua New Guinea, New Caledonia, Vanuatu and north-eastern Australia; Japan and Penghu Archipelago.

Ecology: The species is found in coral reefs (coral slopes or soft bottom habitat) between depths of 5 and 30 m. This solitary species uses its pectoral and pelvic fins for walking on bottom. This species is a master of camouflage and blends in very well with its background and quite often can be seen hiding under plate corals and ledges on coral reefs. It also remains well disguised when hiding in amongst crinoids as it blends in with the fine fronds of the feathery invertebrate. Surprisingly, the weedy scorpionfish is easily overlooked by divers even though it sits on the coral, often in full view.

Biology

Identification: This species is recognized chiefly by its distinctive head shape, the large upturned mouth, dangly tentacles on the snout and skin flaps on the body. Other identifying features of this species include presence of two black spots in the middle of the membrane between the sixth and eighth dorsal-fin soft rays; membrane of the spinous portion of the dorsal fin is not fleshy; firm dorsal-fin spines with strongly pointed tips; 16 or 17 pectoral-fin rays; tips of each caudal-fin ray is divided into two branches; distal margins of the soft-rayed portions of the dorsal, pelvic, anal, and caudal fins are strongly notched; lateral surface of the lacrimal is usually with a pointed spine; suborbital ridge is usually with three-pointed spines; head and body are with elongate black-margined white markings each with a central region of yellow, green, or brown; presence of 12–18 tentacles on the underside of the lower jaw; numerous distinct tentacles on the lateral surface of the body above the lateral line; and long tentacles, with distinct branches along distal margins. The species grows to a maximum length of 25 cm. Body coloration varies from yellow to brown, purple, green, or black and is covered by a maze-like pattern (Motomura and Johnson, 2006).

Fisheries/aquarium/food/pharmaceutical values: It is an aquarium fish.

Rhinopias argoliba (Eschmeyer et al., 1973)

Image not available

Common name(s): Japanese scorpionfish
Distribution: Northwest Pacific: known only from Sagami Bay, Japan.
Ecology: It is a demersal species inhabiting rocky places.

Biology

Identification: These fishes are characterized by the presence of deep, laterally compressed bodies and special shape of the head with an upturned mouth. The eyes are on top of the head and have a special kind of eyebrows. The mouth includes strange protrusions. Dermal appendages are seen above eyes, on the jaws and the body surface. It grows to a maximum length of 15.0 cm (TL).

Food: It feeds on fishes and crabs.

Fisheries/aquarium/food/pharmaceutical values: All the species of *Rhinopia* are aquarium fishes and they readily adapt to captivity. As these fishes are relatively inactive, they can be kept in smaller aquaria. However, smaller *Rhinopias* may be housed in an aquarium of at least 20 gal, while larger adults will do fine in a standard 30-gal tank. Although some décor is needed to help these *Rhinopias*, it is better to keep them in an environment similar to that of their original habitat. Because these fishes tend to spend most of their time in the open (unlike other scorpionfishes, which are more cryptic), they make an interesting addition to the reef aquarium.

Rhinopias cea (Randall and DiSalvo, 1997)

Photo credit: John E. Randall and Louis H. DiSalvo.

Common name(s): Cea's scorpionfish
Distribution: Eastern Pacific: it is known only from Easter Island.
Ecology: This demersal species inhabiting rocky bottoms; depth range 0–5 m.

Biology

Identification: Body of this species is moderately deep (depth 2.15 in SL) and is compressed. Dorsal profile of snout is initially convex, then deeply concave. Eye is protruding well above dorsal profile of head. Interorbital space is deeply de-pressed. Mouth is large and maxilla is extending to below middle of eye. Mouth is oblique, forming angle of about 45° to horizontal axis of body. Lower jaw is projecting. Jaws are with villiform teeth. There are no teeth on palatines. Body is covered with cycloid scales. There are no scales on head except for upper part of opercular membrane. Predorsal scales are extending to ridge linking nuchal spines. Prepectoral and thoracic scales are very small. There are no scales on fins. In fresh fishes, coloration of body is brownish orange-red and is finely mottled with pale and blotched with dark brown. Most conspicuous markings, viz., two dark blotches are seen on body above pectoral fin. This species grows to a length of 20.2 cm.

Fisheries/aquarium/food/pharmaceutical values: Not reported.

Rhinopias eschmeyeri (Condé, 1977)

Photo credit: Wikipedia.

Common name(s): Eschmeyer's scorpionfish, paddle flap scorpionfish

Distribution: Indo-West Pacific.; East Indian Ocean, West Indian Ocean, Australia, Japan, Indonesia, New Zealand, Central/West Pacific; Mauritius to Sri Lanka; Philippines and Papua New Guinea.

Ecology: It occurs on open sand bottoms at depths from 2 to 45 m.

Biology

Identification: This species is usually uniform in color and can be lilac, white, pink, purple yellow, orange, or brick red. That is, its body comprises only one single color and without any markings. Further, it has no additional appendages on the body. However, an important feature is that the fins have almost no notches. Supraorbital appendages are unbranched, flattened and are leaf-like. Other characters of this species: presence of one small black spot, slightly greater than pupil diameter, in the middle of the membrane between the seventh and eighth dorsal-fin soft rays; the absence of scales on the opercle between the tips of the upper and lower opercular spines and the opercular margin; presence of two tentacles on the underside of the lower jaw; lacking tentacles on the frontal below the eyes in anterior view; lacking distinct tentacles on the lateral surface of the body above the lateral line; having short tentacles, without distinct branches along distal margins, on the supraocular and posterior lacrimal spines; having head, body, fins, and tentacles usually without distinct pigmentation or markings; fleshy membrane of the spinous portion of the dorsal fin; distal margins of the soft-rayed portions of the dorsal, pelvic, anal, and caudal fins without notches or very weakly notched; tips of each caudal-fin ray are divided into four branches; dorsal-fin spines are relatively soft and tips are bending easily under the slight pressure; lateral surface of the lacrimal usually smooth or with a bump; suborbital ridge is usually with three bumps, and the soft-rayed portion of the dorsal fin is with a single black spot. It grows to a maximum length of 2 3 cm.

Shedding of skin: The *Rhinopias* spp. are known to shed their skin, or the outer epidermal layer (cuticle). This helps rid the body of algae, parasites, or other encrusting organisms. In some individuals shedding can occur quite often. This species has been reported to shed about every 12 days. The frequency of shedding may also increase if the fish is suffering from parasitic infections. These scorpionfish may also position themselves in front of strong water flow to facilitate the shedding process.

Feeding: These fish rarely swim but move about by crutching along the bottom on their pectoral and pelvic fins. When hunting, they remain motionless and wait for their prey to approach within striking distance or they slowly stalk their quarry. When a specimen is very hungry, it may "hop" rapidly toward a prey item, rather than approaching it slowly. When they get close enough to their prey, the *Rhinopias* will lunge forward and suck in its victim. *Rhinopias* will also rock forward and backward in order to mimic a piece of debris sitting on the bottom.

Fisheries/aquarium/food/pharmaceutical values: It is suitable for aquarium tanks.

Rhinopias filamentosus (Fowler, 1938) (=*Hipposcorpaena filamentosa*)

Photo courtesy: Serge abourjeily.

Common name(s): Filamentous scorpionfish, false rhinopias

Distribution: Western Central Pacific: known from the Philippines, Indonesia, and Papua New Guinea.

Ecology: This demersal, reef associated species is found in silty inshore reefs; depth range 10–95 m.

Biology

Identification: This species is characterized by having variable coloration, from whitish to red-brown; red-brown variety with whitish snout and broad white band on posterior portion of spinous dorsal fin; elevated spinous portion of fin; membrane between first and second spines is moderately incised; strongly incised membranes between lower seven rays (one-third to one-half length of each ray); and 54 longitudinal scale series; large head with concave snout profile; and presence of skin flaps on snout, chin, lower cheek, edge of preoperculum and above eye. It grows to a maximum length of 5.0 cm (TL).

Shedding skin: This species is also capable of shedding its skin.

Fisheries/aquarium/food/pharmaceutical values: Not reported.

Rhinopias frondosa (Günther, 1892)

Photo credit: Wikipedia.

Common name(s): Weedy scorpionfish

Distribution: It is a wide-ranging species that occurs from East Africa to the Caroline Islands, north to Japan and south to Mauritius.

Ecology: It has been reported at depths from 10 to 300 m on bottoms sandy, macroalgae, rubble and rocky substrates.

Biology

Identification: This species appears in two kinds. The first kind has a lot of similarities with the *Rhinopias eschmeyeri* and its dorsal fin is wit out notches. The second type which is largely resembling *Rhinopias aphanes* has a kind of appendages over the body and is therefore also better called the weedy scorpionfish. It is however, distinguished from the *R. aphanes* by its body which is full of pale spots and blotches compared to the continuous stripes of *R. aphanes*. Further, it is also distinguished from *Rhinopias argoliba*, *Rhinopias cea*, and *Rhinopias xenops* by the presence of one or two small black spots, usually less than orbit diameter, in the middle of the membrane between the sixth and ninth dorsal-fin soft rays; absence of scales on the opercle between the tips of the upper and lower opercular spines and the opercular margin, and 16 pectoral-fin rays. It also differs from *R. aphanes* in having fleshy tentacles on the supraocular and posterior lacrimal spines; distal margins of the soft-rayed portions of the dorsal, pelvic, anal, and caudal fins without notches or very weakly notched; tips of each caudal-fin ray is divided into four branches; dorsal-fin spines are relatively soft and tips are bending easily under the slight pressure; and the soft-rayed portion of the dorsal fin is with one or two black spots. It differs from *R. eschmeyeri* in having 9–24 tentacles on the underside of the lower jaw; 2–4 tentacles on the frontal below the eyes in anterior view; presence of numerous distinct tentacles on the lateral surface of the body above the lateral line; and presence of long tentacles, with distinct branches along distal margins, on the supraocular and posterior lacrimal spines. It also differs from *R. aphanes* and *R. eschmeyeri* in having head and body with numerous distinct circular dark-margined spots. This scorpaenid fishes attains a maximum length of 22.5 cm.

Fisheries/aquarium/food/pharmaceutical values: This species is a popular aquarium fish.

Rhinopias xenops (Gilbert, 1905)

Photo credit: John E. Randall.

Common name(s): High-eye scorpionfish, Hawaiian rhinopias, blue spotted scorpionfish

Distribution: Pacific Ocean: Hawaii and the Kii Peninsula, Japan.

Ecology: This demersal species is found on rocks or coral (atolls); depth range 10–124 m. It occurs more shallowly (~25 m) in the cooler waters of the Northeastern Hawaiian Islands and deeper in the larger islands (<60 m) of Hawaii.

Biology

Identification: This species has the following characteristic features: 70–75 vertical scale rows; lachrymal bone with 2 rounded lobes; suborbital ridge with 3 or 4 spines; compressed head and body; elevated rostrum; and long snout. Hawaiian specimens seem to lack the white markings beneath the eye and dorsal fin which are common to these Asian populations. Furthermore, Hawaiian specimens have less of the reddish hue than their Asian counterparts. It can be grown to a maximum length of 15.3 cm (SL).

Fisheries/aquarium/food/pharmaceutical values: Not known.

Scorpaena afuerae (Hildebrand, 1946)

Photo courtesy: Philippe Béarez.

Common name(s): Peruvian scorpionfish

Distribution: Southeast Pacific: Ecuador and Peru.

Ecology: This is a demersal species found in rocky and rubble substrata, at depths of 35–100 m.

Biology

Identification: This species has spine-laden compressed rectangular-shaped body that tapers posteriorly. Body depth is 33%–37% of SL. Body coloration is scarlet red with brown mottling. With the exception of the anterior portion of the dorsal fin, fins have irregular bars and darker spotting. Head is enlarged, depressed, bulbous, and very bony with numerous spines. They have small eyes. There is a deep depression (pit) before and after its eyes and the ridge below its eyes has 3 or 4 spines. Gill cover has 5 spines; the first is the largest and the second is well developed. There is a row of 3 spines behind the eyes and two spines at the upper edge of their gill cover. There are skin flaps above and in front of its upper jaw, which are absent from their lower jaw and body. Fins are large with the anal fin having three spines and 5 rays. Caudal fin is rounded with a ragged margin. Dorsal fin has 12 spines and 9 or 10 rays. Pectoral fin has 19–21 rays. There are

15 or 16 gill rakers. Bodies are covered with smooth scales. It grows to a maximum size of 35.0 cm (TL).

Food and feeding: It dwells on the bottom during daylight hours but become voracious predator during the night, feeding on small crabs, small fish, octopus, and shrimp.

Fisheries/aquarium/food/pharmaceutical values: The Peruvian scorpionfishes are retained and consumed by subsistence fishermen but are generally too small and difficult to handle, thus are mostly a "catch and release."

Scorpaena agassizii (Goode and Bean, 1896)

Photo courtesy: Keiichi Matsuura.

Common name(s): Longfin scorpionfish

Distribution: Western Atlantic: North Carolina, USA and northern Gulf of Mexico to northern South America.

Ecology: This demersal species is found on soft bottoms on the continental shelf and offshore waters between 46 and 300 m depth.

Biology

Identification: Head of this species is very bony with numerous spines. Adults have a very small depression on top of head behind eyes. Eyes are large (150%–220% of snout length). First (top most) preopercular spine is the largest and second is small. Bone over front of mouth is with 2 spines which are pointing down and forwards. Ridge under eye is with 3 spines. Roof of mouth with teeth on sides but not at center. There is no slit behind last gill arch. Dorsal fin has 12 spines and 10 rays; anal fin is with 3 spines and 5 rays; pectoral fin is with 19 rays, with some upper rays branched in adult. Scales are large and smooth. Lateral line is complete. Body coloration is bright red and darker to brownish spots are seen on front lateral line scales, on base of dorsal fin spines and on and above pectoral fin. Its maximum length reaches 25 cm (TL).

Food: It is a mobile benthic invertivore, feeding mainly on crabs, molluscs, and other benthic mobile invertebrates.

Fisheries/aquarium/food/pharmaceutical values: This species commonly occurs as bycatch in industrial shrimp and fin-fish trawl fisheries but is not usually marketed. In the Gulf of Mexico off Mexico, *Scorpaena* spp. are taken incidentally in fisheries targeting rays using gillnets and/or castnets along the coast.

Scorpaena albifimbria (Evermann and Marsh, 1900)

Photo credit: Wikipedia.

Common name(s): Coral scorpionfish

Distribution: Western Atlantic: southern Florida, USA and the Bahamas to Curaçao; northern South America.

Ecology: This shallow-dwelling species is found mostly inshore and associated with reef structure; inhabits small patch reefs; depth range 1–36 m.

Biology

Identification: It has a shallow occipital pit and 3 spines on suborbital ridge. Head is relatively large and snout is of moderate length. Eye is large. Teeth are seen on lower limb. Slit behind fourth gill arch is absent. Pectoral fin rays number 19–21 with the upper most are unbranched following 4–6 branched and remaining 12–15 are unbranched. Dorsal fin has 12 spines and 8–9 rays. Anal fin has 3 spines and 5 rays. Scales are cycloid and lateral line scales numbering 23 are pored. Gas bladder is absent. This species is mottled brown and red. Saddle of dark pigment is located above base of pectoral fin; distal part of pelvic fin is dark; and pectoral fin is dusky. This species changes its color with age and reaches a maximum size of 9.0 cm (SL).

Food: It feeds primarily on small crabs, molluscs, and other benthic mobile invertebrates.

Fisheries/aquarium/food/pharmaceutical values: Not reported.

Scorpaena angolensis (Norman, 1935)

Common name(s): Angola rockfish

Distribution: Eastern Atlantic: Mauritania to Angola, including Cape Verde.

Ecology: It is a demersal species found on sand or mud bottom of coastal waters; depth range 20–311 m.

Biology

Identification: It grows to a maximum length of 25.0 cm (TL). No other information is available on its biology.

Food: It feeds on small fishes and crustaceans.

Fisheries/aquarium/food/pharmaceutical values: It is not a targeted species but may be consumed on a small scale from bycatch. It is caught by fishing gears such as bottom trawls and bottom set longlines.

Scorpaena annobonae (Eschmeyer, 1969)

Image not available

Common name(s): Annobon scorpionfish

Distribution: Eastern Central Atlantic: Annobon Islan.

Ecology: It is a demersal species found on rough bottom; depth range 9–48 m. There is no further information concerning the habitat or ecology of this species except that it was collected from 9 to 47 m depth.

Biology

Identification: It grows to a maximum length of 4.4 cm (SL). No other information is available on the biology of this species.

Major threats: There is current oil drilling in the southeast Atlantic, primarily in the Gulf of Guinea, which may not be an immediate threat. There is also exploration for future oil mining closer to Annobon Island, which may also pose a threat to this species. However, there may be a threat from shipping.

Conservation action: A marine protected area (MPA) that covers 230 km^2 has been created around the Annobon Island to conserve this species.

Fisheries/aquarium/food/pharmaceutical values: Not known.

Scorpaena ascensionis (Eschmeyer, 1971)

Image not available

Common name(s): Ascension scorpionfish

Distribution: Southeast Atlantic: Known only from Ascension Island.

Ecology: It is a bottom-dwelling demersal species and is often found around the continental slope in areas that have rocky or soft bottoms; depth range 16–27 m.

Biology

Identification: It grows to a maximum length of 3.8 cm (SL). No other aspects of its biology have been reported.

Major threats: This species is likely threatened by continued degradation of habitat on Ascension Island, including military development, industrial development, and the recent development of tourism and fishing.

Conservation measure: The Govt. of Ascension has promulgated the Wildlife Protection (Amendment) Ordinance, 2016, which came into force on December 21, 2016 to prevent the catch of this species (http://www.ascension-island.gov.ac/wp-content/uploads/2017/03/ORD-8-Wildlife-Protection-Ord-Schedule-Amendment-2016-Asc.pdf).

Fisheries/aquarium/food/pharmaceutical values: There is no specific fishery for this species but it can be caught as by-catch in bottom trawls.

Scorpaena azorica (Eschmeyer, 1969)

Image not available

Common name(s): Azores scorpionfish

Distribution: Northeast Atlantic (European waters): known only from Azores.

Ecology: This fish is found in demersal and marine environments in subtropical waters, generally on hard bottoms. Its depth range remains poorly known but is mainly distributed on the shelf no deeper than 100 m depth.

Biology

Identification: Head of this species is broad with a short snout. There is a short tentacle just above the eye and several short spines on the head but no flaps of skin decorating the lower jaw though there are a few small ones on the body. Dorsal fin has 12 spines and 9 soft rays and the anal fin has 2 spines and 6 soft rays. Pectoral fins are large and oval and have 18 rays, of which rays numbering two to seven are branched. There are 44 vertical rows of scales on the body but the head, chest, and the base of the pectoral fins are naked. Color of this fish is generally whitish to pale brown, with darker brown spots and a band of darker pigmentation is seen between the dorsal soft rays and the anal fins. Fins are spotted with brown and dorsal spines lack a black spot and caudal fin is unbarred. The maximum recorded length of this species is 9.8 cm (SL).

Little is known of the behavior, diet, and reproduction of this species.

Major threats: There are growing problems around the Azores with marine sand extraction, agricultural run-off and discharge of untreated sewage from tourist resorts and towns. Though fishing limitations are in place, enforcement is lacking. Although the full impact of these threats on the marine environment is unknown, *Scorpaena azorica* is likely to suffer from significant habitat quality declines.

Conservation actions: There are no species-specific conservation measures in force for this species. However, its distribution may overlap with a number of MPAs around the Azores. Further research and monitoring of the habitat status and population size of this species is therefore needed.

Fisheries/aquarium/food/pharmaceutical values: It is unlikely that this species is of commercial importance, although like a large number of congeners, it may be consumed locally.

Scorpaena bergii (Evermann and Marsh, 1900)

Common name(s): Goosehead scorpionfish

Distribution: Western Atlantic: Bermuda, New York to Florida (USA), Bahamas, and Mexico to northern South America.

Ecology: This reef-associated species inhabits shallow clear waters (in turtle grass beds and reefs mainly off islands) from shore to 75 m depth.

Biology

Identification: This species is mottled dark brown with a well-developed occipital pit and a single spine at the end of suborbital stay. Head and snout are short and eyes are of moderate size. Teeth are small and arranged in narrow bands in jaws, palatine and vomer. Preorbital bone has 2 spinous points over maxilla. Preoperculum has 3 spines with first having supplement spine and second is larger than third and closer to first. Gill rakers (3–4) are present on upper limb and 7–10 gill rakers are seen in lower limb. Slit behind fourth gill arch is lacking. Pectoral fin has 16–17 rays with some branched. Dorsal fin has 12 spines and 9 rays; anal fin has 3 spines and 5 rays. Scales are cycloid and lateral line has 23 pored scales. Gas bladder is absent. Dorsal fin has a dark spot between third, fourth, or fifth and seventh or eighth fin spines. Posterior part of dorsal fin and caudal fin bear 3 dark bands. Pectoral fin is irregularly pigmented and anal fin has 3 transverse dark bands. Its maximum recorded length is 10 cm (TL).

Fisheries/aquarium/food/pharmaceutical values: This species is not utilized. In the Gulf of Mexico off Mexico, *Scorpaena* spp. are taken incidentally in fisheries targeting rays using gillnets and/or castnets along the coast.

Scorpaena brachyptera (Eschmeyer, 1965)

Photo courtesy: Smithsonian Tropical Research Institute.

Common name(s): Shortfin scorpionfish

Distribution: Western Atlantic: southern Florida in USA and Panama to Venezuela.

Ecology: This species occurs over rock-rubble bottoms, sometimes preferring areas rich in sponges; depth range 45–120 m.

Biology

Identification: Head of this species is very bony with numerous spines. There is no spine behind front nostril. A shallow pit is seen on top of head behind eyes. A pair of spines is present before this pit. Eyes are moderate. First (top most) preopercular bone is the largest and nasal bones are freely movable. There is no silt behind last gill arch. Roof of mouth is with teeth on sides and not on center. There is no slit behind last gill arch. Dorsal fin is with 12 spines and 9 rays; anal with 3 spines and 5 rays; and pectorals with 19–20 rays (some rays are branched) they are short and not reaching first anal fin spine. Scales are large and smooth. Lateral line is complete. Body is pale, reddish and head is mottled with dark and yellow. Iris is yellow. Inner face of pectoral is bright yellow. Anal is red, mottled and darkest at front. Other fins are red. Caudal fin is with 2 obscure bars. Its maximum length is 7.5 cm (TL) (http://biogeodb.stri.si.edu/caribbean/en/thefishes/species/3421).

Food: It feeds on small crustaceans, crabs and molluscs, and other invertebrates.

Reproduction: It takes place in inshore only.

Fisheries/aquarium/food/pharmaceutical values: Not known.

Scorpaena brasiliensis (Cuvier, 1829)

Photo courtesy: Teodoro Vaske Junior.

Common name(s): Barbfish

Distribution: Western Atlantic: Virginia, USA and northern Gulf of Mexico to Brazil.

Ecology: This reef-associated, demersal species is commonly found in bays, harbors, and on the continental shelf; inhabits soft bottoms; depth range 1–100 m. It is tolerant of low salinity.

Biology

Identification: Head and the snout of this species are moderately short. Occipital pit is well developed and a small to absent posttemporal spine. Preorbital bone has two spinous points over the maxilla. Suborbital ridge has three spinous points, of which the first two are blunt. Preoperculum has five spines—the first bearing a supplemental spine, the second equal or subequal to the third and the fourth and fifth spines moderate to small. The teeth of the barbfishes are small, arranged in narrow bands on the jaws, palatine, and vomer. Barbfishes are mottled red, orange, or blue in color with a paler underside. There is a large dark spot on the upper flank behind the operculum and many spots located on the pale axil of the pectoral fin. There are also two dark bands on the tail. Caudal peduncle is somewhat paler than the remainder of the body. Small brown spots are seen on the pectoral axil in adults. Juveniles are mottled in brown over a tan or pale background. It grows to a maximum length of 35.0 cm (TL).

Food and feeding: These venomous bottom-dwellers are excellent at camouflage, sneaking up on invertebrates or small fish, and sucking them down in one swift gulp. Barbfish feed by camouflaging to blend into its surroundings and lying in wait until an unsuspecting prey items comes within reach. The barbfish then quickly opens its large mouth, creating a vacuum and instantly sucking in the prey. Prey items include crabs, shrimps, and bony fish.

Reproduction: Scorpionfishes are oviparous with females producing eggs that are transparent.

Predators: Larger marine fishes are potential predators. They have the venomous barbs to contend with when attempting to prey on the barbfish.

Parasites: *Holobomolochus serratus*, a parasitic copepod, has been reported to occur in the gill chambers of the barbfish. The microbial parasitic dinoflagellate, *Amyloodinium ocellatum* has also been reported in this fish species (http://www.aquasymbio.fr/en/amyloodinium-ocellatum-scorpaena-brasiliensis).

Fisheries/aquarium/food/pharmaceutical values: This species supports a small artisanal fishery in the Caribbean. It is obtained by mainly using beach nets and as bycatch of shrimp trawl fisheries. Its flesh is of good quality but is usually not marketed commercially. This species is also often seen displayed in aquariums if there are plenty of hiding spaces and appropriate food items. It is found in the aquarium trade at Ceará, Brazil.

Scorpaena brevispina (Motomura and Senou, 2008)

Photo courtesy: Motomura, Hiroyuki.

Common name(s): Japanese shortspined scorpionfish
Distribution: Northwest Pacific: Japan.
Ecology: It is a bottom dweller (demersal) having a depth range 30–45 m.

Biology

Identification: This species has the following characters: pectoral is with fin rays 17; 44 longitudinal scale rows; 13 scales below lateral line; 6 scale rows between last dorsal-fin spine base and lateral line; 17 gill rakers; anteroventral surface of body and base of pectoral fin are covered with exposed scales; no tentacles on the anteroventral surface of lower jaw; anterior surface of preocular spine is with 3 vertical or slightly oblique ridges; lateral surface of maxilla has no distinct longitudinal ridge; occipital pit is extremely deep; dorsal view of occipital pit is rectangular; presence of many small papillae on surface of occipital pit; presence of a long pectoral fin with its posterior tip extending beyond verticals through last dorsal-fin spine base, posterior tip of depressed pelvic fin and second anal-fin spine base; and presence of short third to fifth dorsal-fin spines. It grows to a maximum length of 11.6 cm (SL).

Fisheries/aquarium/food/pharmaceutical values: Not known.

Scorpaena bulacephala (Motomura et al., 2005c)

Image not available

Common name(s): Bullhead scorpionfish

Distribution: West Pacific: off Norfolk Island and off Lord Howe Island, in the northern Tasman Sea; New Caledonia.

Ecology: This demersal species is found at depths of 86–150 m.

Biology

Identification: Body of this species is moderately compressed anteriorly and is progressively more compressed posteriorly. Nape and anterior body are arched. Body depth is relatively deep, slightly less than head length. Very few small papillae are seen on head (small papillae scattered on upper half of head, especially dorsal surface of head). A short, fleshy, slender tentacle is present on posterior end of supraocular spine base; this tentacle is extending beyond tip of tympanic spine. Supraocular tentacle is with several rounded branches along its lateral surface. A short, slender tentacle is also seen on posterior end of preocular spine base. Anterior lacrimal spine is associated with a short, slender tentacle. Posterior lacrimal spine is associated with a short, fleshy tentacle. A short tentacle is on posterior edge of low membranous tube associated with anterior nostril. Five rounded, thin skin flaps are seen along preopercular margin and each flap is occurring near tip of each preocular spine. A short, thin tentacle is on cheek. Pectoral-fin axil is without skin flaps. Body is strongly variegated, mainly brilliant red, suffused with irregular purplish, yellowish brown, and white blotches. Females with translucent unpaired fins, covered with reddish patches and spots. Edge of first dorsal-fin membrane is slightly darker reddish brown and remaining fin is with irregular reddish blotches. Second dorsal fin is pale edged with two broad, poorly defined reddish stripes. Caudal fin is reddish basally, alternating irregular pale and reddish areas distally over fin rays, membranes translucent. Anal fin is similar to dorsal fins, reddish areas most pronounced anteriorly. Pectoral fin is similar to side of body. Pelvic fin is weakly variegated. Pupil is black and iris is reddish with yellowish inner margin. Large, weak purplish oval spot are seen between dorsal-fin spines 7 and 10. This species can reach a maximum length of 8.8 cm (SL) (Motomura et al., 2005a–c).

Fisheries/aquarium/food/pharmaceutical values: Not known.

Scorpaena calcarata (Goode and Bean, 1882)

Common name(s): Smooth-head scorpionfish, red scorpionfish

Distribution: Western Atlantic: Canada to North Carolina; USA and northern Gulf of Mexico to Brazil.

Ecology: This species is mainly found on soft, muddy bottoms and is sometimes associated with coral reefs; depth range 1–216 m.

Biology

Identification: Diagnostic characters of this species include dorsal spine is with 12 spines and 9 soft rays; pectoral fin is with 18–21 rays (some branched); lacrimal bone is with 2 spines ventrally, both pointing forward; infraorbital bones 2 and 3 with 1 and 2 spines respectively; interorbit is narrow and is only 5%–6% SL; occipital pit is absent, except in smallest specimens; and 42–49 vertical scale rows. Body coloration is variable and is dark above and is, cream-colored below. A spot is seen behind head on or below lateral line. Body and the fins have no distinctive markings except for a dark shadow on the upper part of the pectoral fin. It reaches a maximum length of 13.0 cm (TL).

Food: It eats primarily crabs, molluscs and other benthic mobile invertebrates.

Fisheries/aquarium/food/pharmaceutical values: In the Gulf of Mexico off Mexico, *Scorpaena* spp. are taken incidentally in fisheries targeting rays using gillnets and/or castnets along the coast.

Scorpaena canariensis (Sauvage, 1878)

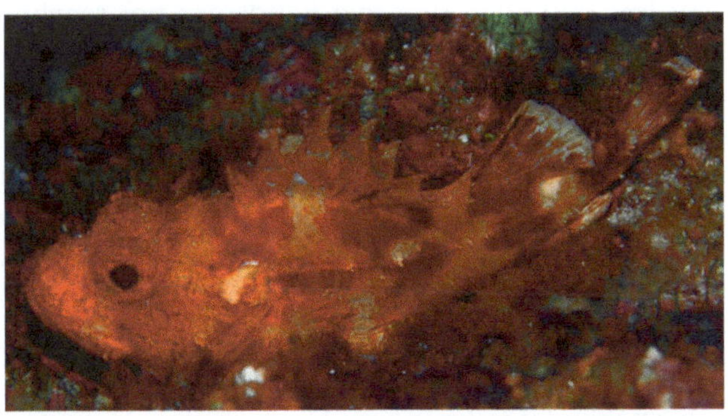

Photo courtesy: Wirtz, Peter.

Common name(s): Canary Scorpionfish, rockfish, goblinfish

Distribution: Eastern Atlantic: Canary Islands and Madeira Island.

Ecology: It is a littoral benthic (demersal) species, frequently found on rocky, sand and mu bottoms from 20 to 130 m depth. It is often found singly.

Biology

Identification: Body of this species IS moderately compressed anteriorly and is progressively more compressed posteriorly. Nape and anterior body ARE not highly arched and body is relatively deep. Head is large, and its length is greater than body depth. Eyes are moderately large and is oriented somewhat dorsolaterally. A short tentacle, slightly longer than nasal spine, is seen behind supraocular spine base. Tiny tentacles are seen along posterior margin of preopercle and central cheek. Mouth is moderately large, slightly oblique and is forming an angle of about 20° to horizontal axis of head and body. Posterior margin of maxilla is reaching level between posterior margins of pupil and orbit. Lateral surface of maxilla is smooth, without ridges. Upper-jaw lip is narrow anteriorly and is becoming broader posteriorly. Dorsal profile of snout is steep, forming an angle of about 40° to horizontal axis of head and body. Nasal spine is simple, conical, directed upward, and its length much greater than anterior nostril diameter. Anterior nostril is with a long tentacle; without distinct branches distally and its length greater than that of nasal spine. Origin of first dorsal-fin spine is above posterior end of lower posttemporal spine base. Posterior margin of opercular membrane is reaching level at origin of fourth dorsal-fin spine. Posterior tip of pectoral fin is not reaching level at origin of last dorsal-fin spine. Posterior tip of pelvic fin is extending slightly beyond anus when fin is depressed. Distal margin of membrane of dorsal-fin spines is not strongly notched. Origin of first anal-fin spine is slightly posterior to origin of last dorsal-fin spine. Head is yellowish white, with brown blotches and spots. Jaws, lips and underside of mandible are white. Body is yellowish white dorsally and is becoming white ventrally. There are 4 vertical (or slightly oblique) irregular gray bands (or elliptical blotches) are seen dorsally; first band is below first to third dorsal-spine bases; second is below middle of spinous portion of dorsal fin, third below soft-rayed portion, and fourth is on caudal peduncle. These bands or blotches are extending below lateral line, although first two bands are not reaching ventral surface of body. Spinous portion of dorsal fin is transparent with 2–6 indistinct gray blotches. Soft-rayed portion of dorsal fin is transparent with 2 distinct gray blotches. Of these blotches, one (larger than orbit diameter) is between first and fifth soft rays, and the other is smaller than pupil diameter and is located between seventh and ninth soft rays. Pectoral fin is white without melanophores. There are no distinct marks on inner surface or axil of pectoral fin. Pelvic and anal fins are white and are mottled with indistinct gray blotches. Caudal fin is white with a large black blotch centrally and several small white spots are on the blotch. It reaches a maximum length of 14 cm (TL) (Motomura, et al., 2005a–c).

Food: It feeds on small fish and swimming invertebrates. Like other scorpionfish, it is also a master of camouflage, enabling it to lie in wait for its victims to come close, before lunging forward and inhaling its prey with its large mouth.

Fisheries/aquarium/food/pharmaceutical values: The species is subject to fishing for local consumption.

Scorpaena cardinalis (Solander and Richardson, 1842)

Photo credit: John E. Randall.

Common name(s): Cook's scorpionfish, Cook's rockcod, Kermadec scorpionfish, sandy-bay cod, Eastern red scorpionfish

Distribution: It is common on NSW coastal reefs; Southwest Pacific: northern New Zealand and offshore islands of the Tasman Sea (Norfolk and Lord Howe Islands, Middleton and Elizabeth Reefs, New South Wales, Australia).

Ecology: It inhabits coral reef, rocky reef, soft sediment, crevices, and caverns at all depths of 0–154 m; occasionally found in large rock pools in the lower littoral zone.

Biology

Identification: This species is distinguished by the following set of characters: dorsal fin rays 9; pectoral-fin rays 16–18; scale rows in longitudinal series 60–74; pored lateral-line scales 23–25; scales above lateral line 6–9, below 21–26; scale rows between sixth dorsal-fin spine base and lateral line 8 or 9; scale rows between last dorsal-fin spine base and lateral line 8–11; predorsal scale rows 4–8; total gill rakers 14–18; exposed cycloid scales covering pectoral-fin base and anteroventral surface of body; no longitudinal ridge on lateral surface of maxilla; presence of median interorbital ridge; presence of occipital pit and supplement preopercular spine; space between upper and lower opercular spines covered by thin skin without sensory pores or canals; a few skin flaps or tentacles on anterodorsal surface of body; absence of a large black blotch on spinous portion of dorsal fin in both sexes; and 2 large white blotches, on caudal-fin base. The coloration of this species is highly variable from light gray/brown to bright red with blotches or may also be a mottling of all these colors and blotches. They have skin flaps on the head, lateral line, and other parts of the body, 12 venomous spines on the dorsal fin and a very large mouth. It can reach a maximum SL of 47 cm.

Age and growth: In this species, growth was slow, with fish attaining, 22 cm after 5 years and 26 cm after 10 years. The oldest fish sampled was estimated to be 33 years old. In contrast to similar-sized congeneric species, males of this species increased in length significantly faster and attained larger sizes than females. Males of *Scorpaena cardinalis* had mean gonadosomatic index (GSI) levels (0.09%) an order of magnitude smaller than did the females (2.8%–4%) during the March spawning peak. Further, this species possesses a cystovarian Type II-3 ovary, a highly specialized form associated with external fertilization and the production of eggs in a floating gelatinous mass. Traits such as slow growth, substantial longevity, variable recruitment patterns and a highly specialized reproductive strategy may make this species vulnerable to overexploitation (Stewart and Hughes, 2010).

Food and feeding: It is nocturnally active and feeds mainly on small mobile benthic animals; and on fish but also crabs, shrimps, and octopus. It stalks its prey and swallows it whole.

Fisheries/aquarium/food/pharmaceutical values: This species is commonly exploited off New Zealand. It is good for eating and is seen in some fisheries. It is often taken by divers and fishermen over rocky reefs.

Scorpaena cocosensis (Motomura, 2004)

Photo courtesy: Hiroyuki Motomura.

Common name(s): Cocos scorpionfish

Distribution: Eastern Pacific: Cocos Island and Galápagos Arch.

Ecology: It is a demersal species found on coralline rubble bottoms; depth range 57–92 m.

Biology

Identification: This species has the following combination of characters: 8 dorsal-fin soft rays; 22 pectoral-fin rays; well-exposed scales covering anteroventral body surface; well-developed interorbital ridges; lateral margins of frontals diverging posteriorly in dorsal view; shallow occipital pit with distinct transverse ridge in front of pit; posterior lacrimal spine directed ventroanteriorly; suborbital ridge is with a small spine; absence of supplemental preopercular spine; reduced second preopercular spine; upper posttemporal spine directed upward; posterior margin of maxilla is just reaching level with posterior margin of pupil; posterior tip of pectoral fin is reaching level with origin of third dorsal-fin soft ray; 48.8% (SL) head length large; large orbit diameter (16.9% SL); and wide interorbital space between supraocular spine bases (9.2% SL). Body and head are red and are mottled with whitish blotches. Dark red spots are seen on operculum. Pectoral fin is red with blackish tip. There are 2 oblique dark red bars across the soft dorsal fin and tail is with 2 dark red bars. It has a maximum length of 6.2 cm (SL) (Motomura, 2004).

Fisheries/aquarium/food/pharmaceutical values: Not reported.

Scorpaena colorata (Gilbert, 1905)

Courtesy image: John E. Randall.

Common name(s): Marbled rosefish

Distribution: Eastern Central Pacific: Hawaiian Islands.

Ecology: It is a demersal, benthic species found on seamounts and knolat depths of 95–219 m.

Biology

Identification: This species has the following diagnostic features. Pectoral fin rays usually 17; absence of coronal spines; shallow to moderate occipital pit; 45 vertical scale rows; lachrymal bone is usually with 3 spines; suborbital ridge is usually with 3 spines, first under eye, second under rear of eye, and third at end of ridge; a dark blotch is often seen on distal part of spinous dorsal between spines 7–9. This species can reach a maximum length of 9.5 cm (SL).

Fisheries/aquarium/food/pharmaceutical values: This species is not generally utilized. It is sometimes caught as bycatch in shrimp trawls.

Scorpaena dispar (Longley and Hildebrand, 1940)

Photo courtesy: Keiichi Matsuura; Teodoro Vaske Junior.

Common name(s): Hunchback scorpionfish

Distribution: Western Atlantic: Florida and northern Gulf of Mexico in USA to Brazil.

Ecology: This species inhabits shallow to deep rocky reefs; depth range 36–125 m.

Biology

Identification: Head of this species is very bony with numerous spines. Adults have a well-developed pit on top of head behind eyes. Eyes are moderate. First (top most) preopercular spine is the largest and is short with a small accessory spine. A slender cirrus is present over eye. The spines (3–4) present in front of mouth are pointing down. Ridge under eye is with 2–4 spines. Roof of mouth is with teeth on sides but teeth are absent at front. There is no slit behind last gill arch. Dorsal fin has 12 spines and 9 rays; anal fin has 3 spines and 5 rays; and pectorals have 17–19 rays and some of them are branched. Scales are large and smooth. Lateral line is incomplete. Body is bright red to orange. There may be 1–2 dark blotches on body behind head (http://biogeodb.stri.si.edu/caribbean/en/thefishes/species/3424). Its maximum size is 23.0 cm (TL).

Food: Its diet is believed to be primarily crabs, molluscs and other benthic mobile invertebrates.

Fisheries/aquarium/food/pharmaceutical values: In the Gulf of Mexico off Mexico, *Scorpaena* spp. are taken incidentally in fisheries targeting rays using gillnets and/or castnets along the coast.

Scorpaena elachys (Eschmeyer, 1965)

Common name(s): Dwarf scorpionfish

Distribution: Western Atlantic: Florida in USA and Antilles.

Ecology: This species is associated with deeper rocky reef; depth range 36–100 m.

Biology

Identification: Body of this species is deep (depth, 38.4% of SL). Head is very bony with numerous spines. Adults have a shallow pit on top of head behind eyes. Snout is very short. Eyes are moderate. First (top most) preopercular spine is the largest and is short with a small accessory spine. The spines (2) present in front of mouth are pointing down; and another at rear on its side. Bone ridge under eye is smooth except for a spine at the rear end. Roof of mouth is with teeth on sides but teeth are absent at center. There is no slit behind last gill arch. Dorsal fin has 12 spines and 9 rays; anal fin has 3 spines and 5 rays; and pectorals have 16–17 rays and some of them are branched. Scales are large and smooth. Lateral line is complete. Body is bright red. A network of dark and pale lines is seen under dorsal fin. It is dusky under and between eyes. Inside gill chamber is black. Pectoral fin is with yellow on inner face and it is dusky on sides. From half of anal fin, it is red, middle is speckled and dark red behind. Its maximum length is 6.4 cm (TL).

Fisheries/aquarium/food/pharmaceutical values: Not reported.

Scorpaena elongata (Cadenat, 1943)

Photo courtesy: A.M. Arias.

Common name(s): Slender rockfish

Distribution: Eastern Atlantic: Mediterranean Sea and Morocco to off northern Namibia.

Ecology: This offshore continental species is demersal and sedentary occurring in rocky areas and on soft substrata; depth range 75–800 m.

Biology

Identification: Head of this species is large and its snout is about equal to orbit diameter (11%–12% of SL). Its preorbital bone is usually with 2 spines over maxilla. Suborbital ridge has 3 or 4 spinous points. Occipital pit is fairly shallow. Pores at symphysis of lower jaw are small and separate. Supraocular tentacle is small. A large flap is associated with posterior preorbital spine and another large flap is at fourth or fifth preopercular spines. Body skin appendages are fairly well developed. Dorsal fin is with 12 spines and 9 rays; and pectoral fin is with 18–20 rays and not reaching over the first anal spine. Scales on sides are ctenoid and there are 45–50 vertical rows of scales on the body. Most of head is naked without scales. Pectoral fin base and chest are naked. There are 13–16 gill rakers. General coloration of body is pink-yellow and most of the head is brownish. Pigment on body is usually concentrated in several large brownish blotches. Posterior part of spinous dorsal fin may have brown pigment but not in a conspicuous spot. It grows to a maximum length of 50.0 cm (TL).

Food: It feeds on fishes, shrimps, and other benthic invertebrates.

Reproduction: Its spawning season is August and September.

Life history traits: Trawl surveys carried out on this species between 1985 and 1998 in the Strait of Sicily (Mediterranean Sea) showed that the females of this species dominated (52%–69%), but the two sexes showed the same length range (9–57 cm). The length at maturity was found to be optimal The above study also indicated that *Scorpaena elongata* of the Strait of Sicily is a slow growing and long living (up to 20–30 years) species (Ragonese et al., 2002).

Parasite: A new species of parasitic copepod, *Prohatschekia mediterranea* n. sp. (Siphonostomatoida: Hatschekiidae), has been described from this species collected off Algeria (Hamza et al., 2007).

Fisheries/aquarium/food/pharmaceutical values: This species is not likely extensively utilized in west Africa but is a species with minor commercial interest in the Mediterranean.

Scorpaena fernandeziana (Steindachner, 1875)

Image not available

Common name(s): Not designated

Distribution: Southeast Pacific: Desventuradas Island and Juan Fernández Island.

Ecology: It is a demersal species.

Biology

Identification: Nothing is known about its biology

Fisheries/aquarium/food/pharmaceutical values: Not reported.

Scorpaena gasta (Motomura et al., 2006)

Photo courtesy: H. Motomura.

Common name(s): Ghostly scorpionfish

Distribution: Eastern Indian Ocean: Western Australia.

Ecology: It is a reef-associated, demersal species; depth range 3–17 m.

Biology

Identification: This species has the following set of characters: 15 pectoral-fin rays; 36–40 longitudinal scale rows; 13 or 14 scales below lateral line; 5 scale rows between last dorsal-fin spine base and lateral line; 0–2 exposed predorsal scale rows; 13 gill rakers; anteroventral surface of body and pectoral-fin base are covered with well-exposed scales; absence of median ridge at the anterior surface of preocular spine; a longitudinal ridge on the lateral surface of maxilla; absence of spines on lateral surface of lacrimal; presence of a simple anterior lacrimal spine; deep occipital pit; presence of a thick skin with numerous small sensory pores covering predorsal area from posterior edge of occipital pit to first dorsal-fin spine origin; similar skin with pores between opercular spines and covering upper spine; several tentacles on head and trunk; 1 or 2 pairs of slender tentacles in anteroventral surface of lower jaw; presence of short second, third, and fourth dorsal-fin spines; shallow caudal-peduncle depth; diagonal cluster of black spots on soft-rayed portion of dorsal fin; and uniform yellowish to reddish spinous portion of dorsal fin with narrow, translucent, submarginal stripe. It grows to a maximum length of 8.4 cm (SL) (Motomura et al., 2006).

Fisheries/aquarium/food/pharmaceutical values: Not reported.

Scorpaena grandicornis (Cuvier, 1829)

Photo courtesy: Patrick O'Donnell.

Common name(s): Plumed scorpionfish

Distribution: It is widespread in the West Indies, found from Florida to Bahia, Brazil, including Bermuda, the Bahamas, Cuba, Haiti, the Dominican Republic, Martinique, Jamaica, Antigua, St. Lucia, Guadeloupe, Curaçao, St. Croix, Honduras, Panama, Colombia, and Venezuela.

Ecology: This reef-associated species inhabits shallow coastal waters over soft bottom and seagrass (Thalassia) beds in bays and channels. It tends to remain stationary for periods of time, moving only when disturbed. It hosts algal and diatom growth on its body in order to enhance its camouflaging capability; depth range 1–15 m.

Biology

Identification: Diagnostic characters of this species include dorsal fin is with 12 spines and 9 soft rays; and pectoral fin is with 18 or 19 rays, some of which are branched. Lacrimal bone is 2 spines on ventral margin of bone. Suborbital ridge is formed by infraorbial bones 2 and 3 with 2 or 3 spines. First preopercular spine is short, extending less than 1/2 distance to posterior margin of operculum. Occipital pit is well developed. Scales are cycloid (smooth to touch), in 42–47 vertical rows. Coloration of body and fins is dark brown; caudal fin is with 3 dark bars, 1 at base, 1 near middle, and 1 at distal margin; upper part of pectoral axil is with small white specks each surrounded by darker pigment; white spots are often extending onto medial surface of fin and on body adjacent to pectoral-fin axil. Coloration of body is dark brown. Tail fin is with 3 brown bars (base, middle and end). Upper part of inside of pectoral base is brown with small white dots and each of which is surrounded by dark pigment. White spots are extending onto fins and side of body next to fin. Lower head and chest are white. It grows to a maximum length of 30.0 cm (TL).

Fisheries/aquarium/food/pharmaceutical values: In the Gulf of Mexico off Mexico, *Scorpaena* spp. are taken incidentally in fisheries targeting rays using gillnets and/or castnets along the coast.

Scorpaena grandisquamis (Ogilby, 1910)

Common name(s): Bigscale scorpionfish

Distribution: Western Pacific: endemic to Australia.

Ecology: It is a reef-associated inshore and benthic species.

Biology

Identification: Nothing is known about its biology

Fisheries/aquarium/food/pharmaceutical values: Not known.

Scorpaena grattanica (Trunov, 2006)

Image not available

Common name(s): Not designated

Distribution: Southeast Atlantic: Grattan Bank, Ascension Island.

Ecology: It is a pelagic-neritic species; depth range 120–170 m.

Biology

Identification: This species is distinguished by the following characters: dorsal fin is with 12 spines and 12 soft rays; distinct occipital pit; cycloid scales; tubular scales in lateral line, 23 + 1; vertical rows of scales on body 45 + 4; monotonous yellowish, spotless coloration on the inside of the pectoral fins and their insertions. It grows to a maximum length of 22.5 cm (SL).

Fisheries/aquarium/food/pharmaceutical values: Not known.

Scorpaena guttata (Girard, 1854)

Photo courtesy: Milton Love.

Common name(s): California scorpionfish, spotted scorpionfish, scorpion, rattlesnake, bullhead, scorpene, sculpin

Distribution: Eastern Pacific: Santa Cruz in central California, USA to Punta Abreojos, Baja California; including northern Gulf of California and Guadalupe Island in Mexico.

Ecology: This demersal species is inhabiting rocky reefs (often lodged in caves and crevices), kelp beds, eelgrass beds and sandy or muddy substrata of bays and along shallow waters; depth range 1–183 m. It spends most of its time on the ocean floor. It emerges at night and does not stay in one territory and individuals travel up to 200 km at a speed of 2.2 km/day.

Biology

Identification: This species has a stocky and slightly compressed body, with a large (spiny) head and mouth. Nape is concave. There are deep depressions before and behind eyes. Ridge below eye is normally without spines, but sometimes with 2–3 spines. Teeth are on sides and front roof of mouth. There are nosskin flaps on head or body. There are 15–19 gill rakers. Dorsal fin is with 12 spines and 9 rays; pectoral fin with 17–19 rays, of which upper rays are branched and lower rays are thickened and unbranched; and anal is with 3 spines and 5 rays. There are 73–86 vertical rows of smooth scales on body and 24–26 scales on lateral line. Coloration of body, head and fins is red brown. Body is densely covered with dark brown and black spots. Fins are with large dark spots, those on dorsal fin are in 2–3 rows. This fish reaches up to 47 cm long (TL). Its maximum recorded life span is 21 years for females and 15 for males.

Food: This carnivorous fish feeds on mobile benthic crustaceans (crabs, primarily juvenile *Cancer anthonyi*), octopuses, squid, cuttlefish, and bony fishes. Juvenile crabs are strongly favored, especially during fall and winter, when they make up the bulk of the diet. The juvenile yellow rock crab (*Metacarcinus anthonyi*) is a very common prey. Another common crustacean prey item is the ridgeback prawn (*Sicyonia ingentis*). Fish species consumed include Californian anchovy (*Engraulis mordax*) and spotted cusk-eel (*Chilara taylori*).

Life history/reproduction: In this species, females grow faster than males. Both females and males mature at 2 years of age. Males tend to mature at a slightly smaller size. It is oviparous and the eggs are implanted in a single layer on the gelatinous walls of hollow, pear-shaped "balloons" of 5–10 in. in length. The mass floats on the ocean surface. Initially colorless and transparent, the mass eventually turns cloudy white as it floats. Each egg is just over a millimeter long. The egg mass is released on the bottom and rise to the surface, and the eggs hatch within the next 5 days. California scorpionfish start spawning at age 3 or 4. Spawning likely takes place around dawn; the fish has been observed congregating daily at this time at the surface of the water during the breeding season. Mating is polygamous. Spawning occurs mostly at night from May through August, with a peak in July. This species forms large offshore spawning aggregations in waters deeper than its off-season habitat. This fish has been reported to return to the same spawning area annually.

Parasites: This fish species is host to a number of parasites, such as the copepods *Bomolochus spinulus, Lepeophtheirus rotundipes, Naobranchia scorpaenae, Pseudodiocus scorpaenus, Hamaticolax spinulus*, and *Chondracanthus gracilis*. A new species of *trematode Trochopus, Trochopus sprostoni* (Monogenea: Capsalidae) has been isolated from the gills of *Scorpaena guttata* in Santa Catalina Channel, California (Arai et al., 1964).

Predators: Predators of this fish include the California two-spot octopus (*Octopus bimaculatus*), which is apparently rarely injured by the spines because its flesh is so pliable.

Conservation: Sport fishing puts some pressure on this species. There are no major threats. The populations off of Southern California apparently undergo short-term fluctuations in size, and it is considered to be a rare species overall along the California coast. In Mexico, there are a smaller number of MPAs that coincide with this species' range. Given its very restricted range, and preference for shallow water areas, more research is needed to determine this species' population status.

Fisheries/aquarium/food/pharmaceutical values: California scorpionfish comprise a minor part of the California sport and commercial fisheries. Within the Southern California Bight, this species is a minor constituent of the partyboat sportfish catch and it ranked 15th in abundance, comprising about 1.5% of all fishes taken. It is captured by fishing gears like hook and line, gill net, and, rarely, otter trawl (Love et al., http://fishbull.noaa.gov/851/love.pdf). It is considered to be a very good food fish once the spines have been avoided. There is a small commercial fishery, as well. The catch is sold to fish processors and to the public; it is sought after in some Asian communities in Southern California. The fish is caught by angling, gillnetting, and otter trawling. The extent of the impact of fishing on populations is unknown.

Scorpaena hatizyoensis (Matsubara, 1943)

Image not available

Common name(s): Not reported

Distribution: Pacific Northwest, including Taiwan, China, Japan.

Ecology: It is a demersal species inhabiting deep water near the rocky reef or sand; its depth range remains unknown.

Biology

Identification: Body of this species is extended with flat side. Head is large, spines with obvious jagged. Eyes are large and are slightly out of the head. Supraorbital bone flap is either small or missing. Mouth is large, and oblique; upper and lower jaws are long; and jaw has no serrated edge. Lacrimal bone is large and square, with several small spines on the outside and with a short protrusion on the upper edge. Orbital is with 5 orbital bone. Gill cover is with a flat spine. Pectoral fins and pelvic fins are usually reddish or reddish brown, with brown or blue round spots; and dorsal fins, anal fins and caudal fins are all pale. The maximum size of this species are unknown.

Food: It feeds on small fish and crustaceans.

Major threats: *Scorpaena hatizyoensis* may be threatened by pollution and intensive fishing, which occur in parts of its range. This species is also taken as by-catch in some areas of its range. However, at present, these threats are not thought to be causing any significant population declines.

Fisheries/aquarium/food/pharmaceutical values: Not reported.

Scorpaena hemilepidota (Fowler, 1938)

Image not available

Common name(s): Not reported

Distribution: Western Central Pacific: Philippines.

Ecology: It is a demersal species inhabiting depths up to 247 m.

Biology

Identification: Nothing is known about its biology.

Fisheries/aquarium/food/pharmaceutical values: Not reported.

Scorpaena histrio (Jenyns, 1840)

Photo courtesy: Allen, Gerald R., Western Australian Museum.

Common name(s): Player scorpionfish, bandfin scorpionfish

Distribution: Eastern Pacific: Mazatlán, Sonora; Mexico to Chile; Cabo San Lucas, Mexico and Galapagos Islands.

Ecology: This is a demersal species that inhabits crevices and ledges of weed-covered rocky and coral reefs; depth range 5–200 m.

Biology

Identification: This species is diagnosed by its dorsal fin which has 12 spines and 9 soft rays; and anal fin is with 3 spines and 5 soft rays. Body is less broad than deep. Head of this species is very bony with numerous spines. A shallow pit is seen behind eyes. Longitudinal ridge under eye is without spine. There are 5 well-developed preopercular spines and of them the top spine is the longest. Roof of mouth is with teeth on sides. There is no slit behind last gill arch. There are several skin flaps on head and body. Dorsal fin has 12 spines and 10 rays;

anal fin is with 3 spines and 5 rays; pectoral fin is with 19–20 rays and lower rays are unbranched and thickened. Scales on body are large and smooth. Lateral line is complete with 24–26 scales. Body is with 43–54 vertical rows of scales. Coloration of body is highly variable and is strongly mottled brown to reddish. A blackish to dark brown blotch is seen on lateral line above middle of pectoral fin. There are red or orange skin flaps within the blotch. It grows to a maximum length of 27.3 cm (TL).

Food: It feeds on mobile benthic crustaceans, octopuses, cuttlefish, squid, and bony fishes.

Fisheries/aquarium/food/pharmaceutical values: Not reported.

Scorpaena inermis (Cuvier, 1829)

Photo courtesy: Smithsonian Institution, National Museum of Natural History.

Common name(s): Mushroom scorpionfish

Distribution: Western Atlantic: Florida (USA), Bahamas, and Yucatan (Mexico) to Curaçao and northern South America.

Ecology: This reef-associated species inhabits primarily offshore islands; also found in clear water over sand and turtle grass; depth range 1–73 m.

Biology

Identification: Head of this species is very bony with numerous spines. There is no pit behind eyes. Eyes are moderate with small mushroom-shaped cirri on its upper one-fourth. First (top most) preopercular spine is the largest and short with a small accessory spine. Bone over front of mouth is with 2 spines which are pointing down. Bone ridge under eye is with 3 spines. Roof of mouth with teeth on sides but not at center. There is no slit behind last gill arch. Dorsal fin has 12 spines and 7 rays; anal fin is with 3 spines and 5 rays; pectoral fin is with 19–21 rays with some rays are branched. Body coloration is mottled red. Tail fin is with 2 indistinct bars (at middle and end). There are no color marks on inside if pectoral base. There is often a brown spot behind head under lateral line. Outer edge of pectoral, soft dorsal and anal fin are dark. Its maximum length reaches 11 cm (SL).

Food: Its diet mainly consists of crabs, molluscs and other benthic mobile invertebrates.

Fisheries/aquarium/food/pharmaceutical values: In the Gulf of Mexico off Mexico, *Scorpaena* spp. are taken incidentally in fisheries targeting rays using gillnets and/or castnets along the coast. This species is infrequently caught in continental waters.

Scorpaena isthmensis (Meek and Hildebrand, 1928)

Photo courtesy: Teodoro Vaske Junior.

Common name(s): Smooth-cheek scorpionfish

Distribution: Western Atlantic: South Carolina, USA and northern Gulf of Mexico to Brazil.

Ecology: This demersal species inhabits sandy, muddy, and rocky bottoms and occurs from near shore to 110 m depth.

Biology

Identification: Head of this species is very bony with numerous spines. There is a well-developed pit on top of head behind eyes. Eyes are moderate with a large branched cirrus above it at rear. First (top most) preopercular spine is the largest and short with a small accessory spine. Bone over front of mouth is with 2 spines which are pointing down. Bone ridge under eye is prominent but without spines. Roof of mouth with teeth on sides but not at front. There is no slit behind last gill arch. Dorsal fin has 12 spines and 9 rays; anal fin is with 3 spines and 5 rays; third spine of anal is longer than second; pectoral fin is with 18–19 rays with some upper rays are branched in adult. Scales are relatively large and smooth. Lateral line is complete. Body coloration is red to brown to tan with greenish mottling. Head is brown, and dark markings are seen at edge of operculum. Iris is yellow. A prominent dark green blotch is seen on the fin membrane between dorsal spines 3–7. Outer half of pelvic fin is dusky and inner half is pink. Tail fin is with 3 dusky bars i.e., at base, center and before end. Its maximum length reaches 16.0 cm (TL) (http://biogeodb.stri.si.edu/caribbean/en/thefishes/species/3428).

Fisheries/aquarium/food/pharmaceutical values: In the Gulf of Mexico, Mexico, *Scorpaena* sp. are taken incidentally in fisheries targeting rays using gillnets and/or castnets along the coast.

Scorpaena izensis (Jordan and Starks, 1904)

Photo courtesy: CSIRO.

Common name(s): Izu scorpionfish

Distribution: Indo-West Pacific.

Ecology: It is a demersal species inhabiting sandy mud bottom of continental shelf; depth range 80–200 m.

Biology

Identification: It has a maximum length of 44.5 cm (TL) and maximum weight of 1.5 kg. No other information is available about its biology.

Fisheries/aquarium/food/pharmaceutical values: Not reported.

Scorpaena jacksoniensis (Steindachner, 1866) (=*Ruboralga jacksoniensis*)

Ian Shaw, Solitary Islands Underwater Research Group (SURG).

Common name(s): Eastern red scorpionfish, cardinal scorpionfish, coral cod, coral perch, fire cod, northern scorpionfish, ocean perch, prickly heat, red rock cod, red scorpion-cod

Distribution: Western Pacific: Eastern Australia, from southern Queensland to Victoria.

Ecology: It is found in shallow estuaries to deep offshore reef; benthic species camouflaged amongst rocks, sponges, corals and algae; depth range 1–73 m.

Biology

Identification: This species is distinguished by the following set of characters: Dorsal fin is with 9 spines (fourth is the longest); pectoral-fin rays 16–18; scale rows in longitudinal series 52–61; pored lateral-line scales 22–24; scales above lateral line 4–6, below 20–25; gill rakers on upper limb 4–6, lower limb 10–13; total rakers 14–19; embedded cycloid scales covering pectoral-fin base (some scales exposed); embedded cycloid scales on anteroventral surface of body; lateral surface of maxilla is without a longitudinal ridge; median interorbital ridge is present; presence of occipital pit and supplement preopercular spine; space between upper and lower opercular spines is covered by thick skin with sensory pores and canals; numerous skin flaps and tentacles on anterodorsal surface of body; and presence of a large black blotch distally on posterior spinous portion of dorsal fin in males. This common species is variable in color, ranging from bright red to a pale grayish-brown, with darker and lighter blotches and mottling. Juvenile may have a white to purplish blotch across the back of its head. It grows to a maximum length of 40.0 cm (SL).

Food and feeding: It is a predatory fish with a big mouth. It often lies motionless on the bottom waiting for passing prey of small fishes and invertebrates. The species swallows its prey whole, so within reason, the bigger the mouth, the larger the prey item that can be consumed.

Fisheries/aquarium/food/pharmaceutical values: Commercial, recreational fish.

Scorpaena lacrimata (Randall and Greenfield, 2004)

Photo courtesy: John E. Randall.

Common name(s): Not reported

Distribution: Eastern Central Pacific: Tahiti.

Ecology: It is a demersal species found in depths up to 400 m.

Biology

Identification: This species has the following set of characters: Dorsal fin is with 12 spines (third and fourth are the longest) and 9 rays; anal fin is with 3 spines and 5 rays; pectoral rays 17; longitudinal scale series 61; body depth is 2.5 in SL; head length (HL) 2.15 in SL; snout length 3.2 in HL; occipital pit is quadrangular and moderately deep; no coronal spines; no suborbital pit; suborbital ridge is with 4 spines; supraoccipital tentacle is present; small tentacles are associated with ventral spines of lacrimal; no tentacles or cutaneous flaps on chin or on body; and one row of small teeth on palatines. Body coloration is pale yellowish (in alcohol) with scattered small dark brown spots on head, body, and dorsal fin; and a short, narrow, dark brown bar is extending ventrally from middle of edge of orbit. It grows to a maximum length of 19.8 cm (SL).

Fisheries/aquarium/food/pharmaceutical values: Not reported.

Scorpaena laevis (Troschel, 1866)

Photo credit: Theo Modder.

Common name(s): Senegalese rockfish

Distribution: Eastern Atlantic: Mauritania to Pointe Noire, Congo, Azores, and Cape Verde.

Ecology: It is a demersal species inhabiting rocky areas in shallow waters; depth range 1–100 m.

Biology

Identification: Head of this species is large; snout is as long as orbit diameter, about 10% of SL; preorbital bone is with 2 spines over maxilla in juveniles and 3 in adults; upper posttemporal spine is present; occipital pit is deep; pores at symphysis of lower jaw are separate. Supraocular tentacle is variable (sometimes, it is longer than orbit diameter); and other tentacles are also variable but usually well developed, especially under the head and on the lower jaw. Dorsal fin is with 12 spines and 9 rays; and pectoral fin is with 18–20 rays. All scales are cycloid; 40–45 vertical rows of scales in body; and pectoral fin base, cheek, and chest are scaled; and total gill rakers, 13–15. Body coloration is variable; in juveniles the axil of the pectoral fin is dark brown to black with some white blotches; adults have large brown spots on the inside of the pectoral fin. It grows to a maximum length of 35.0 cm (SL).

Food: It has been reported to feed on octopus, and presumably other invertebrates and small fishes.

Fisheries/aquarium/food/pharmaceutical values: The fish is edible and is generally caught with hook.

Scorpaena loppei (Cadenat, 1943)

Photo credit: Xisco Ordinas.

Common name(s): Cadenat's rockfish

Distribution: Eastern Atlantic: Morocco, Mauritania, Portugal, Atlantic coast of Spain, Cyprus and the Mediterranean.

Ecology: This is a demersal species inhabiting muddy sand and gravel bottoms; depth range 50–200 m.

Biology

Identification: Head of this species is large; maxilla is with a characteristic longitudinal crest; snout is shorter than orbit diameter; preorbital bone is with 2 spines over maxilla, first is very blunt and second is long and curved to rear; suborbital ridge is with 4 spinous points; upper posttemporal spine is present; occipital pit is present; pores at symphysis of lower jaw are united; tentacles on head and body are moderate or poorly developed; and supraocular tentacle is shorter than orbit diameter. Dorsal fin is with 12 spines and 9 rays; and pectoral fin is with 18 rays. Scales on body are weakly ctenoid; 35–40 vertical rows of scales; and chest, pectoral fin base and head are naked without scales. Total gill rakers are 14–16. General body color is pink with many black spots on body; a dark spot on spinous dorsal fin between spines 6 and 9; and fins are with dusky bands. It grows to a maximum length of 15.0 cm (TL).

Reproduction: Most females of this species reach early sexual maturity in their first year of life which is thought to be facilitated by reduced interspecific competition as this species is the only small scorpion fish inhabiting the benthic habitat of the Balearic Islands (Ordines et al., 2012).

Food: *Scorpaena loppei* has a specialized feeding strategy with a diet primarily comprised of crustaceans (mysids and decapods) but also includes teleost fishes.

Fisheries/aquarium/food/pharmaceutical values: Not reported.

Scorpaena maderensis (Valenciennes, 1833)

Photo credit: Robert Patzner; Wikipedia.

Common name(s): Madeira rockfish

Distribution: Eastern Atlantic: Azores, Madeira, and Morocco to the Canary Islands, Cape Verde and Senegal; Mediterranean Sea.

Ecology: This marine and brackish water, demersal species inhabits shallow coastal waters particularly in rocky areas and favors habitats (including tide pools) with generous brown algal growth, rocky depressions and ledges and small boulders; depth range 20–40 m.

Biology

Identification: Head of this species is large; snout is as long as orbit diameter, 10% of SL; preorbital bone is with 2 large spines over maxilla; suborbitals are with 2 spinous points; upper posttemporal spine is present; first preopercular spine is short, and second is more slender than third; occipital pit is absent; pores at symphysis of lower jaw is separate and small; supraocular tentacle is short; and a few tentacles are seen on lower preopercular spines, on eye, and on some lateral line scales Dorsal fin is with 12 spines and 9 soft rays; and pectoral fin is with 15–16 rays. Ctenoid scales are seen on sides of body slightly; 52–56 vertical rows of scales in body; chest and pectoral fin base are with mostly cycloid scales; and most of head is naked. Total gill rakers are 15–18. Body coloration is variable; head is brown, mottled with white; body is dark with some pale areas; dorsal fin is blotched with brown pigment; pectoral fin is with elongate dark spots; and other fins are with brown and white spots. It grows to a maximum length of 15.0 cm (TL).

Food and feeding: Studies relating to the diet composition and niche breadth of this species have been made. The stomach content of specimens collected along the eastern coast of Sicily (Central Mediterranean Sea) showed that this species preyed on 50 taxa belonging to five major groups, viz., algae, gastropods, crustaceans, polychaetes, fishes. Benthic or epi-benthic crustaceans, such as decapods, amphipods, and isopods, were the most important prey, whereas algae, gastropods, polychaetes, and fishes were only occasionally found ingested. In terms of composition by species, the diet of this species was characterized by a variety of rare or unimportant prey, which was consumed by few individuals only, although sometimes in large amount. As a result, this species can be considered a generalized and opportunistic feeder. Further, the feeding intensity of this species followed roughly a seasonal trend, with a minimum food intake in summer. The individual fish size was found to be the most important factor affecting its diet. According to the observed ontogenetic shift, small-sized individuals were found to feed primarily on small crustaceans (i.e., amphipods and isopods), whereas large-sized specimens consumed preferably bigger and more vagile prey, such as walking and swimming decapods. Furthermore, no significant differences in diet were observed in relation to the sex of predator and sampling season (La Mesa et al., 2007).

Age and growth: A significative dimorphism in size has been observed in this species, being males larger than females. The estimated age range was between 0 and 5 years for females and 0–4 years for males. Females of this species appeared to live longer and attain their maximum size faster than males. The growth performance index of this species ranged between 1.6 and 1.7, which was at the lower end of the range observed in other Mediterranean scorpaenids (La Mesaa et al., 2005).

Fisheries/aquarium/food/pharmaceutical values: In the Eastern Central Atlantic, this is a not a commercial species, though it may be fished locally. This species is rarely present in markets. It is sold fresh. It is likely sold more frequently in local markets in the west African portion of its range than in the Mediterranean.

Scorpaena melasma (Eschmeyer, 1965)

Image not available

Common name(s): Not designated

Distribution: Western Atlantic: Suriname and off Brazil.

Ecology: It is a demersal species; depth range 37–66 m.

Biology

Identification: It is distinguished by the absence of its supplemental preopercular spine. No other information is available on this species

Fisheries/aquarium/food/pharmaceutical values: Not reported.

Scorpaena mellissii (Günther, 1868)

Photo credit: Brown, Judith-St Helena Government, Environment and Natural Resources Directorate.

Common name(s): Melliss's scorpionfish, deepwater gurnard

Distribution: Southeast Atlantic: known only from St. Helena.

Ecology: This demersal species is a bottom-dweller and is often encountered in shallow water; found on rocks, amongst weed, and in crevices; depth range 10–110 m.

Biology

Identification: Like most scorpionfish species, this fish has a mottled appearance and its coloration is varying between light brown and a deep red, with a pattern of lighter color spots along their sides. It grows to a maximum length of 28.5 cm (SL).

Food: It often eats invertebrates and small fish.

Major threats: St. Helena fisheries primarily target tuna and other large pelagics via longline fisheries and small boats operating at depths of up to 100 m. There is interest in the development of commercial trawling in the vicinity of St. Helena. The island is inhabited, and there has been some development of the tourism industry.

Fisheries/aquarium/food/pharmaceutical values: There is interest in the development of commercial trawling in the vicinity of St. Helena.

Scorpaena miostoma (Günther, 1877)

Photo credit: Suzuki, Toshiyuki.

Common name(s): Smallmouth scorpionfish

Distribution: Northwest Pacific: Chiba Prefecture, Japan to Pusan, South Korea; Taiwan.

Ecology: It is a demersal species and no other information is available relating to its habitat.

Biology

Identification: Body of this species is low and is not protuberant under anterior dorsal-fin base. Lateral line is sloping moderately downward above pectoral fin. Mouth is small, and maxilla is not reaching posterior margin of eye. Small embedded scales are seen in front of pectoral fin base. It attains a length of 20 cm (TL).

Reproductive cycle: Biological minimum size of female and male in this species was found to be 12.5 cm in TL. GSI of female (3.81) and male (0.23) were the highest in October. Reproductive cycle was classified in this species into the following successive stages: in female, growing stage, maturation stage, ripe and spawning stage, recovery and resting stage, and in male, growing stage, maturation stage, ripe and spent stage and recovery and resting stage (Sick et al., 1997).

Fisheries/aquarium/food/pharmaceutical values: Not reported.

Scorpaena mystes (Jordan and Starks, 1895)

Common name(s): Pacific spotted scorpionfish

Distribution: Eastern Pacific: California, USA to northern Chile, including the Galapagos Islands.

Ecology: This demersal species dwells in shallow water and seaweed-covered reefs as well as on open, sandy areas between 1 and 85 m depth.

Biology

Identification: It is a heavy bodied fish and is and covered with cycloid scales; and scales are with fleshy appendages. Head is large and is adorned with numerous skin flaps, spines and barbels. An oval depression seen on top of head, just behind posterior margin of orbit is called occipital pit. Mouth is large with strong jaws. There is a single continuous dorsal fin with 12 spines and 8–10 rays. Anal fin is with 3 spines and 5–6 rays; huge pectoral fins extend forward and across bottom of body; and lower most rays are thickened. Body is colored green-gray, brown, red-brown, black and is mottled with light colors. A distinctive dark pigment patch is hidden on axil of pectoral fin. It is red orange to black and is usually white spotted, flashed when swimming or disturbed. It grows to a maximum length of 45.7 cm (TL).

Food and feeding: An opportunistic predator, this cryptic fish lies unseen on the reef top waiting for small fishes to pass by.

Fisheries/aquarium/food/pharmaceutical values: Although the flesh of this species is of excellent quality, it is not utilized commercially due to its venomous spines.

Scorpaena neglecta (Temminck and Schlegel, 1843)

Common name(s): Izu scorpionfish

Distribution: Indo-West Pacific; Korea.

Ecology: It is a demersal species inhabiting continental shelf; depth range 100–150 m.

Biology

Identification: Body of this species is dark brown to brownish purple; and a white bar is sometimes is present below dorsal fin. Fins are completely dark brown or margins are branded yellow, pink or orange. Inner pectoral fin is bright yellow or orange in axil and along edges; and with broad black band along hind margi. Lower rays are crimson colored. Skin is rosy underneath pectoral fin. Neck is humped before dorsal fin origin. Occipital pit is present on upper part of head behind eyes. It grows to a maximum length of 37.5 cm (TL) and weight of 900 g.

Parasite: A new species of parasitic copepod, *Prohatschekia pseudocremouxi* sp. nov. has been isolated from the gills of this species (Moon et al., 2016).

Fisheries/aquarium/food/pharmaceutical values: Not reported.

Scorpaena normani (Cadenat, 1943)

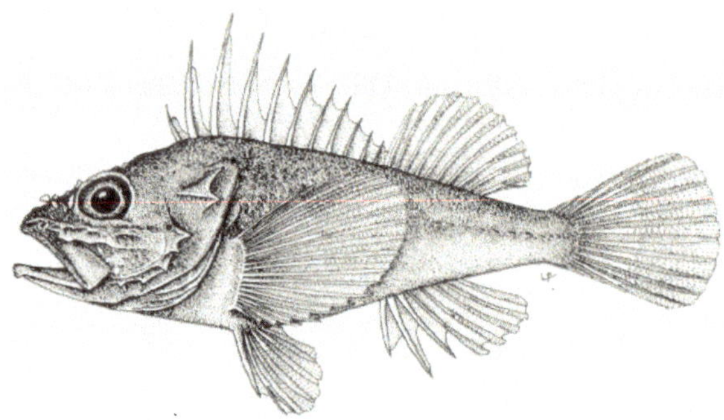

Photo credit: Keiichi Matsuura.

Common name(s): Norman's rockfish

Distribution: Eastern Atlantic: Mauritania to southern Angola.

Ecology: It is a demersal species; depth range 50–300 m.

Biology

Identification: It grows to a maximum length of 20.0 cm (TL). No other data for the biology of this species.

Major threats: While this species is removed as incidental catch and in artisanal fisheries, it is not believed to have major threats.

Fisheries/aquarium/food/pharmaceutical values: This species is caught on bottom trawls, bottom set longlines in local fisheries; however, it is not commercially targeted.

Scorpaena notata (Rafinesque, 1810)

Photo Courtesy: Wikipedia.

Common name(s): Small red scorpionfish

Distribution: Eastern Atlantic: from the Bay of Biscay to Senegal, Madeira, Azores and the Canary Islands, including the Mediterranean and the Black Sea.

Ecology: It is a benthic species and is usually found resting on rocks. It is usually solitary.

Biology

Identification: Head of this species is broad with a short snout and upwardly angled mouth. There is a short tentacle just above the eye and various other shorter tentacles, spines and flaps of skin adorning the head. Dorsal fin has 12 spines and 9 soft rays and the anal fin has 2 spines and 6 soft rays. Pectoral fins are large and oval and have 17–19 rays. There are about 44 vertical rows of scales on the body but the head, chest and the base of the pectoral fins are naked. Color of this fish is reddish-brown and there is a large, semicircular pigmented spot between the sixth and tenth dorsal spines. The fins are spotted with brown and the caudal fin is unbarred. Maximum recorded length of this species is 27.7 cm (TL).

Food and feeding: Studies relating to the feeding ecology and food partitioning of this species in Mediterranean Posidonia seagrass beds showed that *Scorpaena notata* was found to feed mostly on caridean shrimps. Individual size and diel period were the main factors responsible for diet variation and food partitioning that evolved according to both cyclic (temporal) and continuous (ontogenic) trends. A strong interspecific food overlap was observed in juveniles (S.L. <50 mm), whereas intraspecific potential competition was low between juveniles and adults. Among medium and large size classes, intra- and interspecific food overlaps were high and of equal intensity. Narrower food niches at night, due to a stronger specialization on type and size of prey, diminished the intra- and interspecific potential competition during this period of maximum feeding activity. This probably allowed the high number of coexisting scorpaenid populations in Mediterranean seagrass beds. Nevertheless, microhabitat and time partitioning seemed as important as food partitioning in the structuring of fish assemblages (Harmelin-Vivien et al., 1989).

Though it is a predator showing diet specialization for crustaceans, this species favored the thalassinid *Jaxea nocturna* as its main diet. The studies relating to the trophic relationships made in this species further showed that this species seems to rely on soft bottom prey items rather than exploiting resources from the platforms, thus competing with soft-bottom predators for the same resources (Luca et al., 2012).

Growth: The growth rate of *S. notata* individuals is higher in those that inhabit substrata with high amounts of algal biomass off the Balearic Islands; thus, the red algae habitats seem to be an essential habitat for this species. Females of this species dominated in smaller length classes while males prevailed in larger ones, although overall sex ratio did not deviate from 1:1. Females ranged from 10.5 to 18.8 cm TL and from 1 to 6 years; males ranged from 11.4 to 20.7 cm TL and from 1 to 7 years (Neves et al., 2015).

Behavior/food and feeding: It is normally camouflaged against sea fans. It is a solitary, nocturnal fish that hides around rocks and marine plants, and then ambushes its prey of small fish, crustaceans and other bottom-dwelling invertebrates.

Reproduction: Most individuals of this species are sexually mature at 1 year old. It is an oviparous species with external fertilization that deposits its eggs in a gelatinous matrix. In the northwest Mediterranean, the spawning period for this species occurs between July and October. The fecundity of female ranges from 6000 to 33,000 ova and the total number of eggs decreases as the spawning period continues. The fecundity in this species was largely determined by the size and weight of the individuals The maximum egg diameter is ~500 mm. This species exhibits determinate fecundity and is considered to be a multiple spawner a group-synchronous ovarian organization (Muñoz et al. 2005, Ordines et al., 2009).

Parasites: A new species of parasite, viz., *Microcotyle algeriensis* (Monogenea: Microcotylidae) has been isolated from the fishes of this species (Ayadi et al., 2017).

Major threats: The red algae habitats substrates are an essential habitat for this species and are disrupted by trawling activities. A trawling ban on these habitats is highly recommended, as it is an important anthropogenic threat in the Mediterranean Sea.

Fisheries/aquarium/food values: *S. notata* is an important bycatch species for commercial trawlers that operate over the shallow continental shelf off the Balearic Islands. Its annual landings are estimated to be 10–15 t.

Pharmaceutical values: An angiotensin I converting enzyme and antioxidative peptides have been isolated from the head protein hydrolysate of this species by treatment with a protease from the fungus *Penicillium digitatum*. Peptides D-I, E-I, and F-I of this species showed high angiotensin-I converting enzyme inhibitory activity with an IC50 values of 0.98, 1.69 and 1.44 μM, respectively, as well as a synergistic antioxidant activity between the different fractions. This study demonstrated that the peptides produced from the underutilized wastes of this species can be used as potential therapeutic compounds active against oxidative stress and hypertension (Aissaoui et al., 2017).

A new protease was isolated from the intestine of this species. Interestingly, this enzyme presented specificity for casein, egg albumin, bovine serum albumin, and gelatin with obtained degrees of hydrolysis ranging from 3.14% to 6.58%, after 2 h of hydrolysis. These results suggested the potential use of this protease enzyme in the preparation of food protein hydrolysates with interesting properties (Aissaoui et al., 2017).

Scorpaena onaria (Jordan and Snyder, 1900)

Courtesy picture: Wikipedia.

Common name(s): Western scorpionfish, Neglected Scorpionfish

Distribution: Indo-West Pacific: Japan and South Korea; Taiwan; New Caledonia.

Ecology: It is a benthopelagic species found on rocky reefs of continental shelf at depths of 95–500 m; depth range 223–500 m.

Biology

Identification: This species diagnosed by its dorsal fin with 12 spines and 9 soft rays. Body is moderately compressed anteriorly and is progressively more compressed posteriorly. Nape and anterior body are not highly arched in juveniles and are progressively more arched with growth. Body is moderately deep but less than head length. Numerous small papillae are seen on head, especially dorsal surface of head, snout, and lateral surface above suborbital ridge; and numerous tiny tentacles and cirri are present on head. A short tentacle is seen on posterior end of supraocular spine base. A total of 7–12 small tentacles are seen on upper part of outer margin of eye. Mouth is large and is slightly oblique, forming an angle of c. 20–35° to horizontal axis of head and body. Posterior margin of maxilla is extending level with posterior margin of pupil but not extending beyond posterior margin of orbit. Dorsal profile of snout is steep. Origin of first dorsal-fin spine is above lower posttemporal spine. Posterior margin of opercular membrane is extending well beyond level with origin of third dorsal-fin spine. Posterior tip of pectoral fin is not reaching level with tip of depressed pelvic fin or origin of last dorsal-fin spine. Posterior tip of depressed pelvic fin is extending slightly beyond anus but not reaching anal-fin origin. Origin of pelvic-fin spine is posterior to origin of pectoral fin. Origin of first anal-fin spine is slightly posterior to origin of last dorsal-fin spine. Swim bladder is absent. Coloration of body is variegated, mainly red or reddish yellow, suffused with irregular reddish, yellowish, and whitish blotches. It grows to a maximum size of 26.5 cm (SL). This species is sexually dimorphic with only males having a black blotch at the margin of the membrane between the sixth and ninth dorsal fin spines. Further, a mature female differed from an immature female and males in having a longer upper jaw, steeper dorsal profile of the snout, and shorter nasal spine (Motomur et al., 2005).

Fisheries/aquarium/food/pharmaceutical values: *Scorpaena onaria* is taken in both commercial and recreational fisheries. It is taken by hook and line, especially in New Zealand.

Scorpaena orgila (Eschmeyer and Allen, 1971)

Photo credit: John E. Randall.

Common name(s): Bold scorpionfish

Distribution: Southeast Pacific: endemic to Easter Island.

Ecology: It is a demersal species occurring on rocky reefs in shallow waters of 21 m depth.

Biology

Identification: It grows to a maximum length of 38.0 cm. No other information is available on its biology.

Fisheries/aquarium/food/pharmaceutical values: This species is not utilized.

Scorpaena melasma (Eschmeyer, 1965)

Image not available

Common name(s): Not designated

Distribution: Western Atlantic: Suriname and off Brazil.

Ecology: It is a demersal species; depth range 37–66 m.

Biology

Identification: Supplemental preopercular spine absent in this species. No data for other aspects of its biology.

Fisheries/aquarium/food/pharmaceutical values: Not reported.

Scorpaena papillosa (Schneider and Forster, 1801) (=*Ruboralga ergastulorum*)

Photo credit: Wikipedia.

Common name(s): Red rock cod, Southern red scorpionfish, chained scorpionfish, common red gurnard, common red rock cod, red rock cod, Southern red scorpioncod, Southern rockcod

Distribution: Indo-West Pacific: New Zealand and southern Australia.

Ecology: This reef-associated species lives in marine, demersal (on or near sea floor), temperate waters, at depths of 5–50 m. It can be found on rocky bottoms, in bays, shallow estuaries, and in muddy waters as well as other environments, such as in offshore kelp beds and shallow seagrass beds. Juveniles of this species are sometimes found in large rock pools.

Biology

Identification: The chief diagnostic features of this species include dorsal fin with 12 spines and 9–10 soft rays; and anal fin is with 3 spines and 5 soft rays. A small row of spines is seen beneath the eyes, and its gill cover margin is containing 3 spines. Head and body of this species are robust. Its large mouth contains small, thin teeth that form velvety bands (villiform). Spines are present on cheeks. Body is mottled and is reddish to reddish-brown with darker blotches and whitish patches. A black spot is present on the dorsal fin. Some individuals are with a white band behind the eyes and on the tail base. It is the only species in the genus *Scorpaena* that has a pair of coronal spines on the interorbital space. This species grows to a maximum length of approximately 30 cm (TL).

Food: This nocturnal species feeds on a wide variety of invertebrates, mainly on crustaceans and other fishes.

Fisheries/aquarium/food/pharmaceutical values: It is not a commercial species.

Scorpaena pascuensis (Eschmeyer and Allen, 1971)

Photo credit: John E. Randall.

Common name(s): Not designated

Distribution: Southeast Pacific: Easter Island.

Ecology: It is a demersal species.

Biology

Identification: It grows to a maximum length of maximum length of 5.6 cm (SL). Other data on this aspect is not available.

Fisheries/aquarium/food/pharmaceutical values: Not reported.

Scorpaena pele (Eschmeyer and Randall, 1975)

Photo credit: John E. Randall.

Common name(s): No common name
Distribution: Eastern Central Pacific: only from the Hawaiian Islands.
Ecology: It is a demersal species; depth range 176–240 m.

Biology

Identification: This species has the following characteristics: dorsal fin is with 12 spines and 9 soft rays; anal fin has 3 spines and 5 soft rays; 24 vertebrae; occiput is with a shallow pit; lachrymal bone is with 3 spines over maxillary; suborbital ridge is with 4 spines; and coronal spines are absent. Coloration of body is mostly red, marbled and is spotted with white. It grows to a maximum length of 13.5 cm (SL).

Fisheries/aquarium/food/pharmaceutical values: Not reported.

Scorpaena pepo (Motomura et al., 2007)

Photo courtesy: Hiroyuki Motomura.

Common name(s): Pumpkin scorpionfish
Distribution: Northwest Pacific: Taiwan.
Ecology: It is a demersal species occurring up to a depth of 200 m.

Biology

Identification: Body of this species is moderately compressed anteriorly and is progressively more compressed posteriorly; nape and anterior part of body are moderately arched; body is moderately deep but less than head length. Numerous small papillae are present on head, especially on dorsal surface, snout, and lateral surface above suborbital ridge; numerous tiny tentacles and cirri are seen on head, including dorsal margin of eye; and a very short tentacle is seen at posterior end of supraocular spine base. Mouth is large, slightly oblique and is forming an angle of about 40° to horizontal axis of head and body; posterior margin of maxilla is extending well beyond a vertical through posterior margin of orbit; lateral surface of maxilla is smooth, without distinct ridges; and lower jaw is with a symphysial knob. Dorsal profile of snout is steep, forming an angle of about 50° to horizontal axis of head and body. Nasal spine is simple, conical, and directed upwards; and its length is slightly greater than anterior nostril diameter. Anterior nostril is with a distinct tentacle and its length greater than nasal spine. Origin of first dorsal-fin spine is anterior to base of supracleithral spine. Posterior margin of opercular membrane is extending beyond a vertical through origin of fourth dorsal-fin spine. Posterior tip of pectoral fin is not reaching a vertical through tip of depressed pelvic fin and origin of last dorsal-fin spine. Posterior tip of depressed pelvic fin is not reaching anus. Origin of pelvic-fin spine is posterior to origin of pectoral fin. Origin of first anal-fin spine is posterior to origin of last dorsal-fin spine.

Head and body are bright yellowish-orange; mottled with very pale gray blotches dorsally; and small distinct black spots are scattered on head and body, except for ventral surface, median fins and upper 1/2 of pectoral fin. Underside of lower jaw is bright orange and is mottled with distinct white blotches. One or 2 irregular narrow white bands are seen across spinous portion of dorsal fin medially; membrane of spinous portion of dorsal fin is yellowish-orange and is mottled with poorly defined whitish and blackish blotches; and a large black blotch, greater than orbit diameter, is seen on outer margins of membranes between 6th and 11th spines in males. Soft-rayed portion of dorsal fin is orange proximally, and grayish subdistally, with poorly defined whitish blotches and blackish spots over fin; and distal margin of the fin is whitish. Pectoral-fin base is whitish; and pectoral fin is yellowish-orange with small black spots scattered dorsally, and poorly defined whitish blotches ventrally. Pelvic and anal fins are yellowish-orange, with poorly defined whitish blotches. Caudal fin is yellowish-orange basally, gray subdistally, and whitish distally. It grows to a maximum length of 24.5 cm (SL) (Motomura et al., 2007).

Fisheries/aquarium/food/pharmaceutical values: Not reported.

Scorpaena petricola (Eschmeyer, 1965)

Image not available

Common name(s): Not known

Distribution: Southwest Atlantic: off Brazil.

Ecology: It is a demersal species inhabiting up to depths of 73 m.

Biology

Identification: It is a small species with a generally pallid head and body. Pectoral-fin rays are usually 18, sometimes 17; second anal-fin spine is longer than third anal-fin spine; and body depth is 33%–38% SL.

Fisheries/aquarium/food/pharmaceutical values: Not reported.

Scorpaena plumieri (Bloch, 1789)

Photo courtesy: John E. Randall.

Common name(s): Spotted scorpionfish, pacific spotted scorpionfish, stone scorpionfish, and spotted stone scorpionfish

Distribution: Western Atlantic: Bermuda, Massachusetts, and northern Gulf of Mexico to southern Brazil. Eastern Atlantic: Ascension and St. Helena.

Ecology: This demersal species usually dwells in shallow coral reef and rocky areas. It remains motionless and is well camouflaged among rocks and plants at a depth range of 1–60 m.

Biology

Identification: Body of the fish is robust with numerous protrusions and prominent head. It has 12 dorsal, 2 pelvic, and 3 anal short and thick fin spines covered with mucous-rich integumentary sheath. Occipital pit is well developed. Pectoral fins have broad bases and are fan-like in shape. Dorsal fin is continuous and deeply notched. Caudal fin is truncate. Mouth is terminal with the maxilla reaching past the eye. Coloration of the body is very similar to that of a rocky bottom. A pattern of reddish tones, cream, with dark bands (brown to reddish) is seen. Below the pectoral fin, numerous white or dark spots are seen. Tail is with three dark bars. Body is often with large appendages or "plumes" above the eyes and numerous fleshy flaps on chin and head. Skin is rough looking. The typical life span of this species is approximately 10–15 years. Body length in adult fishes is up to 50 cm and maximum weight is 1.6 kg.

Behavior/Adaptations: *Scorpaena plumieri* is not usually an aggressive species. It rests throughout the day by lying motionless and most of its movement takes place during the night. This species is well camouflaged between the fire coral (*Millepora alcicornis*), barnacles, red encrusting sponge, and chicken liver sponge (*Chondrilla nucula*). It displays characteristic bright white spots on dark background of pectoral axil when disturbed. Juveniles after hatching at the water surface, migrate toward shallow waters to take their place amongst coral reefs and other rocky substrate. Because this species remains motionless for long periods of time, considerable loads of crustaceans and algae often live on it. The shedding of its outer layer of the skin is performed by the fish in order to get rid of the fouling organisms.

Food and feeding: It is carnivorous preying mainly on fish and crustaceans. Its method of hunting is unique. It is an ambush predator. Lying motionless and camouflaged by substrate, it is undetected by prey. It waits until the prey is close enough so that it can strike. When it strikes, its large cavernous mouth is used like a vacuum to quickly pull the prey into their mouth and swallow it.

Reproduction: These solitary animals where the sexes are separate and migrate to waters of 40–120 m deep where spawning takes place. Spawning occurs during the period of June to August. The females are oviparous and the eggs are transparent or sometimes can have a green color. During spawning, the eggs are released by the female and the sperm are released by the male. The resulting fertilized eggs float at the surface of the water. The maximum clutch size is about 15,000 eggs. Once the eggs are produced and fertilized, they remain floating at the surface and their color changes from transparent to more white. The females do not remain to care for the offspring. The parents leave after spawning occurs to return to the shallow waters where they normally reside. The eggs are left, where they are susceptible to predation but the transparency of the eggs help to reduce their vulnerability. After just 2 days, the eggs will hatch and baby scorpionfish will remain near the surface until they are big enough to swim down to the reefs where the adults swim.

Predators: The predators of the spotted scorpionfish are snapper including the schoolmaster (*Lutjanus apodus*) and mutton snapper (*Lutjanus analis*) as well as sharks, snappers, rays, and moray eels.

Parasites: Parasites of the spotted scorpionfish include the digenic trematodes *Neopecoelus scorpaenae* and *Helicometrina nimia*.

Fisheries: This species, owing to its venom, has a minor commercial fishery, and it is caught as bycatch mostly on hook-and-line and with traps.

Food value: Spotted scorpionfishes are caught as incidental bycatch throughout their range primarily on hook and line, in traps, and shallow water trawls. The flesh which is said to taste a bit like chicken is marketed fresh for human consumption.

Aquarium value: This species is widely used in home aquaria. It is also found in the aquarium trade at Ceará, Brazil.

Pharmaceutical properties of venom: The antitumoral effect associated with cytotoxicity of the venom extracts of this species has been studied on cultured murine glioblastomas cells (RT2). Inhibition of cell adhesion and proliferation were recorded at concentrations higher than 10 mg/mL. At concentration of 100 mg/mL, the cells were completely lysed. The venom showed gelatinolytic activity which was responsible at least partially, for the said antitumoral effect (Soprani et al., 2007).

Scorpaena porcus (Linnaeus, 1758)

Photo courtesy: Wikipedia.

Common name(s): Black scorpionfish, European scorpionfish, sea pig, sea scorpion, small-scaled scorpionfish

Distribution: Eastern Atlantic: British Isles to the Azores, and the Canary Islands, including Morocco, the Mediterranean Sea, and the Black Sea.

Ecology: It is a demersal, solitary, and sedentary species. It is nonmigratory and is commonly found among rocks and algae up to depths of 800 m. In the Mediterranean Sea, *Scorpaena porcus* was found associated with *Posidonia oceanica* beds and inside marine caves.

Biology

Identification: Head of this species is broad with a short snout and upwardly angled mouth. There is a short tentacle just above the eye and various other shorter tentacles, spines, and flaps of skin adorning the head. Upper jaw is without a conspicuous longitudinal ridge; and two pores are present under the chin behind the junction of the lower jaws. Dorsal fin has 12 spines and 9 soft rays and anal fin has 2 spines and 6 soft rays. Pectoral fins are large and oval and 16–18 rays. Scales on body are small, emarginate; 65–70 vertical rows of scales; and chest, pectoral fin base and head are naked. Gill rakers 16–18, 5–6 on upper arch, 8 on ceratobranchial and 3–5 rudiments on hypobranchial. The black scorpionfish has a maximum length of 40.5 cm (TL) and maximum published weight is 870.00 g.

Biometry analysis: Biometry analysis of this species made in the samples collected from the middle and south Adriatic showed sexual dimorphism in seven morphometric measurements. There were no differences in meristic characters between sexes. Relative growth studied by comparing changes in morphological characters with growth in standard and head length showed the existence of one unique population of *S. porcus* in the Adriatic Sea (Ferri et al., 2010).

Life span: *Scorpaena porcus* can live up to 12–18 years of age. The oldest age estimates for this species vary between regions; in the Adriatic Sea, it can live to 11 years, 6 years in the Mediterranean, and 7 years in the Sea of Marmara.

Food and feeding: It feeds on small fishes such as blennies and gobies, crustaceans, and other bottom-dwelling invertebrates. Studies relating to feeding habits of this species showed that this species fed mainly upon red mullet (*Mullus barbatus*) (35%) and mud shrimp

(*Upogebia pusilla*) (28%) but also upon harbor crab, *Liocarcinus depurator*; brown shrimp, *Crangon crangon*; sea horse, *Hippocampus* sp., and others, which included *Ulva lactuca*, shrimp larvae, *isopods*, hermit crab, and mollusc, *Bittium reticulatum*. It was found that there were significant differences between its dietary habits in summer and winter times. The stomachs of this species were found to be empty (V% = 28.3) in summer samples and it was found to feed on harbor crab most frequently in summer (63.4%) and on red mullet (57.5%) in winter (Başçınar and Sağlam, 2009).

Studies relating to the diet composition; size-related changes in the diet; and changes in feeding habits of this species made showed that this species was found to prey on fish, crustaceans, molluscs and algae. The red mullet (*M. barbatus*) (34.7%) was found as the main diet component followed by the mud shrimp, *U. pusilla* (28.1%). Other important food items included harbor crab, *L. depurator* (27.3%) and brown shrimp [*C. crangon* (3.2%); seahorse *Hippocampus* sp. (5.0%); and other items (1.7%)] which included *Ulva lactuca*, shrimp larvae, isopods, hermit crab, and mollusc, *Bittium reticulatum*. The diet of this species (sizes, 6–12 cm; 12–18 cm) was composed mainly of the red mullet, *M. barbatus* (62.2%). However, for the larger sized fishes (18–24 cm size class), sea horse, *Hippocampus* sp. (32.2%) was found as the main prey species. Depending on season, the stomachs of this species were found to be empty (V% = 28.3) in summer samples. Further, it was found to feed on harbor crab most frequently in summer (63.4%) and on red mullet (57.5%) in winter (Başçınar and Sağlam, 2009).

Studies relating to the feeding ecology and food partitioning of this species in Mediterranean Posidonia seagrass beds showed that *S. porcus* was found to feed mostly on brachyurans. Individual size and diel period were the main factors responsible for diet variation and food partitioning that evolved according to both cyclic (temporal) and continuous (ontogenic) trends. A strong interspecific food overlap was observed in juveniles (S.L. < 50 mm), whereas intraspecific potential competition was low between juveniles and adults. Among medium and large size classes, intra- and interspecific food overlaps were high and of equal intensity. Narrower food niches at night, due to a stronger specialization on type and size of prey, diminished the intra- and interspecific potential competition during this period of maximum feeding activity. This probably allowed the high number of coexisting scorpaenid populations in Mediterranean seagrass beds. Nevertheless, microhabitat and time partitioning seemed as important as food partitioning in the structuring of fish assemblages (Harmelin-Vivien et al., 1989).

Though it is a predator showing diet specialization for crustaceans, burrowing shrimp, *Alpheus glaber*, hairy crab *Pilumnus hirtellus*, bivalve *Corbula gibba*, and European anchovy *Engraulis encrasicolus* were found to be its favored by this species. Further, *E. encrasicolus* was found to be its natural prey. The studies relating to the trophic relationships made in this species further showed that this species seems to rely on soft bottom prey items rather than exploiting resources from the platforms, thus competing with soft-bottom predators for the same resources (Luca et al., 2012).

Reproduction: Little is known about its reproduction. It attains sexual maturity in 3 years old, rarely in 2 years. The spawning season of *S. porcus* in the Black Sea is in summer from

June to August; and eggs are covered with a mucilaginous membrane. It also exhibits sexual dimorphism and the growth rate of females is higher than that of males.

Parasites: Two isopod species, *Nerocila orbigny* and *Ceratothoa parallela* (Crustacea, Isopoda, Cymothoidae), have been reported for the first time in this species collected from the eastern Adriatic Sea (Ferri et al., 2008).

Nutritional values: This fish contained moisture, protein, fat, ash, carbohydrate, 81.8%, 15.9%, 0.2%, 1.6%, 0.6%, respectively, and energy 68 kcal/100 g. The essential amino acids content of this species was 52% of total amino acids. The dominant amino acids were lysine and glutamic acid. The ratio of essential/nonessential amino acids (E/NE) was calculated as 1.1. The fatty acid composition indicated 53.2% saturated fatty acids (SFA), 35.1% monounsaturated fatty acids (MUFA) and 1.8% polyunsaturated fatty acids (PUFA). Predominant SFA, MUFA, and PUFA were palmitic acid, oleic acid, and arachidonic acid, respectively. The major, macro, and microelements were potassium, calcium, and iron zinc, respectively (Kaya and Kocatepe, 2014).

Fisheries/aquarium/food/pharmaceutical values: *S. porcus* is of commercial interest for its meat which is tough and delicious and is traditionally consumed in the Black Sea Area. In 2002 in Turkey, the total catch of this species was 240 t (104 t from Black Sea, 36 t from the Sea of Marmara, 88 t from Aegean Sea, and 12 t from the Mediterranean Sea). This species is also the second most frequently caught scorpaenid species in the MPA of Cap de Creus in the northwestern Mediterranean. Between 2008 and 2010, in the Cap de Creus MPA, catch per unit effort peaked in the summer which was concurrent with the spawning of this species.

Heavy metal content: As this species is widely consumed in the Black Sea area, pollution studies were made to assess the level of heavy metal accumulation in the different tissues of this species. The results of this investigation showed that the highest heavy metal (Al, Cu, Ni, As, Cd, Hg, Pb, U) accumulation was generally found in the liver. The maximum aluminum level and the minimum uranium level were found in the tissues. In terms of heavy metals, Al, Cu, Cd, and Hg showed a statistically significant difference between tissues ($P < 0.05$). It was determined that heavy metal concentrations obtained from the muscle tissues did not exceed the national and international recommended limits; and also, it was found that daily intake amounts did not exceed tolerable daily intake amounts. In terms of public health, it was found out that there was not any risk in consumption of scorpionfish in the study area (Çulha et al., 2016).

Major threats: The reduction in oxygen concentrations and the increase in the organic matter content in littoral waters; and the impoverishment of algal communities.

This species exhibits life history traits that make it vulnerable to exploitation including low mobility, long life span and slow growth. Longevity estimates of this species vary between regions which may be due to geological factors and fishing pressures. If current fishing pressures are not reduced on *S. porcus* and existing regulations not revised, the stock may be in danger.

In Sevastopol Bay, a highly polluted area of the Black Sea, the growth rate of *S. porcus* was lower and over half of the studied females had anomalies of oogenesis. An increase in the activity of superoxide dismutase (SOD) was also observed; SOD is a key enzyme of the antioxidant blood protection. A low growth rate of *S. porcus* was also observed in the Sevastopol coastal zone. Earlier studies indicated that the decreased number of older adult individuals of this species in this region could primarily be due to chronic pollution.

Conservation measures proposed: Developing actions concerning the construction of artificial reefs for the restoration of optimal conditions.

There are no species-specific conservation measures in place for this species. There are no length regulations for catching this species in Turkey; however, a minimum fishing size of 18 cm TL is recommended and that fishing with trammel nets should be prohibited between June and September when this species spawns.

The average size in both male and female *S. porcus* in the Cap de Creus MPA was always found to be greater than their size at sexual maturity. Maximum obtained sizes were much larger inside the MPA than in the nonprotected areas. Improvement of the monitoring and management of this species is therefore suggested.

Scorpaena russula (Jordan and Bollman, 1890)

Common name(s): Reddish scorpionfish, short-spine red scorpionfish

Distribution: Eastern Pacific: Culiacán, Sinaloa, Mexico to Peru.

Ecology: It is a demersal found in coastal waters; depth range 10–160 m. It inhabits sand and rubble substrata.

Biology

Identification: Body of this species is lender with depth of 30–33 of SL. Head is very bony with numerous spines. Snout is short (7%–11% SL). It is narrow between eyes. There are no ridges and weak ridges between eyes. Between eyes, it is flat and is covered with many small cirri and with large tentacles on rear base of spine above rear of eye. Pit behind eyes is a slight depression with a front edge with many cirri. Longitudinal ridge is seen under eye with 2–3 small spines. Top preopercular spine is the largest and second spine is very reduced. Roof of mouth is with teeth on sides. There is no slit behind last gill arch. There are no skin flaps on body. Dorsal fin is with 12 spines and 9 rays; anal fin is with 3 spines and 5 rays; and pectoral fin is with 20–22 rays, of which lower 14 rays are unbranched and thickened. Scales are large and smooth on body; lateral line is complete; with 24–25 scales; and 43–49 vertical rows of scales on lower front body and they are imbedded and not well exposed. Body is mottled red and brown; pectoral fin is with blackish tip; 3–4 irregular rows of small spots on soft dorsal rays. It grows to a maximum length of 15.0 cm (TL).

Fisheries/aquarium/food/pharmaceutical values: Not reported.

Scorpaena scrofa (Linnaeus, 1758)

Photo courtesy: Wikipedia.

Common name(s): Red scorpionfish, bigscale scorpionfish, large-scaled scorpion fish

Distribution: Eastern Atlantic: British Isles to Senegal including Madeira, the Canary Islands, and Cape Verde; throughout the Mediterranean except Black Sea.

Ecology: It is a marine and brackish water demersal species found over rocky, sandy or muddy bottoms; depth range 20–500 m. In the Mediterranean Sea, it is often associated with *P. oceanica* meadows. The juveniles of this species often camouflage between the rocks of the backwash, and when they are more than 20 cm, they progressively go down, looking for a calm place for their sedentary life.

Biology

Identification: Body of this species is oval with an impressing, large head. Occipital pit is moderate. Pores are seen at symphysis of lower jaw small and separate. Supraocular tentacle is usually small or absent; and numerous tentacles are present on lower jaw. Both the jaws are armed by villiform teeth. Dorsal fin is with 12 spines and 9 rays; pectoral fin is with 18–20 rays. Scales on body are ctenoid, and there are 45 vertical rows of scales in body; most of head is naked or with embedded cycloid scales; and pectoral fin base and chest are naked without scales. There are 14–17 gill rakers. Coloration of body is variable from brick red to light pink with dark mottling or blotches; fins are blotched with brown; distal part of pelvic fins are usually dusky; and dorsal fin is with a black spot frequently present between spines 6 and 11. It grows to a maximum length of 50.0 cm (TL) and weight of 3.0 kg.

Behavior: This species is a sedentary, solitary and nonmigratory fish. By day, it lives in burrows and caves. At night it comes out to hunt.

Food and feeding: Studies relating to the feeding ecology and food partitioning of this species in Mediterranean Posidonia seagrass beds showed that, *Scorpaena scrofa*, was found to feed mostly on fishes. Individual size and diel period were the main factors responsible for diet variation and food partitioning that evolved according to both cyclic (temporal) and continuous (ontogenic) trends. A strong interspecific food overlap was observed in juveniles (S.L. <50 mm), whereas intraspecific potential competition was low between juveniles and adults. Among medium and large size classes, intra- and interspecific food overlaps were high and of equal intensity. Narrower food niches at night, due to a stronger specialization on type and size of prey, diminished the intra- and interspecific potential competition during this period of maximum feeding activity. This probably allowed the high number of coexisting scorpaenid populations in Mediterranean seagrass beds. Nevertheless, microhabitat and time partitioning seemed as important as food partitioning in the structuring of fish assemblages (Harmelin-Vivien et al., 1989).

Fisheries/aquarium/food/pharmaceutical values: *S. scrofa* is a much appreciated, commercial species in the Mediterranean and Sicilian markets. This species is not sought after for food or trade in west Africa. *S. scrofa* is a traditional ingredient in Marseille bouillabaisse. It is also widely used in Japanese cuisine.

The high quality of its meats, the excellent palatability and the high market value make this species a good candidate for aquaculture (Mancuso, 2011).

Major threats: In the Eastern Central Atlantic, it is caught primarily as incidental catch but this is not believed to be a major threat. In the Mediterranean, this is a commercial species and its main catching method is trawls, gill nets, trammel nets, and longlines.

Scorpaena sonorae (Jenkins and Evermann, 1889)

Photo courtesy: John Snow.

Common name(s): Sonora scorpionfish

Distribution: Eastern Central Pacific: Mexico.

Ecology: This species is demersal, inhabiting sand-rubble substrata at depths down to 91 m.

Biology

Identification: This species has a spine-laden compressed, elongated, oval-shaped body that tapers toward the tail; its body depth is 28%–32% of SL. Head is very bony with numerous spines and a short snout. Head and body are covered with numerous skin flaps. It has large eyes. Small weakly developed depressions (pits) are seen before and after the eyes and the ridge below its eyes has zero to two small spines. Its dorsal fin has 12 or 13 spines and 8 or 9 rays; anal fin has 3 spines and 5 rays; pectoral fins have 19–21 rays; and caudal fin is straight. Body lacks skin flaps and is covered with smooth scales. It grows to a maximum length of 15.8 cm (SL). It varies significantly in color and features various shades of mottled reds and browns with irregular dark bars on its sides and white on its belly. A key to identification of this species is a large round dark spot on its soft dorsal fin. Anal fin is red with wide bars; caudal fin has black and white bars; pectoral fins are deep red and disproportionately large; and pelvic fins are reddish. Tall these fins have black margins.

Behavior: It is a small and very rare species and virtually nothing is known about its behavioral patterns.

Fisheries/aquarium/food/pharmaceutical values: Not reported.

Scorpaena stephanica (Cadenat, 1943)

Photo courtesy: Info Nero di Sole.

Common name(s): Spotted-fin rockfish, Malhado-fin rockfish

Distribution: Eastern Atlantic: Mauritania to Angola.

Ecology: This is a demersal species found on sandy or muddy bottoms; depth range 46–201 m.

Biology

Identification: This species has a sturdy body and thorny head. Big mouth which is terminal and oblique, with small conical teeth on the maxillary arches, vomer, and palatine (2 rows). Muzzle is equal to the diameter of the eye or slightly more long. Preorbital bone is with two thorns above the jaw; suborbital ridge is with 4–5 spines; preopercular margin is with 4 spines; and opercular is with 2 spines. Occipital fossa is present. Small flaps are seen on the head. Dorsal fin is with 7 spines and 10 soft rays and anal fin is with 3 spines and 5–6 rays.

Pectoral fins are large and round, with some branched rays. Chest and head are naked without scales. It grows to a maximum length of 40.0 cm (TL).

Food: It is carnivorous feeding on invertebrates and small fishes.

Fisheries/aquarium/food/pharmaceutical values: This species is removed as bycatch in bottom trawls and bottom set longlines. It is consumed rarely and mostly thrown back.

Scorpaena sumptuosa (Castelnau, 1875)

Photo courtesy: Philip Good.

Common name(s): Western red scorpionfish, Western red rock cod

Distribution: Eastern Indian Ocean: it is endemic to southwestern Australia.

Ecology: It is an inshore, solitary, benthic species found in rocky reefs; depth range 5–40 m.

Biology

Identification: This species is diagnosed by the following characters: deep-bodied species with a prominent hump on the shoulder; large head with spiny ridges; eyes are elevated on the head.; lateral surface of maxilla is with a distinct longitudinal ridge; relatively thick skin with several small sensory pores covering nearly the entire predorsal area to near the first pored lateral line scale; anteroventral surface of lower jaw is with one or two pairs of slender tentacles in ventral view. It has a mottled red to brown coloration, with two vertical broad bands down the body extending onto the fins. It grows to a maximum length of 40.0 cm (TL).

Food and feeding: Western red scorpioncod are ambush predators. They may lay motionless for hours at a time, amongst corals, algae, and invertebrates before extending the prehensile mouth and sucking in unsuspecting prey.

Fisheries/aquarium/food/pharmaceutical values: Not reported.

Scorpaena thomsoni (Günther, 1880)

Photo courtesy: Wikipedia.

Common name(s): Not designated
Distribution: Southeast Pacific: Desventuradas Islands and Juan Fernández Islands.
Ecology: It is a demersal species inhabiting bottom waters.

Biology

Identification: It grows up to a length of 40 cm. Biology of this species is not studied.
Fisheries/aquarium/food/pharmaceutical values: Not reported.

Scorpaena tierrae (Hildebrand, 1946)

Image not available

Common name(s): No common name
Distribution: Southeast Pacific: Chile.
Ecology: It is a demersal species.

Biology

Identification: No data
Fisheries/aquarium/food/pharmaceutical values: Not reported.

Scorpaena uncinata (de Buen, 1961)

Image not available

Common name(s): No common name
Distribution: Southeast Pacific: Chile.
Ecology: It is a demersal species inhabiting up to a depth of 137 m.

Biology

Identification: No data
Fisheries/aquarium/food/pharmaceutical values: No data.

Scorpaena wellingtoni (Victor, 2013)

Photo courtesy: Benjamin C. Victor.

Common name(s): No common name as it is a new species

Distribution: Tagus Cove on Isla Isabela in the Galápagos Archipelago, endemic to Galápagos Islands.

Ecology: Not known.

Biology

Identification: This species has the following combination of characters: dorsal fin is with 12 spines and 8 rays; anal fin is with 3 spines and 5 rays; pectoral-fin rays is 21, second through fifth ray branched; penultimate dorsal-fin spine is more than half length of ultimate spine; 4 upper-limb gill rakers and 6–7 in lower limb; relatively deep body (depth 36.2%–37% SL); upper body below spinous dorsal fin is arched; large eye and short snout; moderate to shallow occipital pit, no significant suborbital pit; narrow and shallow interorbital space; interorbital ridges are not developed; suborbital ridge is with two small spines, one at midpoint and one at end; of five preopercular spines, first spine is reaching halfway across operculum, second spine is much shorter than third; 23 lateral line scales, some with tentacles; anterior mandibular pores are paired; markings include dark bar on cheek below rear half of eye, dark blotches on operculum above and below first preopercular spine, indistinct dark blotch on anterior midbody, series of dark spots and saddles along base of spinous and soft dorsal fins, dusky caudal peduncle, two dark bands on both soft dorsal and anal fins, clear caudal fin, dark base and tip of pectoral fin, and dark pelvic fins.

Live colors are unknown. Ethanol-preserved specimens are mottled, with a broad dark bar on the cheek below the rear half of eye, dark blotches on the operculum above and below the first preopercular spine, a series of concentrated dark spots and dusky saddles along base of spinous and soft dorsal fins followed by a dusky caudal peduncle, spinous dorsal fin has a dark blotch at the front and broadly over the mid fin, both soft dorsal and anal fins have two dark bands, caudal fin is clear, there are dark patches over the base and at the tip of the pectoral fins, and the pelvic fins are dusky (Victor, 2013).

Fisheries/aquarium/food/pharmaceutical values: Not reported.

Scorpaenodes africanus (Pfaff, 1933)

Image not available

Common name(s): African speckled scorpionfish

Distribution: Eastern Central Atlantic: Senegal and Annobon Island.

Ecology: It is a demersal species inhabiting rocky habitats in inshore areas.

Biology

Identification: It grows to a maximum length of 90.0 cm (TL). No detailed information is available

Food: It feeds on mobile invertebrates

Fisheries/aquarium/food/pharmaceutical values: Not reported.

Scorpaenodes albaiensis (Evermann and Seale, 1907)

Photo credit: John E. Randall.

Common name(s): Longfingered scorpionfish, splitfin scorpionfish

Distribution: Indo-West Pacific: from east coast of Africa to the Philippines, Tonga, and Queensland.

Ecology: It is a rare, solitary species that inhabits coastal coral reefs at depths of 0–35 m. It is very secretive in coastal coral reefs and rarely seen by divers, even at night.

Biology

Identification: Body of this species is moderately compressed anteriorly and is progressively more compressed posteriorly. Nape and anterior body are not arched. Body depth is shallow, less than head length. Slender tentacles are seen on supraocular, posterior lacrimal, parietal, nuchal, and preopercular spines. Short, broad tentacle is present on posterior edge of low membranous tube associated with anterior nostril. Distinct tentacles are absent from membranes of all fins. Pectoral-fin axil is without skin flaps. Scales are covering dorsal and lateral surfaces of head and upper margin of maxilla. Dorsal profile of snout is not steep, forming angle of about 20° to horizontal axis of head and body. Origin of first dorsal-fin spine is above supracleithral spine. Posterior margin of opercular membrane is reaching level of base of third dorsal-fin spine. Posterior tip of pectoral fin is extending beyond level of origin of anal fin. Origin of pelvic-fin spine is just below origin of pectoral fin. Posterior tip of depressed pelvic fin is not reaching origin of anal fin. Origin of first anal-fin spine is slightly posterior to level of last dorsal-fin spine. Gill rakers are relatively short and spinous; and slit is present behind fourth gill arch. Body coloration is variable and is mottled with reddish to grayish. Subopercle is with a large blackish spot, equal to or larger than pupil diameter. It grows to a maximum length of 12.0 cm (TL) (Motomura et al., 2009).

Fisheries/aquarium/food/pharmaceutical values: This species is of no importance to fisheries but sometimes occurs as bycatch in various nearshore fisheries between 3 and 8 m depth.

Scorpaenodes arenai (Torchio, 1962)

Photo courtesy: Costa, Francesco.

Common name(s): Messina rockfish

Distribution: Northeast Atlantic: Azores. Mediterranean; straits of Messina and of Sicily (Italy) and the Ionian Sea (Greece).

Ecology: It is a demersal species inhabiting rocky caves and sandy bottoms along the coast; depth range 14–31 m.

Biology

Identification: It has the following diagnostic features: suborbital ridge is with 4–6 spinous points; a second row of spines may be present below main ridge of spines on suborbitals; interorbital spines and the 2 additional spines on the frontals near the midline are absent; supplemental preopercular spine is usually present; dorsal fin is with 13 spines and 9 or 10 rays; and pectoral rays are 17–19. There are 50 vertical rows of scales on body and 6–8 gill rakers are seen on lower arch. Background coloration is red with bars on the body below the middle and end of the spinous dorsal fin, and between the soft dorsal fin and anal fin and at the base of caudal fin; and a dark blotch is present near the end of the spinous dorsal fin. It grows to a maximum length of 11.0 cm (TL).

Food: No data.
Reproduction: It probably spawns in September.
Fisheries/aquarium/food/pharmaceutical values: Not reported.

Scorpaenodes barrybrowni (Pitassy and Baldwin, 2016)

Photo courtesy: Reef Builders.

Common name(s): Not designated as it is a new species
Distribution: Curaçao, Caribbeans.
Ecology: This reef fish inhabits very deep waters; depth range 95–160 m.

Biology

Identification: This species is extremely colorful, tiny with an adult size of about only 4.0 cm. It is distinguished by the following combination of characters: dorsal-fin soft rays 8; pectoral-fin rays 16–17, rays 11–12 (from uppermost ray) noticeably longer than rest in smallest specimens; anal fin is with 3 spines, second longer, more robust than first or third; caudal-fin rays 25–27; spines on suborbital ridge 4; secondary suborbital ridge spines are absent; two prominent round to oblong pores in suborbital sensory canal immediately ventral to suborbital ridge; coronal, interorbital, upper post temporal and postorbital spines are absent; lateral line is incomplete, 8–10 pored scales extending from behind supracleithral spine to mid body; cirri are associated with nasal, supraocular, and parietal spines and are present on posteroventral projection of lacrimal and upper left quadrant of orbit; and there are no cirri associated with postocular, tympanic, supracleithral, and lower posttemporal spines. Body is relatively elongate and its depth at origin of dorsal fin is 30%–32% SL; and depth at caudal peduncle 9%–10% SL.

Body coloration is mostly orange to pinkish orange, paler on underside of head and belly; lower portion of body is with diffuse areas of translucent yellow pigment; and body lappets are pink to white. Nasal and supraorbital cirri are pink to orange; pupil is black; snout, anterior portions of jaws, and dorsal portion of cheek are pink to orange; posterior portions of jaws and ventral portion of remainder of head are pink to clear; series of indistinct, narrow reddish-brown bars present across dorsal portions of head and nape; and lappets on jaws are creamy yellow. Body is with five orange-brown to brown blotches; posterior most three are forming well-defined bars and anterior most blotch is on nape above rear corner of opercle and is extending posteriorly beneath first three dorsal-fin spines; second blotch is broadest, irregular in shape and is with pale center; third marking is a bar beneath anterior half of soft dorsal fin; fourth is the narrowest and most indistinct, located on caudal peduncle immediately behind bases of dorsal and anal fins; and fifth is strong, narrow, and situated on posterior edge of caudal peduncle. Spinous dorsal fin is mostly orange with scattered pale or pale pink streaks; and soft-dorsal fin is with broad, bright orange stripe at base; base of caudal fin is with strong orange bar; anal fin is with broad orange stripe at base; pelvic fin spine is whitish and pectoral fin base is pale pinkish (Baldwin et al., 2016).

Fisheries/aquarium/food/pharmaceutical values: Not reported.

Scorpaenodes bathycolus (Allen and Erdmann, 2012)

Common name(s): Deepreef scorpionfish

Distribution: Western Pacific: Indonesia.

Ecology: This reef-associated species occurs in relatively flat or gradually sloping sand-rubble bottom with scattered rock and coral outcrops in 60 m depth.

Biology

Identification: This species is diagnosed with the following characters. Dorsal fin is deeply notched and its spines are gradually increasing in length to seventh spine; remaining dorsal spines are gradually decreasing in length to penultimate spine; fourth soft dorsal ray is the longest; pectoral fins are with middle two rays which are greatly prolonged compared to other rays; lowermost pectoral four rays (and two prolonged rays immediately above) are unbranched with free, filamentous distal tips. Coloration of body is blotchy reddish with scattered brown spots and streaks. Dorsal fin is with large reddish vertically elongate patches along base and red distal tips on spinous portion; and a row of reddish spots along outer margin of soft dorsal fin. Caudal fin is with red base, and about four irregular red bars; anal fin is with scattered large red spots; pectoral fin is clear with faint reddish spots and a few scattered small brown spots; and pelvic fins are whitish. It grows to a maximum length of 6.3 cm (SL).

Fisheries/aquarium/food/pharmaceutical values: No data.

Scorpaenodes caribbaeus (Meek and Hildebrand, 1928)

Photo courtesy: Wikipedia.

Common name(s): Reef scorpionfish

Distribution: It is widespread through the western Atlantic. Florida, the Bahamas, Haiti, Jamaica, the Virgin Islands, Arcas Cay, Isla Lobos Mexico, Cozumel, Belize, Panama, Los Roques, Punta de Betín, Playa Brava, and Cabo de La Aguja off Venezuela.

Habitat and Activity. The reef scorpionfish is found in tropical reefs and rocky crevices in relatively shallow water, between 5 and 55 m in depth, where there is high visibility. Camouflage is therefore imperative for this species. They are nocturnal fishes and will spend the majority of the day resting, hidden in between coral. During the nighttime, they will hunt small fishes. This species is solitary in nature.

Population Ecology. The reef scorpionfish is a solitary, nocturnal, and ambush predator. Therefore, the fish will be found by itself and will rarely ever be found in groups. The only period that pairs of fish of this species can be seen is during mating season. Their solitary behavior may be attributed to the way they hunt and acquire food, which is by individually ambushing prey and hiding in coral. The life span of individuals of this species ranges between 10 and 15 years.

Biology

Identification: This species has the following diagnostic characters. Dorsal fin is with 13 spines and 9 soft rays. Pectoral fin is with 18–20 rays, some branched. Lacrimal bone is with 2 rounded lobes but without definite spines on ventral margin and usually with a small spine on its posterolateral face. Suborbital ridge is formed by infraorbital bones 2 and 3 and is with 1–4 spines and with a second row of 1–6 spines below main ridge. Coronal spines are usually present near ends of interorbital ridges. Occipital pit is absent. A slit is present behind the last hemibranch. Scales are ctenoid (rough to touch) in 40–45 rows. Lateral-line scales, 20–24, last on base of caudal fin. It has a terminal mouth with maxilla extending past the eye. It possesses small teeth arranged in narrow bands within the mouth. The tail of the fish is relatively short with respect to its body size; however, it is heavily notched. Body has highly variable coloration but is usually variously spotted and blotched with brown; a poorly defined dark blotch on posterior end of spinous dorsal fin; 1–3 brown bands on pectoral fins in juveniles, which become increasingly spotted with maturity; caudal peduncle is pale in specimens less than about 30 mm SL but spotted in specimens are larger than 30 mm. It is a small species, reaching 8.5 cm (SL).

Behavior: The majority of the reef scorpionfish's juvenile life is spent hiding among coral and rocky crevices as the relative size of the juvenile fish make them vulnerable prey for bigger carnivorous or omnivorous fish. This species uses its poisonous spines to protect itself from predators

Food and feeding: It is an omnivorous species, usually hunting small fish, aquatic snails, and small crustaceans. It is very good at camouflage, and this technique helps as an imperative part of its hunting arsenal. This species will feed mostly on small fish by remaining motionless in between rocky crevices or in coral. Due to its excellent camouflage, the reef scorpionfish is not seen by smaller fish seeking food or shelter making them the ideal food source. Its 65% of the food source is small herbivorous fish, with the remaining constitutes aquatic snails and small crustaceans. The mechanism of feeding employed by this fish is not very intricate. It will remain motionless and camouflaged, unseen by the smaller fishes.

When a prey comes by the predator, it will vacuum suck the prey toward itself, which is then eaten using the narrow row to teeth within the mouth. Due to this pattern of feeding, this species is considered as an ambush predator

Reproduction: Mating season for the reef scorpionfish occurs between May and June, when the waters are warm. This species is oviparous. The female fish will release on average between 2000 and 15,000 eggs directly into the water, immediately after the male reef scorpionfish will release sperm cells into the water, thus fertilizing the eggs. The transparent eggs then float to the surface. Within 2 days, the eggs hatch with the fry remaining close to the surface until they become over 2.5 cm in size after which they return to the coral reef beneath where they will begin their solitary journey, with no further parental help.

Fisheries/aquarium/food/pharmaceutical values: It is captured in shore to about 18 m. It is not infrequently taken as bycatch in seines and nets, where its venomous spines can pose a threat to fishermen, although the fish itself is not marketed.

Major threats: The reef scorpionfish is neither an endangered nor vulnerable species. However, the relative numbers of this fish has seen decline in the last few decades as although the species has no real natural predators. Many fishes are caught by humans for sale as exotic aquarium fish. This is very evident in the species population located between the Coast of Florida and the Bahamas, with heavy fishing and habitat destruction leading to reduced numbers.

Scorpaenodes corallinus (Smith, 1957)

Photo credit: Randall, J.E.

Common name(s): Coral scorpionfish

Distribution: Indo-Pacific: East Africa to French Polynesia; Ouvéa Atoll, Loyalty Islands.

Ecology: It is a benthic, solitary, and secretive (remaining concealed) species living on coral reefs; It is rarely seen at night at scuba depths; depth range 0–18 m.

Biology

Identification: It is distinguished with the following characters. Vertical scale rows in low 40s in body; suborbital ridge is with 3 spinous points in a single row; interorbital spines are absent or present as a lump; small spines are at midline between tympanic spines present;

and white bar on caudal peduncle and diagonal bar from pectoral to dorsal fin. It grows to a maximum length of 10.5 cm (TL).

Major threats: There have been no confirmed population declines in this species. However, because of its affinity with coral reefs, this species may be experiencing population declines due to habitat loss in parts of its range. However, significant global-level population declines are not suspected at this time.

Fisheries/aquarium/food/pharmaceutical values: Not reported.

Scorpaenodes elongatus (Cadenat, 1950)

Image not available

Common name(s): African spotted scorpionfish

Distribution: Eastern Atlantic: Senegal and Pointe Noire.

Ecology: It is a demersal species probably found over rocky bottoms.

Biology

Identification: It grows to a maximum length of 15.0 cm (TL). No other detailed information

Fisheries/aquarium/food/pharmaceutical values: No data.

Scorpaenodes englerti (Eschmeyer and Allen, 1971)

Photo credit: Randall, J.E.

Common name(s): Englert's scorpionfish

Distribution: Southeast Pacific: Easter Island and Desventuradas Islands.

Ecology: It is a demersal species inhabiting rocky intertidal areas; depth range 0–38 m.

Biology

Identification: It grows to a maximum length of 9.0 cm. No detailed information.

Fisheries/aquarium/food/pharmaceutical values: Not reported.

Scorpaenodes evides (Jordan and Thompson, 1914) (=*Scorpaenodes littoralis; Thysanichthys evides; Sebastella littoralis*)

Photo credit: John E. Randall.

Common name(s): Cheekspot Scorpionfish, Little Scorpionfish

Distribution: Indo-West and Central Pacific: Australia, New South Wales, Hawaii, Japan; subtropical locations in the Indian and Pacific Ocean, and the Red Sea.

Ecology: This reef-associated, benthic species occurs primarily in shallow tropical marine waters. It is commonly observed upside down on the roofs of caves; depth range 1–40 m.

Biology

Identification: Body of this species is not strongly compressed anteriorly. Nape and anterior body are not strongly arched. Body depth is moderate, less than head length. Short, broad tentacle is present on posterior edge of low membranous tube associated with anterior nostril. Pectoral-fin axil is without skin flaps. Numerous scales are covering occiput, area behind orbit, and cheek; few scales are in interorbital space; and no scales on lateral surface of maxilla and anteroventral surface of head. Well-exposed ctenoid scales are seen on lateral surface of trunk, and scales are becoming cycloid on abdomen. Exposed cycloid scales are seen on anteroventral surface of body and pectoral-fin base. Body scales are extending onto basal parts of fin rays or fin membranes, except for bases of pelvic fins. Lateral line is nearly straight from upper end of gill opening to caudal-fin base. Underside of dentary is with three distinct sensory pores on each side. There is no distinct longitudinal ridge on lateral surface of maxilla. Upper and lower jaws are with villiform teeth which are of Y-shaped patch on vomer. Underside of lower jaw is without ridges. Gill rakers are short and spinous; and slit is present behind fourth gill arch. Swim bladder is present. Dorsal profile of snout is steep. Origin of first dorsal-fin spine is above supracleithral spine. Posterior tip of pectoral fin is extending slightly beyond a vertical through origin of anal fin. Origin of pelvic-fin spine is slightly posterior to a vertical through origin of pectoral fin. Posterior tip of depressed pelvic fin is not reaching origin of anal fin. Origin of first anal-fin spine is anterior to a vertical through last dorsal-fin spine.

Head is reddish-black dorsally and laterally, reddish ventrally and is with distinct blackish blotch on subopercle. Trunk Island variegated, mottled with brown, red, and/or orange blotches and spots. Spinous portion of dorsal fin is blackish anterobasally. Soft rayed portion of dorsal fin is semitransparent with reddish areas basally and distally. Pectoral, pelvic, and anal fins are with reddish spots. Caudal fin is semitransparent with reddish submarginal band; and reddish spots are seen on caudal-fin rays. It grows to a maximum length of 10.5 cm (Motomura et al., 2010).

Fisheries/aquarium/food/pharmaceutical values: Not reported.

Scorpaenodes guamensis (Quoy and Gaimard, 1824)

Common name(s): Guam scorpionfish

Distribution: Indo-Pacific: from east coast of Africa to the Society Islands, and southern Japan south to Queensland, Australia.

Ecology: This reef-associated, solitary and cryptic, demersal species is found singly well camouflaged around inshore rock walls, crevices, shallow lagoons, channels and sandy bottoms with algae growth or sponges; depth range 0–20 m.

Biology

Identification: This species is distinguished by the following set of characters: pectoral-fin rays 17–20, some rays are branched in individuals over c. 40 mm SL; pelvic fin rays 5; dorsal-fin spines 13 and soft rays 7–9; anal spines, 3 and soft rays 4–5; total gill rakers 11–15; presence of nasal spine; and absence of interorbital and coronal spines. This species is exceedingly scaly with 40–46 body scale rows in longitudinal series; fine ctenoid scales are occurring around the eye, on preorbital, top of snout, about a third the way out on soft dorsal and anal fins, and nearly half way out the pectoral fins; lower jaw, chin, maxillary, and lips are naked; and cirri occur on tips of cranial spines and on lateral line pores. Palatine teeth are absent. Body is oblong, compressed. Adult breeding males appear to be distinguishable from adult females by having the first anal spine nearly as long as the second, whereas in females and immature the first spine is much shorter than the second. This small species reaches sexual maturity at a length of 5.0–6.5 cm. Head and body tan; and with darker gray reticulate markings. A large dark spot (blotch) is present on opercle. It grows to a maximum length of 12.5 cm (SL) (Motomura et al., 2016).

Food: It feeds nocturnally on small fish and swimming invertebrates such as small shrimps, crabs, and polychaetes.

Fisheries/aquarium/food/pharmaceutical values: Not reported.

Scorpaenodes hirsutus (Smith, 1957)

Photo credit: John E. Randall.

Common name(s): Hairy scorpionfish

Distribution: Indo-Pacific: Red Sea and East Africa to the Marquesan and Pitcairn islands; north to Ryukyu and Hawaiian Islands; south to Australia.

Ecology: This benthic species inhabits lagoon and seaward reefs in the euphotic zone at depths of 8–58 m.

Biology

Identification: This species is distinguished by the following characters: body is covered with short hairlike filaments; vertical scale rows of about 30; presence of 4 spines on suborbital ridge, first on lateral face of lachrymal bone; usually 1 additional spine below suborbital ridge at level between second and third spines; presence of interorbital spines; and absence of small spines at midline between tympanic spines. It grows to a maximum length of 5.6 cm (SL).

Major threats: The hairy scorpionfish may be undergoing localized declines due to habitat degradation. However, owing to its broad distribution of this species, this habitat loss is not considered to be a major threat for this species at present.

Fisheries/aquarium/food/pharmaceutical values: No data.

Scorpaenodes immaculatus (Poss and Collette, 1990)

Photo courtesy: Stuart G Poss, Bruce B Collette.

Common name(s): No common names found

Distribution: Western Indian Ocean: known only from Walters Shoals.

Ecology: This shallow water, demersal species is nonmigratory and has a depth range 40–49 m.

Biology

Identification: This species is diagnosed by the following characters: dorsal fin has 14 spines and 8 rays; anal fin is with 3 spines and 5 rays; pectoral fin is with 19 rays, 2–9 rays are branched; lateral line is with 27 scales; with last scale is exceeding over base of caudal fin; gill rakers 16 (left) and 17 (right) and caudal fin is with 12 branched rays. Head is moderately large and deep posteriorly. Interorbital one is with suborbital ridge but without spine. Nasal spine is stout and slightly curved. Preocular spine is sharp; supraocular spine is stout; and postocular spine is strong. Coronal spines are absent. Operculum is with 2 prominent strong spines extending posteriorly beyond edge of bone. Maxilla is extending almost to vertical with posterior border of orbit. Operculum is covered with scales except posteriorly. Body scales are of moderate size with about 50 vertical rows. Underside of head is without scales. Scales on posterior part of head and on body are ctenoid. Cirri are absent on body. Dorsal fin origin is above middle of opercle. Anterior dorsal fin spines are tipped with simple cirri. Caudal fin is somewhat rounded. Body coloration is uniformly red with small red spots over soft fin rays. Spinous dorsal fin is with more red pigment proximally. It grows to a maximum length of 8.9 cm (SL) (Poss and Collette, 1990).

Fisheries/aquarium/food/pharmaceutical values: Not reported.

Scorpaenodes insularis (Eschmeyer, 1971)

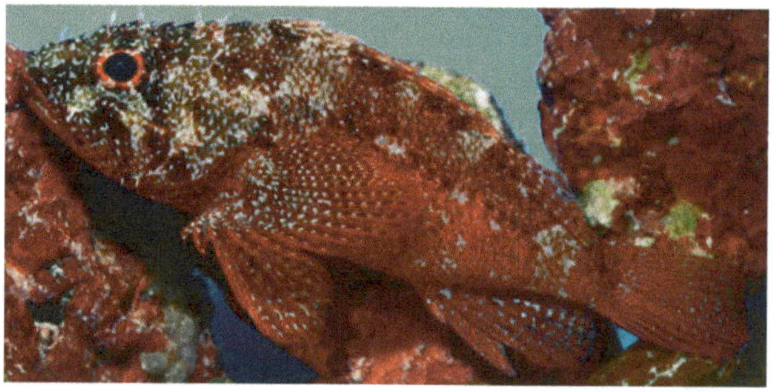

Photo courtesy: Osmar Luiz Jr.

Common name(s): Red scorpionfish.

Distribution: Eastern Atlantic: St. Paul's Rocks, St. Helena, and Ascension Islands.

Ecology: It is a demersal fish occurring in inshore, and it has been collected from rocks, caves and crevices; small individuals may inhabit rubble; depth range 3–35 m.

Biology

Identification: It grows to a maximum length of 10.0 cm (SL). No further information.

Food: Its diet consists mainly of smaller crustaceans such as (mostly juvenile) shrimp and crabs.

Fisheries/aquarium/food/pharmaceutical values: It is an aquarium fish and a tank of at least 800 L is recommended for its maintenance.

Scorpaenodes investigatoris (Eschmeyer and Rama-Rao, 1972)

Image not available

Common name(s): Not designated

Distribution: Western Indian Ocean: Pakistan.

Ecology: It is a demersal species.

Biology

Identification: It grows to a maximum length of 7.3 cm (SL). No further information.

Fisheries/aquarium/food/pharmaceutical values: No data.

Scorpaenodes kelloggi (Jenkins, 1903)

Photo credit: John E. Randall.

Common name(s): Kellogg's scorpionfish

Distribution: Indo-Pacific: East Africa to the Hawaiian, Line, and Society islands, north to Taiwan, south to northern Australia; throughout Micronesia.

Ecology: It is a benthic species that lives on shallow coral reefs; occurs in coralline areas from near shore to the outer reef slope; common but rarely seen under rocks or rubble where it hides during the day; depth range 6–24 m.

Biology

Identification: It is distinguished by its dorsal fin which is with 13 spines and 7–8 soft rays; anal fin is with 3 spines and 5 soft rays; 3–4 spinous points on suborbital ridge; presence of interorbital spines; and absence of no small spines at midline between tympanic spines. Body coloration is white mottled with dark brown, four bars; and last bar is on the caudal peduncle. This species can reach a maximum length of 5.0 cm (TL).

Major threats: There have been no confirmed population declines for this species. However, because of its affinity with coral reefs, it may be experiencing population declines due to habitat loss in parts of its range.

Fisheries/aquarium/food/pharmaceutical values: Not reported.

Scorpaenodes minor (Smith, 1958)

Photo credit: John E. Randall.

Common name(s): Minor Scorpionfish

Distribution: It is native to the Indian Ocean and the western Pacific Ocean; Indo-Pacific: East Africa to Samoa and the Austral Islands, north to the Ryukyu Islands.

Ecology: This species mainly inhabits coral reefs or reefs along the coastal environment; with camouflage ability, it is usually found hiding in coral reefs, rock crevices, or reef flats of mixed sand and rubble bottom and is not easy to be found; depth range 1–52 m.

Biology

Identification: This is a solitary species with cryptic coloring. It is distinguished by its dorsal fin which has 12–14 spines and 7–9 rays; pectoral fin is with 14–16 rays; and circular caudal fin. Body of this species is extended and its sides are flat. Head is large and is sharp jagged and serrated. Eye is large, is in lateral position, and is up to the dorsal margin. Mouth is large. Body coloration is brown or reddish brown; anal and caudal fins are light colored; and dark red stripes are seen on fins. It grows to a maximum length of 5.2 cm (TL).

Food: It feeds on other fishes and crustaceans.

Fisheries/aquarium/food/pharmaceutical values: It has no economic value and this small fish may be of use only for academic research.

Scorpaenodes muciparus (Alcock, 1889)

Photo credit: Osmany, Hamid Badar.

Common name(s): Not designated

Distribution: Indo-West Pacific: Bay of Bengal, Arabian Sea, Gulf of Oman, Pakistan.

Ecology: It is a demersal species; depth range 42–320 m.

Biology

Identification: This species is distinguished by its dorsal fin with 13 spines and 9 rays; anal fin with 3 spines and 5 rays; and pectoral fin with 19 rays. Pectoral fin is reaching to over soft rays. Spines on head are well developed. First preopercular spine is long; second is small and close to first than to third; and fourth is broad. Tentacle and other fleshy appendages are few. Dorsal and anal fins are fairly long. Scales on flanks are ctenoid; vertical scale rows, 50; and lateral line scales are 23 plus 2 in caudal fin. Scales on chest and pectoral base are cycloid. Dorsal parts of head are scaled; and underside of head is unscaled or with cycloid scales. There are 18–20 gill rakers including rudiments. Upper arch has 6 short spiny rakers and lower arch has 9–10 rakers. There is a small slit behind fourth gill ach. Dark pigments are most intense on pectoral fin and as a spot on spinous dorsal fin. It grows to a maximum length of 8.7 cm (SL) (Eschmeyer, 1969).

Fisheries/aquarium/food/pharmaceutical values: Not reported.

Scorpaenodes parvipinnis (Garrett, 1864)

Photo credit: John E. Randall.

Common name(s): Lowfin scorpionfish, coral scorpionfish, shortfinned scorpionfish

Distribution: Indo-Pacific: Red Sea to the Marquesan and Tuamotu Islands, north to the Ryukyu and Hawaiian Islands, south to Lord Howe Island.

Ecology: This benthic species inhabits areas of rich coral growth from near shore to the outer reef slopes; depth range 1–49 m.

Biology

Identification: This species has the following features: interorbital area and snout scaled; presence of small spines at midline between tympanic spines; presence of interorbital spines; extra spines are usually present on upper rear margin of eye after supraocular spine; dorsal spine is not usually longer than orbit diameter; and body is often covered with small skin flaps. It is also distinguished from other species of *Scorpaenodes* by having suborbital ridge with more than 5 spinous points, usually with 10 or more in adults. It is a mottled reddish to brown scorpionfish often with a large whitish area on the anterior body and a pale bar on the caudal peduncle. It can reach a maximum length of 14 cm (TL).

Major threats: There have been no confirmed population declines for this species. However, because of its affinity with coral reefs, it may be experiencing population declines due to habitat loss in parts of its range.

Fisheries/aquarium/food/pharmaceutical values: Not reported.

Scorpaenodes quadrispinosus (Greenfield and Matsuura, 2002)

Photo credit: David Greenfield.

Common name(s): Fourspine scorpionfish

Distribution: It is found distributed in the eastern Indian Ocean (Christmas Island, Australia) and in the Pacific Ocean (East Asia, Micronesia, Melanesia, Timor Sea, and Polynesia).

Ecology: This demersal species occurs in high-energy habitats, on outer reef flats, and surge channels of exposed reefs; depth range 0–12 m.

Biology

Identification: This species is distinguished by its pectoral fin rays which are usually 17 and 39–43 vertical scale rows in body. Body of this species is not strongly compressed anteriorly and is progressively compressed posteriorly. Nape and anterior body are not strongly arched. Body depth is relatively deep and is less than head length. Numerous tiny, slender tentacles are present on head and body. Pectoral fin axil is without skin flaps. Scales are covering dorsal and lateral surfaces of head and midlateral surface of maxilla. Well-exposed ctenoid scales are seen on lateral surface of trunk. Scales are becoming cycloid on abdomen. Exposed cycloid scales are present on anteroventral surface of body. Body scales are extending onto basal fin rays or membranes, except for bases of pelvic fins. Lateral line is nearly straight from upper end of gill opening to caudal-fin base. Origin of first dorsal-fin spine is above supracleithral spine. Posterior margin of opercular membrane is reaching a vertical through base of third dorsal-fin spine. Posterior tip of pectoral fin is not reaching a vertical through origin of anal fin. Origin of pelvic-fin spine is slightly posterior to a vertical through origin of pectoral fin. Posterior tip of depressed pelvic fin is not reaching origin of anal fin. Origin of first anal-fin spine is anterior to a vertical through last dorsal-fin spine. Coloration of body is strongly variegated, mainly from pale pink to dense red, with 4 indistinct, vertical, dark bands on lateral surface of body; first band is above opercle, second between middle of spinous portion of dorsal fin and pectoral fin, third below 9th–12th spines of dorsal fin, and fourth is below soft rayed portion of dorsal fin. All fins, except for spinous portion of dorsal fin, are pinkish to reddish, with small reddish to blackish spots scattered on rays. Caudal-fin base is with a vertical, reddish band, its width less than pupil diameter. There are no distinct black blotches on opercle or subopercle. It grows to a maximum length of 10.0 cm (SL).

Fisheries/aquarium/food/pharmaceutical values: Not reported.

Scorpaenodes rubrivinctus (Poss et al., 2010)

Image not available

Common name(s): Not designated

Distribution: Eastern Pacific: Galápagos and Cocos Islands (Costa Rica).

Ecology: This bathydemersal species is often observed from rocky reefs, often in association with sponges, upon the platform and along the margins of the Galápagos archipelago; depth range 160–412 m.

Biology

Identification: Head of this species is large (45%–47% SL). Snout is notably extended (13%–15% SL). All head spines are sharp, well developed, except nasal spines, which are relatively short when present Body is laterally compressed, deepest between third dorsal spine and pelvic fin. Dorsal fin is with 13 spines and 8–10 soft rays; it is originating above supracleithral spine; third or fourth spine is the longest; penultimate is shortest; and fin membranes between spines are notably incised. Anal fin which is with 3 spines and 5 soft rays; spines are prominent and are slightly curved; and second spine is the longest and thicker than first or third. Pectoral fin is with 8–10, usually 9 ventral most unbranched rays, 9–10 branched rays in middle of fin, and a single unbranched ray dorsally. Longest ray of pectoral fin (dorsal most unbranched ray) is reaching to base of third anal spine. Pelvic-fin spine is relatively large, reaching more than half the distance from pelvic-fin insertion to the anal-fin origin. Caudal fin is with 5 or 6 branched rays in ventral half and 6 or rarely 7 branched rays in dorsal half. Coloration of body is pink with orange bars; first bar is faint and is behind head between first three dorsal spines. Pectoral-fin rays are checkered with orange, membrane clear. Caudal-fin rays are orange. Belly and chest are pale pink, becoming orange on ventral 40% of body. Posterior portion of caudal peduncle immediately anterior to caudal-fin base is often with five to nine distinct but relatively minute melanic spots. It grows to a maximum length of 10.1 cm (SL) (Poss et al., 2010).

Fisheries/aquarium/food/pharmaceutical values: Not reported.

Scorpaenodes scaber (Ramsay and Ogilby, 1886)

Photo credit: John E. Randall.

Common name(s): Pygmy scorpionfish, cheekspot scorpionfish, pygmy rockcod

Distribution: Indo-Pacific.

Ecology: It is a marine and brackish water species as it is found from estuary reefs to offshore reef habitats. It is predominantly found in shallow coral, gravel, or rock formations; sheltered lagoons and caves in the shoreline to the outer reefs; and depth range is 3–30 m. With camouflage ability, it often hides its body.

Biology

Identification: The cheekspot scorpionfish can be distinguished from other-related species by the dark blotch on the operculum and the red-spotted fins. It has small ctenoid scales and 13 venomous dorsal fin spines. There are small leaf-like appendages on the head and mouth. Its lateral line is prominent and almost straight. Its coloration is variable from brown to pink or red. The maximum recorded length for this species is 12 cm (TL).

Food: It feeds on other fishes and crustaceans

Fisheries/aquarium/food/pharmaceutical values: It has no economic value, and it is only for academic research.

Scorpaenodes smithi (Eschmeyer and Rama-Rao, 1972)

Photo courtesy: CSIRO.

Common name(s): Little scorpionfish, Smith's scorpionfish

Distribution: Indo-West Pacific.

Ecology: This demersal species usually occurs deep along drop-offs in caves and on rock walls but in some locations with strong currents in shallow water amongst sponges. It is also found in sand and mud bottoms in 42–110 m.

Biology

Identification: It grows to a maximum length of 7.7 cm (TL). No other information is available.

Fisheries/aquarium/food/pharmaceutical values: Not reported.

Scorpaenodes steenei (Allen, 1977)

Common name(s): Steene's scorpionfish

Distribution: Eastern Indian Ocean: endemic to southwestern Australia.

Ecology: This reef-associated species is found inshore on reef areas; depth range 1–17 m.

Biology

Identification: It is a pinkish scorpionfish with blackish markings on the head and a black spot on the middle of the gill cover, a red snout and lips, dusky reddish-gray bars and blotches on body, reddish bars on caudal peduncle, and red and whitish bands on the fins. It grows to a maximum size of 10 cm (SL).

Fisheries/aquarium/food/pharmaceutical values: Not reported.

Scorpaenodes steinitzi (Klausewitz and Frøiland, 1970)

Photo credit: John E. Randall.

Common name(s): Not reported

Distribution: Western Indian Ocean: Red Sea.

Ecology: It is a demersal species.

Biology

Identification: It grows to a maximum size of 5.5 cm (SL). No other information is available.

Fisheries/aquarium/food/pharmaceutical values: Not reported.

Scorpaenodes tredecimspinosus (Metzelaar, 1919)

Photo courtesy: Carvalho Filho, Alfredo.

Common name(s): Deepreef scorpionfish, thorny scorpionfish

Distribution: Western Atlantic: North Carolina, USA, and the Bahamas to São Paulo, Brazil.

Ecology: This reef-associated species occurs mostly on deep coral reefs and steep, coral-covered slopes; depth range 7–82 m.

Identification: Head of this species is very bony with numerous spines. There is no pit on top of head behind eyes. There are no small spines below main row of 1–3 spines on bony ridge under eye. Bone above front of mouth is with 2 rounded lobes on lower edge and a small spine at rear side. Teeth are seen on center of roof of mouth and not on sides. There is no small slit behind last gill arch. Dorsal fin is with 13 spines and 9 rays and anal fin is with 3 spines and 4–5 rays. Last dorsal and anal rays are branched. Pectoral fin is with 17 rays which are long and pointed with longest rays on center. While some of these pectoral rays are branched, lowest rays are unbranched. Scales are rough and head is scaly. Body coloration is red-brown to yellowish. There is a black blotch is at end of spiny dorsal fin; and reminder of fins are with faint brown spots or transparent. Rear of tail base is white followed by a red bar which extends onto base of tail fin. It grows to a maximum length of 6.5 cm (TL).

Food: It tends to eat primarily crabs, molluscs, and other benthic mobile invertebrates.

Fisheries/aquarium/food/pharmaceutical values: This species is not utilized. In the Gulf of Mexico off Mexico, *Scorpaena* spp. are taken incidentally in fisheries targeting rays using gillnets and/or castnets along the coast.

Scorpaenodes tribulosus (Eschmeyer, 1969)

Photo courtesy: David Catania.

Common name(s): Not designated

Distribution: Western Indian Ocean: Somalia and Kenya.

Ecology: It is a demersal species; depth range 73–176 m.

Biology

Identification: Its diagnostic features are dorsal fin is with 13 spines and 8 soft rays, last soft ray is double; anal fin has 3 spines and 5 rays, last soft ray is double; pectoral rays 19, some rays branched; and pectoral fin is reaching to level of first anal spine. Spines on head are fragile. Underside of the head is covered with ctenoid scales rather than being naked or covered with cycloid scales. It grows to a maximum size of 5.1 cm (SL) (Eschmeyer, 1969).

Fisheries/aquarium/food/pharmaceutical values: Not reported.

Scorpaenodes varipinnis (Smith, 1957)

Photo credit: John E. Randall.

Common name(s): Blotchfin scorpionfish

Distribution: Indo-Pacific: East Africa to Micronesia, north to Taiwan, south to Australia; including Society Island.

Ecology: It is reef-associated species living in and around sponges and coralline areas; depth range 0–200 m.

Biology

Identification: This species is diagnosed by its dorsal fin which has 13 spines and 8–9 soft rays; and anal fin is with 3 spines and 5 soft rays. Body of this species is compressed; relatively deep; and body depth is 2.6–3.0 in SL. Interorbital and coronal are spines absent. There are three suborbital spines. Coloration of body is with poorly defined large reddish brown blotches. There are no distinct dark blotches on opercle or subopercle. A blackish arc is seen at base of pectoral fins. Caudal fin peduncle is with a white bar followed by a narrow reddish bar. This species can reach a maximum length of 13 cm (TL).

Fisheries/aquarium/food/pharmaceutical values: Not reported.

Scorpaenodes xyris (Jordan and Gilbert, 1882)

Photo courtesy: Robertson, Ross.

Common name(s): Rainbow scorpionfish

Distribution: Eastern Pacific: California, USA to Peru and the Galapagos Islands.

Ecology: This reef-associated common but inconspicuous species is found in cracks, crevices, under ledge overhangs and other recesses on rocky reefs, steep slopes, and especially on walls. Juvenile fishes normally stay near the protective spines of long-spined urchin, *Diadema mexicanum*; depth range 0–30 m.

Biology

Identification: There is a longitudinal ridge under eye is with 2–3 spines. Absence of pit on top of head behind eyes. This small fish has a large mouth and is covered with weakly ctenoid scales. Teeth are seen on center of roof of mouth and not on sides. There are no teeth on the palatine. A small slit may or may not be present behind last gill arch. Dorsal fin is with 13 spines and 10 rays and anal fin is with 3 spines and 5 rays. Last dorsal and anal rays are branched. Pectoral is with 18 rays; some rays are branched; lower 10 rays are unbranched and thickened; and longest rays are at center. Scales are rough and head is scaly. Skin flaps are absent on scales of body. Small spiny rays at top and bottom of base of tail fin. This cryptic and rare species is small, with variable, mottled red coloration. A dark rounded spot is seen on lower corner of gill cover. A lateral stripe of alternating red and white dashes is found, and a white blotch is on back at meeting point of first and second dorsal fins. It grows to a maximum length of 15.0 cm (TL). Larvae of this species are oceanic and are widely distributed in tropical eastern Pacific (TEP). They have distinct and prominent bladelike ridges arising from the parietal bones with 2 posterior spines.

Fisheries/aquarium/food/pharmaceutical values: Not reported.

Scorpaenopsis altirostris (Gilbert, 1905)

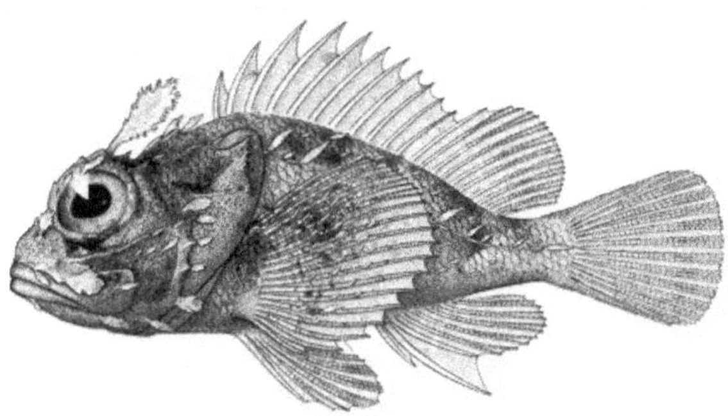

Image credit: Smithsonian Institution, National Museum of Natural History, Department of Vertebrate Zoology, Division of Fishes.

Common name(s): Not designated

Distribution: Eastern Central Pacific: Hawaii.

Ecology: This demersal species inhabits coarse sand bottom with broken shells and coral; depth range 79–134 m.

Biology

Identification: This species is diagnosed by its dorsal fin which has 12 spines and 9 soft rays; and anal fin is with 3 spines and 5 soft rays. In this species, about 1/4 of orbit is extending above dorsal profile of head. Spines seen dorsally on head are relatively low; supraorbital is very broad-based; postorbital is more blade-like and is angling laterally. Interorbital ridges are originating medial to rear edge of posterior nostrils. Occipital pit is moderately deep; front edge is incurved and with no distinct ridges at margins. Suborbital pit is small but relatively deep, below anterior half of orbit. Suborbital ridge is with 4 spines, the first on lacrimal below ridge. Upper opercular spine is divided into 2–4 points. Lacrimal is with 2 spines over maxilla, the anterior is angling forward and the posterior is projecting backward. Supraocular tentacle is varying from shorter than supraocular spine to longer than orbit diameter. Pectoral fins are long and are extending to a vertical base of first and second anal soft rays. It grows to a maximum length of 4.6 cm.

Fisheries/aquarium/food/pharmaceutical values: Not reported.

Scorpaenopsis barbata (Rüppell, 1838)

Photo courtesy: Wikipedia.

Common name(s): Bearded scorpionfish

Distribution: Western Indian Ocean, from the Red Sea to the coasts of Somalia, of Madagascar and South Africa.

Ecology: This reef-associated species lives along sandy beaches, rocky coastlines and coral reefs and in shallow waters at depth up to 30 m. The feathery fins which cover the scorpion fish body help in the camouflage of the animals in the surrounding coral. Also, the coral-like frilly, fleshy flaps of skin protruding from its chin add to the species camouflage ability and help the fish to blend with its environment.

Biology

Identification: It has a broad spiny head with a large mouth and feathery fins. Many fleshy outgrowths of skin protrude from their chin (hence the name of bearded scorpionfish). Its dorsal fin has twelve spines and nine soft rays. Anal fin has three spines and five soft rays. Upper opercular spine is broad with a strong median ridge. Space between opercular spines is with 1 or more ridges. Paired interorbital ridges are prominent, separated by a channel more than twice width of 1 ridge, and ending at a ridge extending medially from postocular spine along incurved edge of moderately deep occipital pit. An oblique ridge is lateral to each interorbital ridge from above middle of orbit, converging posteriorly with interorbital ridge. Tympanic spines are about as long as postocular spines, close behind but separated from ridge extending medially from postocular spines. Median interorbital ridge is prominent, reaching well into anterior interorbital space. First and second suborbital spines are oblique, sharply ridged, and with supplemental ridges. This fish is brightly colored and heavily camouflaged. Body is heavily pigmented with dark brown or reddish brown, mottled with whitish and blackish blotches; often with a dark brown bar extending ventrally from posterior half of eye and broadening onto lower cheek. Maximum length female is 25 cm and male 17 cm (TL).

Food and feeding: This ambush hunter usually rests during the daylight in crevices in the corals and from its position it suddenly leaps at its unsuspecting prey which could be small fish and crustaceans.

Reproduction: Eggs hatch in about 2 days into tiny fry that remain near the water surface until they grow bigger and they reach about 5 cm in length, they swim down into the ocean to join the reef community.

Fisheries/aquarium/food/pharmaceutical values: Not reported.

Scorpaenopsis brevifrons (Eschmeyer and Randall, 1975)

Photo credit: John E. Randall.

Common name(s): Bigmouth scorpionfish, shortsnout scorpionfish

Distribution: Eastern Central Pacific: Midway and Hawaiian Islands.

Ecology: This reef-associated, nonmigratory, and benthic species is usually found on rocky bottom or coral reefs but also on sand and rubble bottom; depth range 1–38 m. It is uncommon on rubble or barren reefs at scuba depths.

Biology

Identification: Body of this species is stout with a round head and protruding eyes. It has the following identifying features: pectoral rays usually 19; a blunt head profile, long jaw (21%–27% SL); no hump behind head; longest dorsal spine 4–6; 5 spines on the suborbital ridge and 3 on the lachrymal bone overlying the maxillary. Body coloration is variable. There is no black band across anal and pelvic fins. This species can grow to a maximum SL of 13.3 cm.

Fisheries/aquarium/food/pharmaceutical values: No data.

Scorpaenopsis cacopsis (Jenkins, 1901)

Photo credit: John E. Randall.

Common name(s): Jenkin's scorpionfish

Distribution: Eastern Central Pacific: Hawaii.

Ecology: This reef-associated, solitary species is benthic, and tends to live beneath ledges or near cave entrances on the outer edge of reefs; depth range 4–60 m.

Biology

Identification: Body of this species has a tapering head with coloration of orange with ornate pattern of black lines and multicolored flecks. Other identifying features of this species include vertical scale rows 50–55; lachrymal bone with 2 spines over maxillary, first points down and forward, second points down and back, sometimes second split distally into 2 points; suborbital ridge usually with 5 spines, sometimes some spines split in larger specimens. It grows to a maximum length of 51.0 cm (TL).

Food: It normally feeds on small fishes (holocentrid, acanthurid, and *Aulostomus chinensis*).

Major threats: Its population has been greatly reduced due to fishing pressure by spear-fishermen. This anecdotal information is not however supported by quantitative data and is therefore not fully understood.

Fisheries/aquarium/food/pharmaceutical values: This species is of no importance to commercial aquariums but is commonly sought after in fisheries as a prized food fish. Scorpaenopsis cacopsis is caught in small-scale commercial and recreational fisheries via spear fishing.

Scorpaenopsis cirrosa (Thunberg, 1793a)

Photo credit: John E. Randall.

Common name(s): Weedy stingfish, weedy scorpionfish; frilly scorpionfish

Distribution: Northwest Pacific: southern Japan and China south to Hong Kong and northern Taiwan.

Ecology: This is a demersal species; depth range 3–91 m.

Biology

Identification: This species has the following diagnostic features: dorsal fin is with 12 spines and 8–10 soft rays; anal fin is with 3 spines and 5 soft rays; pectoral fin is with 17–18 soft rays; and gill rakers 13–16. About one-fourth of eye is projecting above dorsal profile of head; interorbital space is hemispherical and not deep; and fourth or fifth dorsal spine is usually the longest. Body color will vary with depth and environment. It grows to a maximum length of 23.1 cm (SL).

Fisheries/aquarium/food/pharmaceutical values: It is not a commercial food species.

Scorpaenopsis cotticeps (Fowler, 1938)

Photo credit: John E. Randall.

Common name(s): Sculpin scorpionfish

Distribution: Indo-West Pacific: Japan, Philippines, South China Sea, Queensland (Australia), Seychelles, Somalia, and the Gulf of Aden.

Ecology: This demersal species inhabits shallow rocky reefs; over rubble or rubble and sand bottom; depth range 15–70 m. With camouflage ability, it often hides the body and therefore it is not easy to be found.

Biology

Identification: Body of this species is long and oval with slightly flat side. Head profile is blunt; and eyes are large and protruded a little from the dorsal contour of head. There is no hump behind head. Head length is of 3.6–4.2 times the eye diameter. Mouth is large and is in top position. Eye tentacles are with different shapes and length. Dorsal fin is with 12 hard spines and 9 soft rays; anal fin is with 3 spines and 5 soft rays; and pelvic fin is with 18–19 soft rays. There are 33–37 vertical scale rows in body; and 4 predorsal scales. Gill flap is with one dark spot. It is a mottled reddish-brown scorpionfish, often with a broad pale bar on the body behind the soft dorsal and anal fins, and sometimes a pale "saddle" on the back below the middle of the dorsal fin. It grows to a maximum size of 7.5 cm (TL).

Food: Its food consists of other fishes and crustaceans.

Fisheries/aquarium/food/pharmaceutical values: This small fish, in addition to academic research, has no economic value. Occasionally, it is caught at the end of trawling.

Scorpaenopsis crenulata (Motomura and Causse, 2011)

Photo courtesy: H. Motomura.

Common name(s): Serrated deepwater scorpionfish
Distribution: Southwestern Pacific: Wallis and Futuna Islands.
Ecology: It is a small deepwater (benthopelagic) species inhabiting depths up to 600 m.

Biology

Identification: This species is distinguished by the following characters: pectoral-fin rays 17; scale rows in longitudinal series 37; and pored lateral-line scales 19. Body of this species is moderately compressed anteriorly and is progressively more compressed posteriorly. Nape and anterior body are moderately arched. Body depth is slightly less than head length. Numerous small papillae and tiny tentacles are seen on head. Posterior lacrimal spine tentacle is linked posteriorly to head by skin. A tentacle on posterior edge of low membranous tube is associated with anterior nostril. Supraocular tentacle is minute and its length is less than supraocular spine length. Underside of lower jaw is with a pair of tentacles. Mouth is large, slightly oblique; and posterior margin of maxilla is extending well beyond a vertical through posterior margin of orbit. Lower jaw is with a symphysial knob. Upper and lower jaws ARE with a band of conical teeth. There are no teeth on middle of vomer; and two rows of small teeth are on each anterior side of vomer, becoming one row posteriorly, forming a V-shaped patch on vomer. There are no palatine teeth. Underside of lower jaw is smooth, without ridges. Dorsal profile of snout is steep, forming an angle of about 50° to horizontal axis of head and body. Nasal spine is simple, sharp, directed dorsally, and its length greater than anterior nostril diameter.

Origin of first dorsal-fin spine is above supracleithral spine. Posterior margin of opercular membrane is reaching the vertical through fourth dorsal-fin spine base. Posterior tip of pectoral fin is reaching a vertical through second anal-fin soft ray base. Origin of pelvic-fin spine is just below origin of pectoral fin. Posterior tip of depressed pelvic fin is extending beyond anus but not reaching first anal-fin spine base. Origin of first anal-fin spine is just below origin of last dorsal-fin spine. Head and body are mostly yellowish-white. Maxilla is with poorly defined dark blotches posteriorly. Inside of mouth is white; and there is no black mark at front of upper jaw behind teeth. All fins, except for pectoral fin, are whitish, without distinct dark markings. Outer surface of pectoral fin is without distinct marking; and there is a large black blotch distally between third and seventh rays (Motomura and Causse, 2011). It can reach a maximum length of 4.9 cm (SL).

Fisheries/aquarium/food/pharmaceutical values: Not reported.

Scorpaenopsis diabolus (Cuvier, 1829)

Photo credit: John E. Randall.

Common name(s): False stonefish, devil scorpionfish

Distribution: Indo-Pacific: Red Sea and East Africa to the Hawaiian Islands and French Polynesia, north to southern Japan, south to the Great Barrier Reef and New Caledonia.

Ecology: This reef-associated benthic species is found on the seabed among rubble, seaweed-covered rocks or on rocks encrusted with coralline algae on reef flats, lagoons, and the seaward side of reefs. It is sometimes partially covered with sediment; depth range 1–70 m.

Biology

Identification: The false stonefish has a large rather flattened head with a wide, horizontal mouth; a humped back, and a tapering body. Its back is arched and is frequently coated with filamentous or coralline algae Its dorsal fin has 12 spines and 8–10 soft rays; anal fin has 3 spines and 5–6 soft rays; and pectoral fin rays are usually 18. There are about 45 vertical scale rows in body; lachrymal bone is with 2 or 3 spines over maxillary, first points forward, followed by 1 or by 2 close-set spines which point down and back; suborbital ridge is with 4 or more spinous points, usually more than 8 or 10 points, not in a row and of various sizes; and a shallow pit is seen below front corner of eye. It has a large head, pointed snout and short dorsal fin spines. Small eyes that don't stick out of the head, with small shallow depression near the eyes. Skin is rough with low conical projections, spines, and tassels. Coloration of body is a combination of mottled gray and white with reddish-brown blotches. Inner sides of the broad pectoral fins have orange, black, and white blotches and the fins can be "flashed" as a warning. It can reach a maximum length of 30.0 cm (TL).

Behavior: This species exhibits biofluorescence, that is, when illuminated by blue or ultraviolet light, it reemits it as red, and appears differently than under white light illumination. Biofluorescence may assist in intraspecific communication and camouflage. Further, when threatened, the fish will flare its pectoral fins exposing the bright colors on the inner surface. This is believed to startle potential predators as a warning.

Food and feeding: This carnivorous fish feeds mainly on fishes. It is a predator that relies on its camouflage to catch passing prey.

Fisheries/aquarium/food/pharmaceutical values: Not reported.

Scorpaenopsis eschmeyeri (Randall and Greenfield, 2004)

Photo credit: John E. Randall.

Common name(s): Not reported

Distribution: Southwestern Pacific: Fiji and New Caledonia; Great Barrier Reef in north Queensland.

Ecology: It is a benthic species frequenting rocky reef areas; depth range 1–24 m.

Biology

Identification: This species has the following diagnostic features: dorsal fin has 12 spines and 9 rays; anal fin has 3 spines and 5 rays; pectoral fin rays 17–18 (rarely 18); longitudinal scale series 44—48; body depth is 2.8–3.05 in SL; head length is 2.2–2.3 in SL; and snout length is 3.0–3.15 in head length. Nearly one-half of orbit extending above dorsal profile of head. A pair of interorbital ridges is flaring posteriorly, then curving medially to join slight incurved ridge at front of occipital pit; and median interorbital ridge is extending half way back in interorbital space. Occipital pit is shallow and not flat. Coronal spines and pretympanic spines are absent. Suborbital pit is not well developed; suborbital ridge is with 4 spines, the first on lacrimal; two ventral spines on lacrimal, one directed mainly anteriorly and the other is curving posteriorly. Supraoccipital tentacles are absent. Body color is variable. Darkest blotches are seen usually above tip of upper opercular spine; two on lateral line; two obliquely above; two in soft portion of dorsal fin, and one in anal fin. It can reach a maximum length of 11.5 cm (SL).

Fisheries/aquarium/food/pharmaceutical values: Not reported.

Scorpaenopsis furneauxi (Whitley, 1959)

Photo courtesy: Australian National Fish Collection, CSIRO.

Common name(s): Furneaux scorpionfish

Distribution: Western Pacific: northeastern coast of Queensland to vicinity of Pine Peak Island; extending to Torres Strait and eastern Arafura Sea.

Ecology: This benthic species is found inshore on reef areas; 10–102 m.

Biology

Identification: It is diagnosed by its longest dorsal spine which is 1.9–2.2 in head length. There is no distinct median ridge extending from between posterior nostrils to anterior fourth or more of interorbital space. Tympanic spines are small and are about half length of postocular spines. It grows to a maximum length of 9.6 cm (SL).

Fisheries/aquarium/food/pharmaceutical values: No data.

Scorpaenopsis gibbosa (Bloch and Schneider, 1801)

Photo credit: Cook, David C.

Common name(s): Humpbacked scorpionfish, false stonefish

Distribution: Western Indian Ocean: Kenya and Mozambique to islands in the western Indian Ocean.

Ecology: This reef-associated, benthic species is usually found immobile at the bottom camouflaged among rocks and corals.

Biology

Identification: It is a robust fish, with a markedly humped back (size of hump increases with age). Pectoral fin is with 17 or 18 rays (usually 17); and there are 43–48 scales above lateral line. Ascending process of premaxilla is narrow, and its maximum width 1.8–2.2 in orbit diameter; a series of papillae or nodules is seen across interorbital space between supraocular spines; and nasal spine is single. Body coloration is variable and is usually reddish and marbled and mottled with green, yellow, maroon, and white. Dorsal fin is green or grading to red, with orange mottling; anal fin is brown or maroon, spotted and marbled, often with a broad oblique brown bar; caudal fin is with a pink to brown crossbar at base, followed by a pinkish bar and a wide maroon bar that is streaked with green and white; pectoral fins are grading from maroon with green mottling at base to dark maroon brown, and yellowish-orange terminally; inner surface of pectoral fins is blue at base, bright orange yellow and is with black spots; and a black subterminal bar is over body of fin and this bar is yellowish-white terminally. Maximum recorded size for this species is 21 cm (SL).

Food and feeding: This predatory species hunts by ambushing prey. Its food consists of other fishes and crustaceans.

Fisheries/aquarium/food/pharmaceutical values: This fish is sold fresh in small quantities in market of its range.

Scorpaenopsis gilchristi (Smith, 1957)

Image not available

Common name(s): Gilchrist's scorpionfish

Distribution: Western Indian Ocean: off the Tugela River, Natal, South Africa.

Ecology: It is a demersal, reef-associated species inhabiting up to depths of 26 m.

Biology

Identification: It is diagnosed by the following features: lachrymal is with 5 spines; and 2 predorsal scales. Suborbital ridge is seen directly below eye. There is no suborbital pit. Pores at symphysis of lower jaw open medially to a deep midventral pit. It grows to a maximum length of 6.0 cm (TL).

Fisheries/aquarium/food/pharmaceutical values: Not reported.

Scorpaenopsis insperatus (Motomura, 2004)

Photo courtesy: H. Motomura.

Common name(s): Sydney scorpionfish

Distribution: Southwestern Pacific: Chowder Bay, Sydney Harbour, Australia.

Ecology: This reef-associated, demersal, nonmigratory species is found on coarse sandy bottom at depths of 4–14 m.

Biology

Identification: This species has the following diagnostic features: no median interorbital ridge; well-developed interorbital ridges conjoined at anterior edge of occipital pit; upper opercular and posterior lacrimal spines simple, not divided; 30–34 longitudinal scale rows; 3 suborbital spines; fifth dorsal fin spine the longest; large orbit diameter (5.4–5.6 in HL); short snout length (3.7–3.9 in HL) less than orbit diameter; and narrow interior. It grows to a maximum length of 5.2 cm (SL). It differs from the other species in the genus Scorpaenopsis by the presence of only one spine on the upper operculum and 30–34 scale rows along the side of the body.

Food and feeding: Unknown. It is presumed to be a predator like other scorpionfishes.

Fisheries/aquarium/food/pharmaceutical values: No data.

Scorpaenopsis lactomaculata (Herre, 1945)

Common name(s): Whiteblotched scorpionfish

Distribution: Western Indian Ocean.

Ecology: It is a demersal species inhabiting sandy or muddy bottoms.

Biology

Identification: This species is diagnosed by the following characters: fourth or fifth dorsal spine is usually longest in head; occipital pit is absent or very shallow; snout is very long (2.7–3.0 in head length); space between opercular spines is naked without scales; first dorsal spine is short; supraocular, and postocular spines are broadly joined in adults and are flaring outward to form a shelf over posterior half of eye. It grows to a maximum length of 28.0 cm (TL).

Fisheries/aquarium/food/pharmaceutical values: It is taken occasionally by trawl.

Scorpaenopsis longispina (Randall and Eschmeyer, 2001)

Photo credit: John E. Randall.

Common name(s): Not reported

Distribution: Western Indian Ocean: Mauritius to Kwazulu-Natal (South Africa); including Réunion and Madagascar.

Ecology: It is a demersal, reef-associated species; depth range 6–27 m.

Biology

Identification: This species is identified by its dorsal fin which has 12 spines and 9 soft rays; anal fin is with 3 spines and 5 soft rays; and 24 vertebrae. Second anal spine is long (1.55–1.75 in head length). It grows to a maximum length of 9.2 cm (SL). No other information is available on this aspect.

Fisheries/aquarium/food/pharmaceutical values: Not reported.

Scorpaenopsis macrochir (Ogilby, 1910)

Photo credit: John E. Randall.

Common name(s): Humpback Scorpionfish; rough humpbacked scorpionfish, flasher scorpionfish

Distribution: This species is widely distributed in tropical and subtropical waters of the Indo-Pacific; western and northwest Australia, the Moluccas, and the Philippines to the Marquesan and Society Islands, north to Ryukyu Islands, south to Rowley Shoals, Tonga, and the Mariana and Caroline Islands in Micronesia.

Ecology: This species occurs in depths less than 80 m. It is typically found on soft substrates, where it sometimes buries itself, camouflaged among rubble where it lies in wait for passing prey.

Biology

Identification: This species has a broad, spiny head with a wide space between the eyes, a highly arched back, and a divided upper opercular spine. It has 12 spines in its dorsal fin and 9 soft rays and anal fin has 3 spines and 5 soft rays. Coloration of body of this fish is brown mottled with white. Its large pectoral fin is flushed with yellow and orange on its inside and has a complete, broad black band near its margin and there are no large back spots. *Scorpaenopsis macrochir* is very similar to *Scorpaena neglecta* but differs by the presence of two to six points on the nasal spine (versus serrations in *S. neglecta*). It can also be distinguished from *S. neglecta* by the relative positions of the bulge on the snout and the posterior nostrils. The posterior margin of the bulge in *S. macrochir* extends well beyond the level of the posterior nostrils, whereas in *S. neglecta*, the bulge does not extend beyond the level of the posterior nostrils. This species grows to 13 cm in length.

Behavior: As its common name suggests, this species possesses bright orange and yellow "flash colors" on the inner surfaces of its pectoral fins to ward off and confuse attackers. These bright colors also serve as a warning that this fish is venomous.

Fisheries/aquarium/food/pharmaceutical values: It is an aquarium fish, and it is occasionally found in the aquarium trade.

Scorpaenopsis neglecta (Heckel, 1837)

Photo courtesy: Wikipedia.

Common name(s): Yellowfin scorpionfish, bandtail scorpionfish

Distribution: Indo-West Pacific: Eastern Indian Ocean and the western Pacific Ocean, ranging from India east to Japan, Indonesia, and Australia.

Ecology: This demersal species is found on the continental shelf; occurs in shallow rocky and coral reefs; open sand and mud bottoms; harbors and lagoons; it is found singly well camouflaged; depth range is 0–40 m.

Biology

Identification: In this species, nape and anterior body is highly arched, giving a humpback appearance. Supraorbital ridge is serrated. Pterotic, lower posttemporal, and second suborbital spines are replaced by serrate ridges. Nuchal spine is with 1 or 2 supplemental spines. Anterior ridge of lacrimal is serrated. Orbit diameter is 1.70–1.95 in snout length. Body color is highly variable and is strongly mottled with yellowish to reddish and grayish blotches. Axil of pectoral fins is with black spots. It can reach a maximum length of 19.0 cm (SL).

Food: This species feeds on small fish and swimming invertebrates.

Fisheries/aquarium/food/pharmaceutical values: This species is caught by trawls. Its commercial importance is not known.

Scorpaenopsis obtusa (Randall and Eschmeyer, 2001)

Photo courtesy: Smithsonian National Museum of Natural History.

Common name(s): Shortsnout scorpionfish
Distribution: Western Central Pacific: Philippines south to Australia.
Ecology: It is demersal species inhabiting coral reefs; depth range 0–15 m.

Biology

Identification: This species is diagnosed by the following characters. Dorsal fin is with 12 spines and 9 soft rays; anal fin is with 3 spines and 5 soft rays; 18 pectoral-fin rays on each side of body; 39 scale rows in longitudinal series; 21 pored lateral-line scales; 6 scales above lateral line; 19 scales below lateral line; 5 scale rows between sixth dorsal-fin spine base and lateral line; 6 scale rows between last dorsal-fin spine and lateral line; 4 predorsal-fin scale rows; and 12 total gill rakers.

Body of this species is not compressed anteriorly but is progressively more compressed posteriorly. Mouth is large, oblique and is forming an angle of about 60° to horizontal axis of head and body. Posterior margin of maxilla is reaching a straight line linking posterior margin of orbit with edge of retroarticular. Lower jaw is projecting. Vomerine tooth plate is distinct, forming a V shaped patch but with few teeth. There is no palatine teeth. Nasal spine is simple and is directed upward. Anterior nostril is with a tentacle. Origin of first dorsal spine is just above origin of supracleithral spine. Posterior margin of opercular membrane is extending slightly beyond level of origin of fourth dorsal spine. Posterior tip of pectoral fin is reaching or extending beyond level of origin of first anal fin soft ray. Posterior tip of pelvic fin is not reaching or extending slightly beyond anus when depressed. Origin of last dorsal spine is just above origin of first anal spine. In alcohol, its head and body are brown dorsally and whitish ventrally. Dorsal fin membrane is with scattered melanophores. Pectoral fin is whitish with a broad black band near base and a semicircular black submarginal spot. Axil of pectoral fin is white with 20–25 small black spots. Inner surface of pectoral fin is white with a semicircular black spot near base of first 4–6 rays and associated membranes, and a broad submarginal black band is extending over first 6 rays and associated membranes. Pelvic fin is black, except white distal margin. Anal fin is whitish with a broad black band. Caudal fin is white. It grows to a maximum length of 10.0 cm (TL) (Motomura and Shinohara, 2005).

Fisheries/aquarium/food/pharmaceutical values: Not reported.

Scorpaenopsis orientalis (Randall and Eschmeyer, 2001)

Image not available

Common name(s): Not designated

Distribution: Northwest Pacific: Japan.

Ecology: It is a demersal, reef-associated species; depth range 8–20 m.

Biology

Identification: This species has the following diagnostic features. About one-half of eye is projecting above dorsal profile of head; interorbital space is V-shaped and deep; maximum depth is about 2.5 in orbit diameter; third dorsal spine is usually longest (2.15–2.5 in head length); first dorsal spine is 2.0–2.5 in second spine; and snout length is 2.8–3.05 in head length. It grows to a maximum length of 27.8 cm (SL).

Fisheries/aquarium/food/pharmaceutical values: No data.

Scorpaenopsis oxycephala (Bleeker, 1849)

Photo credit: John E. Randall.

Common name(s): Smallscale scorpionfish, tasseled scorpionfish,

Distribution: Indo-West Pacific: Red Sea and Sodwana Bay, South Africa to the Mariana Islands, north to Taiwan; Palau and Guam in Micronesia.

Ecology: It is a reef-associated, solitary species inhabiting clear-water outer reef slopes and channels from depths of 1 to 400 m, in areas with mixed rock and coral substrates.

Biology

Identification: Body of this species is oblong and is slightly compressed. Dorsal profile is more convex than ventral; and dorsal profile of head is with slight slope. Occipital pit is very shallow; and deep concavity is seen before eyes. Mouth is large, oblique; lower jaw is projecting with symphysial knob; and maxilla is long and distally broad. Eyes are directed slightly upward, with elevated supraorbital ridges. Snout is very long and nostrils are close to each other; and anterior nostril is with a multiciliate flap. Villiform bands of teeth are seen in jaws; inverted "v" shaped band of teeth is present on vomer; palatine teeth are absent; and patches of teeth are also seen on roof and floor of pharynx. Tongue is smooth, free and is rounded apically. Gill openings are wide, and gill membrane is free from isthmus. Pseudobranchiae are present; and there is no slit below fourth gill arch. Gill rakers are reduced to spiny knobs. Spines on head well are developed. Filaments are seen on head. Thread-like tentacles are present on pectoral fin, body surface, and inner side of pectoral fin axil. Small filaments are seen on dorsal fin. Lateral line running is high, straightly to caudal peduncle. Ctenoid scales are seen on most part of the body; and small cycloid embedded ctenoid scales are present on chest, belly and bases of pectorals, pelvic, and anal. Minute small embedded scales are present below eye. There are no scales on pectoral axil. Origin of dorsal fin is above second half of opercle. Origin of dorsal and pelvic is on same line where as origin of pectoral is slightly anterior. Dorsal fin is continuous. Soft dorsal is with one spine. First dorsal spine is very shortest. Fifth to eleventh dorsal spines are gradually smaller. Pectorals are with rounded edge and are reaching to above or vent; upper two to six rays of pectoral are divided and eight to last lower rays are fleshy, with pointed tips. Pelvic fin is shorter than pectorals. First anal spine is shortest, and second is the longest, and robust. Rays are divided. Caudal finis slightly rounded.

Coloration of body is reddish, clouded with brownish to gray, pale red mottling and having a few blackish spots, lower flanks and ventral side pale. Ventrally, it is pale red with gray mottling. A narrow blackish mark is often present anteriorly on snout to upper lip. Maxilla and lower jaw are with gray to red-brown streaks. Dorsal fin is dusky, and soft rays are red-brown distally. Caudal fin is with two to three black bands radiating. Pectoral is with black bands, pink and red-brown streaks and undivided pectoral rays are spotted. Ventral fin is with brown spots. Anal fin is with three narrow dark horizontal bands across lower half. Caudal fin is with two or three black bands radiating. It can reach a maximum length of 36 cm (TL).

Fisheries/aquarium/food/pharmaceutical values: It is taken as a food fish by subsistence fisheries. It may also be taken as bycatch in trawls and is also caught by spear.

Scorpaenopsis papuensis (Cuvier, 1829)

Photo credit: John E. Randall.

Common name(s): Papuan scorpionfish

Distribution: Pacific Ocean: Indonesia and the Philippines to the Society Islands, north to the Ryukyu Islands, south to the Great Barrier Reef and Fiji.

Ecology: It is a reef-associated, solitary species and is found both in marine and brackish water in coral and rocky areas; also in coastal and slopes, estuaries, and lagoons, often in algae covered rubble zones; depth range 1–40 m. It is difficult to see this fish due to its camouflaged colors.

Biology

Identification: This species is diagnosed by its dorsal fin which has 12 spines and 9 soft rays; anal fin is with 3 spines and 5 soft rays; and its snout length is more than its orbit diameter. It has a moderately-humped back and prominent skin tassels on underside of head. Snout length is higher than orbit diameter. Juveniles are with large, forward pointing tentacles above the eyes. Space on opercle between opercular spines is scaled. There are 12 dorsal spines. It has a dark purplish blotch on the rear of the spinous dorsal fin. It can grow to a maximum length of 25 cm

Behavior: The Papuan scorpionfish exhibits biofluorescence, that is, when illuminated by blue or ultraviolet light, it reemits it as red, and appears differently than under white light illumination. This biofluorescence may assist this fish in intraspecific communication and camouflage.

Food: It feeds on fishes.

Fisheries/aquarium/food/pharmaceutical values: It occasionally makes its way into the aquarium trade.

Scorpaenopsis pluralis (Randall and Eschmeyer, 2001)

Image not available

Common name(s): Not designated

Distribution: Eastern Central Pacific: Hawaii.

Ecology: It is a demersal species; depth range 100–110 m. Very little is known on its habitats.

Biology

Identification: This species has the following diagnostic features. Occipital pit is deep. Mouth is oblique and its gape is forming an angle of about 35°. Maxilla is not reaching a vertical at rear edge of orbit. Lower opercular spine is divided into 2–4 points. Posterior lacrimal spine is divided into 3 points. It reaches a length of 7.4 cm (SL).

Food: Apparently, this species feeds on fish and crustaceans.

Fisheries/aquarium/food/pharmaceutical values: Not reported.

Scorpaenopsis possi (Randall and Eschmeyer, 2001)

Photo courtesy: Winterbottom, Richard.

Common name(s): Poss's scorpionfish

Distribution: Indo-Pacific: Red Sea and East Africa to Polynesia.

Ecology: It is a solitary, demersal species found on rock or coral bottoms; depth range 1–55 m.

Biology

Identification: This species has the following diagnostic features: dorsal fin is with 12 spines and 9 soft rays; anal fin has 3 spines and 5 soft rays; and 24 vertebrae. Upper opercular spine is single. Supraocular tentacles of adults are absent or very small. Adult individuals possess various intensities of coloration (with a light orange or bright crimson ventral part of the body and pectoral fins in different fishes) irrespective of the sex. It grows to a maximum length of 19.4 cm (SL).

Sexual dimorphism is almost totally absent in this species. The males have urogenital papillae up to 3 mm in length. Recently, spawned females differ from the males in opened genital pores.

Food: It feeds on fishes.

Fisheries/aquarium/food/pharmaceutical values: It occasionally makes its way into the aquarium trade.

Scorpaenopsis pusilla (Randall and Eschmeyer, 2001)

Image not available

Common name(s): Not designated

Distribution: Eastern Central Pacific.

Ecology: It is a demersal species and is found in sand and rubble bottom with scattered, small coral heads.

Biology

Identification: This species has the following diagnostic features: dorsal fin is with 12 spines and 9 soft rays; anal fin has 3 spines and 5 soft rays; 24 vertebrae; 43 longitudinal scale series; interorbital ridges are not elevated posteriorly to join tip of tympanic spines; and base of parietal spines is curving strongly into occipital pit. It grows to a maximum length of 4.1 cm (SL).

Fisheries/aquarium/food/pharmaceutical values: This species is not utilized.

Scorpaenopsis ramaraoi (Randall and Eschmeyer, 2001)

Photo credit: John E. Randall.

Common name(s): Rama Rao's scorpionfish

Distribution: Indo-West Pacific: Pakistan and Sri Lanka to New Britain, Papua New Guinea; Taiwan, Gulf of Thailand, Sri Lanka and India.

Ecology: It is a demersal species and is found in silty sand and rocky/coral bottoms; depth range 1–60 m. Most species from the Scorpaenidae family are found near or on the bottom and are known to lead solitary lives and only aggregate for reproduction.

Biology

Identification: Body of this species is oblong and is slightly compressed. Dorsal profile is more convex than ventral; dorsal profile of head is with gentle slope and with transverse quadrangular depression across occiput; deep concavity is seen on snout before eyes; interorbital space is narrow; and coronal ridges are low. Mouth is large, and oblique; lower jaw is slightly projecting, with symphysial knob; maxilla is long; jaws have a band of slender, conical, incurved, inwardly depressible teeth in about 6–7 rows at front of upper jaw and 6 in lower, narrowing to 1–2 rows at posterior end; teeth of vomer are small and closely set; conical teeth are in about 4 rows forming an inverted V-shaped patch of teeth; and palatine teeth are absent. Tongue is thick and broadly rounded. Gill opening is wide and gill rakers are short and stout with small spinules. Preopercle is slightly rounded, and opercular flap is pointed, reaching origin of pectoral fin base. Spines on head are well developed and moderate in size. Maxilla is with one long and two to five short filaments; four to five broad filaments are seen on preopercular margin; small filaments are present on lower opercular edge; and three long and two to four short filaments, on each side of lower jaw.

First dorsal spine is shortest and half long as second. This spine is shorter than second and fourth; fourth spine is the longest; fifth to eleventh spines are gradually smaller; and soft rays are divided. Pectorals are with rounded edge, reaching to slightly beyond vent; its upper 2–6 rays are divided, and eighth to last ray are fleshy, with pointed tips. Pelvics are shorter than pectorals and their origin is behind that of pectorals. First anal spine is very short, the second is the longest and strongest; and anal rays are also longer than dorsal rays and are divided. Caudal is slightly rounded. Body coloration is variable; body and head are usually brownish to gray, with bluish green, green or pale red mottling; lower flanks and ventral side are pale; chest is pale red with white and gray mottling; and maxilla and lower jaw are with gray to dark brown streaks. Dorsal fin is dusky, and its spinous part is with irregular gray to brown patches; and soft dorsal is mottled dark green. Pectorals are brown, with vertical gray bars. Pelvics are red with white streaks. Anal fin is greenish brown, with gray streaks. Caudal is mottled red and brown with irregular gray bars. Filaments on body are gray to dark brown. Filaments are bluish or green, those above eyes reddish; and iris is scarlet. It grows to a maximum length of 20.8 cm (TL) (Naranji and Velamala, 2016).

Major threats: It is unlikely that *Scorpaenopsis ramaraoi* is being impacted by any major threat processes. Most Scorpaenidae species in the western central Pacific are small and dangerous to handle, and therefore, they do not form the basis of large fisheries.

Conservation: There are no known species-specific conservation measures in place, or needed, for *S. ramaraoi*. However, its distribution may coincide with a number of MPA designations.

Fisheries/aquarium/food/pharmaceutical values: Not reported.

Scorpaenopsis rubrimarginata (Fricke et al., 2013)

Common name(s): Not designated as it is a new species

Distribution: This species is only known from the holotype, which was collected at 80 m depth off Saint-Gilles, Réunion, Mascarene Islands, southwestern Indian Ocean.

Ecology: It is a pelagic-neritic species.

Biology

Identification: Scales are ctenoid on most of the body, becoming cycloid on abdomen; and 42 in longitudinal series. Gill rakers are broad and spiny; and eight rakers are on lower arch. Upper jaw is with three series of small teeth, which are slightly recurved. Teeth are similar in the lower jaw. Palatine teeth are absent. One-fourth of eye is extending above dorsal profile of head. Occipital pit is present but is not deep; anterior edge of pit is incurved; and there is no transverse ridge across posterior end of occipital pit. Nasal spines are small. Preocular spine is strong and is slightly curved backward. Supraorbital ridge is barely expanded over posterior half of eye. A pair of small supraocular spines is present. Interorbital ridges are moderately developed and are separated at narrowest place by less that width of a single ridge. Tympanic spines are longer than postocular spines. There is no ridge connecting tympanic and parietal spines. Parietal and nuchal spines are elevated; and parietal spines are shorter than nuchal spines. Suborbital pit is present and is shallow. Suborbital ridge is with five spines. Upper opercular spine is double. Lower opercular spine is preceded by a ridge. Postocular region and basal part of opercle are scaled but no scales are seen on opercle between spines. There are 4 predorsal scale rows. Preopercle is with a total of three spines. Upper preopercular spine is strong. A small, triangular supraocular tentacle is present. No other tentacles are seen on head or body. Anterior end of each mandibular canal is with two pores which are not opening to a common midventral pit. Body of this species pale yellowish; and dorsal parts of head, opercle, and anterior dorsal parts of body are with oblique, brownish yellow to rose gray bars. Eye is brownish yellow to rose gray, dorsally with crimson blotches and streaks. Dorsal fin is pale whitish to yellow, with broad, crimson bars. Anal fin is bright crimson. Caudal fin is pale whitish to yellow, with a basal and a distal vertical crimson bar. Pectoral fins are pale whitish to brownish yellow. Male is with a large black spot basally on 8th–10th pectoral fin rays. Pelvic fins are bright crimson. It grows to a maximum length of 2.4 cm (SL) (Fricke et al., 2013).

Fisheries/aquarium/food/pharmaceutical values: Not reported.

Scorpaenopsis venosa (Cuvier, 1829)

Photo credit: Randall, J.E.

Common name(s): Raggy scorpionfish

Distribution: Indo-West Pacific: East Africa to Papua New Guinea, north to Taiwan and the Philippines, south to the Great Barrier Reef; Saya de Malha Bank.

Ecology: This reef-associated species is seen in protected coral reef crevices, and sandy and muddy areas of bays or lagoons, often where the fish can be well camouflaged in areas of algae, seagrass or sponge; occurs along continental margins or islands associated with continents, not at true oceanic islands; depth range 2–100 m.

Biology

Identification: Body of this species is oblong and is slightly compressed. Predorsal profile is with deep slope from nape to snout; and snout is with deep groove depression before eyes. Mouth is large and oblique; lower jaw is slightly projecting s and is with small symphysial knob. Maxilla is broad. Interorbital space is deeply very concave with an elevated smooth coronal ridge on either side; and a deep quadrangular pit across occiput is seen. Eyes are moderate, elevated and are slightly upward. Nostrils are round, of equal size; and posterior nostril is situated in deep depression before eye. Teeth are villiform and are in bands on jaws. Tongue is smooth, free and is rounded apically. Gill opening is wide and gill membrane is free from isthmus; pseudobranchiae are present; and there is no slit below fourth gill arch. Gill rakers are reduced to spiny knobs. Preopercle is slightly rounded, and opercular flap is pointed. Spines on head are well developed. Preopercle is with five spines and first two spines well developed. Post orbital spines are short. Filaments are present; two on snout, and four on preorbital. Short filaments are present on preocular spine; and six filaments are present on over maxilla. Origin of dorsal and pelvic is in same line where as origin of pectoral is slightly anterior. Origin of dorsal fin is above hind border of opercle; and dorsal spines are moderately strong. Posterior part of soft dorsal is rounded. Pectorals are rounded and their upper rays are long; two to five rays are divided and lower rays are fleshy with pointed tips. Pelvic fin reaches vent and pelvics are shorter than pectorals, Anal fin is with three spines; second spine is strong, stout and longest; first anal spine is shortest; and its rays are longer than longest anal spine height. Caudal cut is nearly truncate. Coloration of body is dark, dull red; marbled with grayish or tiny light-blue ocelli scattered over the body; and lower side of throat and belly are pink. Head is dark red brown with paler, three dark bars below eye; and upper and lower jaws are with pale and dark gray mottling. A dark triangle is seen below the eye. Dorsal fin is dark brown with irregular white and gray streaks; dark blotches are present between seventh and tenth dorsal spines; one or two irregular vertical gray bands are present on the caudal; and pectorals are with numerous dark spots. Anal fin is with three white bars. Filaments on body are gray to dark brown. The maximum recorded TL is 25.0 cm (Krishna, 2016).

Antimicrobial activity of venom and gonad: The crude and purified fractions of venom gland and gonad extracts of this species have been reported to show antimicrobial activity and produce zone of inhibition against pathogens viz., *Pseudomonas* sp., *Staphylococcus aureus, Vibrio harveyi, Vibrio parahaemolyticus, Proteaus* sp., and *Escherichia coli*. The purified extracts of spine formed maximum inhibition (15.5 mm) against *V. harveyi* in fraction II and minimum inhibition (6.5 mm) against *S. aureus* in fraction II. Fraction III of gonad produced maximum inhibition (18 mm) against *V. harveyi* and Fraction III produced minimum (6.0 mm) against *S. aureus* (Sasikala and Ravindran, 2013).

Fisheries/aquarium/food/pharmaceutical values: Not reported.

Scorpaenopsis vittapinna (Randall and Eschmeyer, 2001)

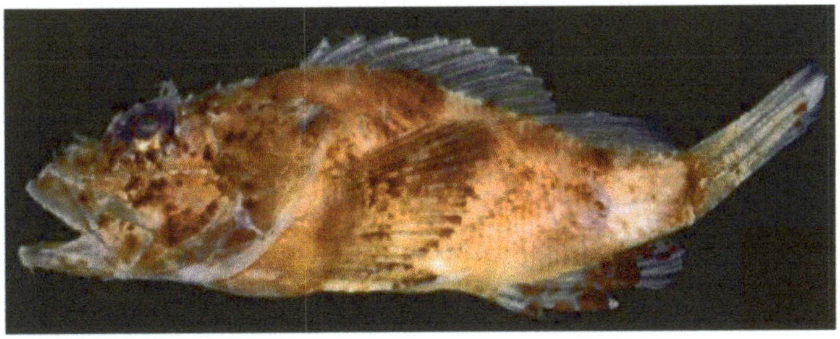

Photo courtesy: Hiroyuki Motomura, Mizuki Matsunuma and Hsuan-Ching Ho.

Common name(s): Bandfin scorpionfish, bigmouth Scorpionfish, bandedfin stingfish

Distribution: Indo-Pacific: Red Sea and East Africa to the Society Islands; Japan, Taiwan, or Australia.

Ecology: This species lives in coral reefs and areas that have a rubble bottom; depth range 2–28 m.

Biology

Identification: This species is diagnosed by the following characters. Dorsal fin is with 12 spines and 8 soft rays; anal fin is with 3 spines and 5 soft rays; 18 pectoral-fin rays on each side of body; 41 scale rows in longitudinal series; 19 pored lateral-line scales;7 scales above lateral line; 13 scales below lateral line; 5 scale rows between sixth dorsal-fin spine base and lateral line; 5 scale rows between last dorsal-fin spine and lateral line; 6 predorsal-fin scale rows; and 11 total gill rakers. Body is moderately compressed anteriorly and is progressively more compressed posteriorly. Mouth is large, oblique and is forming an angle of about 20° to horizontal axis of head and body; and its posterior margin of maxilla is just reaching a vertical through posterior margin of orbit. Vomerine tooth plate is forming a V-shaped patch, with few teeth; and there are no palatine teeth. Nasal spine is simple and is directed dorsally. Ascending process of premaxilla is not intruding into interorbital space; and posterior margin of ascending process is not extending beyond level of posterior margin of posterior nostril. There is no median interorbital ridge. Interorbital ridges are poorly developed; and interorbital space is shallow and narrow. Preocular and supraocular spines are small. Postocular spine is simple and is larger than supraocular spine. There are no coronal or pretympanic spines. Occipital pit is shallow. Suborbital ridge is with 3 spines (2 spines are on left side of head). Suborbital pit is present. Preopercle is with 5 spines; uppermost spine is the largest and bearing supplemental preopercular spine; and lower 4 spines are simple, blunt. Upper opercular spine is with 2 points; and lower opercular spine is simple with a strong median ridge. Origin of first dorsal-fin spine is just above supracleithral spine. Posterior margin of opercular membrane is reaching a vertical through fourth dorsal-fin spine base. Posterior tip of pectoral fin is just reaching a vertical through third anal-fin spine base. Posterior tip of depressed pelvic fin is reaching origin of anal fin. Origin of last dorsal-fin spine is just above origin of first anal-fin spine. In preserved specimens, head and body are gray dorsally and whitish ventrally. Distinct melanophores are scattered on head; and dark blotches are present on body. There is no black blotch on spinous portion of dorsal fin. There are no distinct markings on axil and inner surface of pectoral fin. Pelvic fin is whitish with a black band. Anal fin is white with irregular black bands. It can grow to a maximum length of 8.5 cm (TL) (Motomura et al., 2011a,b).

Fisheries/aquarium/food/pharmaceutical values: Not reported.

Sebastapistes ballieui (Sauvage, 1875)

Picture courtesy: John E. Randall.

Common name(s): Spotfin scorpionfish

Distribution: Eastern Central Pacific: Hawaiian Islands; Midway Islands.

Ecology: It is a reef-associated; nonmigratory species Inhabiting crevices or under rocks on shallow reefs; depth range 1–11 m.

Biology

Identification: This species has the following characteristics: dorsal fin has 12 spines and 9 soft rays; anal fin is with 3 spines and 5 soft rays; coronal spines are present; no occipital pit; present of 40–45 vertical scale rows; lachrymal bone is with 3 spines over maxillary, posterior 2 spines are close set and are pointing down and to rear and first spine points mostly forward; and suborbital ridge is with one spine at posterior end of ridge. Body coloration is olive green or beige mottled with red and white; a prominent branched flap above each eye, and an obscure dark spot on the spiny dorsal fin. It grows to a maximum length of 11.7 cm (TL).

Food and feeding: It emerges at night from its habitats to feed. Its diet consists of other fishes and crustaceans

Fisheries/aquarium/food/pharmaceutical values: Not reported.

Sebastapistes coniorta (Jenkins, 1903)

Photo credit: John E. Randall.

Common name(s): Humpback nohu, speckled scorpionfish

Distribution: Eastern Central Pacific: Hawaiian Islands and Johnston Island.

Ecology: It is a reef-associated, benthic and nonmigratory species usually hiding among branches of cauliflower and antler corals (especially *Pocillopora meandrina*); depth range 1–25 m.

Biology

Identification: Lachrymal spines usually 5. Coronal spines absent; no occipital pit; about 50–55 vertical scale rows; a small spine sometimes present at base of first spine over maxillary in large specimens; suborbital ridge more or less a double ridge, a spine at end, sometimes a spine under eye, fourth spine on lachrymal bone is far below level of suborbital ridge and not counted as part of suborbital ridge; supplemental preopercular spine well above first preopercular spine; well-marked spine at upper rear of orbit in area of sphenotic. Body and fins with scattered small dark spots, no dark spot on spinous portion of dorsal fin. Greenish-yellow with reddish-brown blotches and peppered with small brown dots during the day, red or brown with white blotches and scattered brown dots at night It grows to a maximum length of 10.0 cm (TL).

Food and feeding: It feeds mainly on crabs and shrimps and is a nocturnal species.

Major threats: Speckled scorpionfish is associated with coral reefs, where it is specifically found among the branches of the coral *P. meandrina*. Globally, coral reefs are in decline; local and global pressures, including overfishing, climate change, pollution, and coastal development are affecting more than 75% of reefs. More than 65% of the reefs in the Pacific Ocean were classified as threatened, with the highest concentrations of threatened reefs in Melanesia, Hawaii, Samoa, and the Society Islands. Since 1999, no statistically significant trend in coral cover across the Main Hawaiian Islands could be detected but local declines approaching 20% have occurred in some areas, most often in the central archipelago. Although a substantial portion of this species' distribution overlaps with the Papahanaumokuakea Marine National Monument, some areas within the national monument are experiencing bleaching events that are likely to negatively affect the habitat quality of this species

Fisheries/aquarium/food/pharmaceutical values: Not reported.

Sebastapistes cyanostigma (Bleeker, 1856)

Photo courtesy: Ken Clifton, Flickr.

Common name(s): Yellowspotted scorpionfish, coral scorpionfish, pink spotted scorpionfish, stingfish

Distribution: Indo-Pacific: Red Sea south to Port Alfred, South Africa and east to the Line Islands, north to Ryukyu Islands, south to Samoa and Australia.

Ecology: This reef-associated benthic species is found often in shallow water (less than 15 m) among the branches of *Pocillopora eydouxii* corals in surge areas of seaward reefs. Also observed between the branches of the fire coral Millepora and Styllophora. It is solitary or in small groups; depth range 2–30 m.

Biology

Identification: This species has the following diagnostic features; dorsal fin is with 12 spines and 9–10 soft rays; anal fin is with 3 spines and 5–6 soft rays; and yellow and white spots on body; no dark bars on underside of head. It can reach a maximum length of 10 cm (TL).

Food and feeding: These fishes are nocturnal feeders, dieting mainly on small fishes, crustaceans and other planktonic organisms. Surge areas of the reef that provide them with passing prey are their favorite spots. It is a miniature nocturnal ambush predator. During the day, it hides mainly in *Pocillopora* coral heads and it can be seen hiding deep in the coral. At night it emerges and positions itself in a suitable spot to ambush its prey. These fishes will often be three to four in a small area within 10–20 cm of each other. Pairs are commonly seen.

Major threats: There have been no confirmed population declines as far as this species is concerned. However, because of its affinity with coral reefs, this species may be experiencing population declines due to habitat loss in parts of its range. However, significant global-level population declines are not suspected at this time. As of 2008, 15% of the world's coral reefs were considered under imminent threat of being "Effectively Lost" (with 90% of the corals lost and unlikely to recover soon), with regions in East Africa, South and South-east Asia, and the wider Caribbean being the most highly threatened. Of 704 zooxanthellate reef-building coral species which were assessed by using the International Union for Conservation of Nature Red List Criteria, 32.8% are in categories with elevated risk of extinction.

Fisheries/aquarium/food/pharmaceutical values: It is of no commercial importance but occasionally makes its way into the aquarium trade, because of its small size and coloration.

Aquarium habit: These fishes are reclusive by nature and spend the daylight hours hiding. However they can easily be fed with live shrimp initially. Usually, they will take to being stick fed. As they are smaller than the average scorpionfish, medium size fishes cannot fit in their mouth. They can therefore safely be maintained in a tank with medium size fishes.

Sebastapistes fowleri (Pietschmann, 1934)

Photo courtesy: Adams, Mary Jane.

Common name(s): Pigmy scorpionfish, dwarf Scorpionfish

Distribution: Indo-Pacific: Mauritius and the Comoro Islands to Pitcairn; north to the Hawaiian Islands.

Ecology: This reef-associated, solitary benthic species is found in areas with rubble or mixed sand and rubble, within or near coral reefs; depth range 2–61 m.

Biology

Identification: This species has the following combination of diagnostic characters: dorsal fin has 12 spines and 9 soft rays; anal fin is with 3 spines and 5 soft rays; 16 pectoral-fin rays; 30–34 scale rows in longitudinal series; no palatine teeth; posterior lacrimal spine directed ventrally or anteroventrally; two suborbital spines; no coronal spine; lower opercular spine with median ridge, not covering by scales; and ctenoid scales on dorsal and lateral surface of body. Eye is not extending above the dorsal profile of the head. Head and body are reddish, mottled with white or pink. Spinous portion and basal soft-rayed portion of dorsal fin, and basal parts of pectoral, pelvic, anal, and caudal fins are reddish. This is the smallest of all the scorpionfishes as it can reach a maximum length of only 4.7 cm (TL).

Food: It feeds on small fishes.

Major threats: There have been no confirmed population declines of this species. However, because of its affinity with coral reefs, it may be experiencing population declines due to habitat loss in parts of its range. However, it is also found on rocky reefs, and significant global population declines are not suspected. As of 2008, 15% of the world's coral reefs were considered under imminent threat of being "Effectively Lost" (with 90% of the corals lost and unlikely to recover soon), with regions in East Africa, South and South-east Asia, and the wider Caribbean being the most highly threatened. Of 704 zooxanthellate reef-building coral species which were assessed by using the International Union for Conservation of Nature Red List Criteria, 32.8% are in categories with elevated risk of extinction.

Fisheries/aquarium/food/pharmaceutical values: Though this ornamental fish species is edible, it is not commercially important.

Sebastapistes galactacma (Jenkins, 1903)

Photo credit: John E. Randall.

Common name(s): Galactama scorpionfish, milky scorpionfish

Distribution: Pacific Ocean: Guam, Pohnpei, Hawaii, and Rapa.

Ecology: This reef-associated, benthic species is inhabiting coral and rubble areas of seaward reefs in depths up to 200 m.

Biology

Identification: This species t has the following combination of diagnostic characters: dorsal fin has 12 spines and 9–10 soft rays; and anal fin is with 3 spines and 5 soft rays; It has two long "horns" which arise over the eye. It is often with darker mottlings on the back and upper sides. It can grow to a maximum length of 10 cm (TL).

The entire fish is densely covered with a multitude of fine white spots

Major threats: There have been no confirmed population declines as far as tis species is concerned. However, because of its affinity with coral reefs, it may be experiencing population declines due to habitat loss in parts of its range. However, it is also found on rocky reefs, and significant global population declines are not suspected. As of 2008, 15% of the world's coral reefs were considered under imminent threat of being "Effectively Lost" (with 90% of the corals lost and unlikely to recover soon), with regions in East Africa, South and Southeast Asia, and the wider Caribbean being the most highly threatened. Of 704 zooxanthellate reef-building coral species which were assessed by using the International Union for Conservation of Nature Red List Criteria, 32.8% are in categories with elevated risk of extinction.

Fisheries/aquarium/food/pharmaceutical values: Not reported.

Sebastapistes mauritiana (Cuvier, 1829)

Photo credit: John E. Randall.

Common name(s): Spineblotch scorpionfish, Mauritius scorpionfish

Distribution: Indo-Pacific: East Africa to the Marquesan and Rapa Islands, north to the Ryukyu Islands, south to the Great Barrier Reef and Samoa.

Ecology: This reef-associated, benthic species is found singly and well camouflaged on sheltered reefs, harbors, under rubble, etc. but at night on sandy bottoms; it is also inhabiting outer intertidal reef flats and shallow lagoon reefs exposed to wave action; solitary and cryptic; depth range 0–10 m.

Biology

Identification: This species t has the following combination of diagnostic characters: dorsal fin has 12 spines and 9 soft rays; and anal fin is with 3 spines and 5 soft rays. It has very short snout; underside of head is with weak dark bars; spinous dorsal is normally with dark spot; strong coronal ridges with spine; and distinct occipital pit which is deepening with age; Body coloration is dark brown to reddish brown mottled with white and darker brown. This species can reach a maximum length of 9.5 cm (TL).

Food and feeding: It feeds nocturnally on small fish and swimming invertebrates.

Major threats: There have been no confirmed population declines about this species. However, because of its affinity with coral reefs, it may be experiencing population declines due to habitat loss in parts of its range. However, it is also found on rocky reefs, and significant global population declines are not suspected. As of 2008, 15% of the world's coral reefs were considered under imminent threat of being "Effectively Lost" (with 90% of the corals lost and unlikely to recover soon), with regions in East Africa, South and South-east Asia, and the wider Caribbean being the most highly threatened. Of 704 zooxanthellate reef-building coral species which were assessed by using the International Union for Conservation of Nature Red List Criteria, 32.8% are in categories with elevated risk of extinction.

Fisheries/aquarium/food/pharmaceutical values: Although there is no commercial use for this species, it is commonly taken as bycatch.

Sebastapistes nuchalis (Günther, 1874)

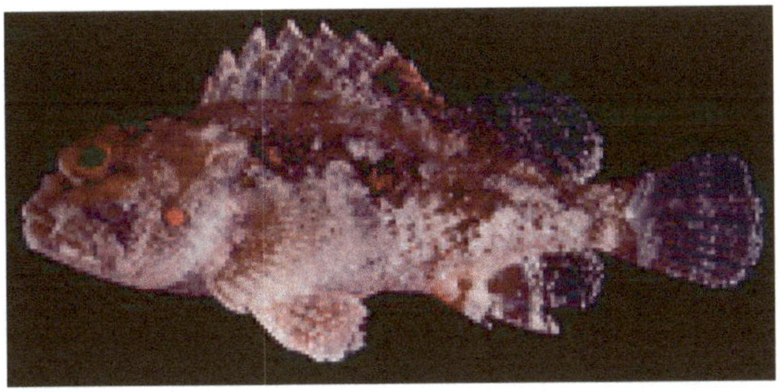

Photo courtesy: John E. Randall.

Common name(s): Not designated

Distribution: Indo-Pacific.

Ecology: This benthic, inshore species is found on rocky and sandy bottoms.

Biology

Identification: It has the following diagnostic features: large head; snout is shorter than eye diameter; preorbital bone is with 2 spines over maxilla; suborbitals with 2 spinous points; absence of occipital pit; pores at symphysis of lower jaw are separate; supraocular tentacle is absent; and absence of skin flaps on head and body; dorsal fin is with 12 spines and 9 rays; pectoral finis with 15 rays; 24 tubular scales on lateral line; and there are 5 gill rakers on lower arch. Coloration of body is red-brown, mottled, and marbled; and fins are barred with oblique, brown, narrow bands. It grows to a maximum length of 10.2 cm (TL).

Food: It mainly feeds on crustaceans.

Fisheries/aquarium/food/pharmaceutical values: Not reported.

Sebastapistes perplexa (Motomura et al., 2014)

Photo courtesy: Hiroyuki Motomura.

Common name(s): Not designated as it is a new species

Distribution: Pacific coast of Japan, where it ranges from the Boso Peninsula to Shikoku, including the Izu Islands.

Ecology: This benthic species is found in moderately shallow water areas of depth range, 2–48 m.

Biology

Identification: This species t has the following combination of diagnostic characters: dorsal fin has 12 spines and 9 soft rays; and anal fin is with 3 spines and 5 soft rays; pectoral-fin rays 14; and exposed cycloid scales on opercle between tips of opercular spines. Body scales are not extending onto rays or membranes of fins, except on base of caudal fin. Embedded cycloid scales are also seen on cheek. Lateral line is sloping downward from second pored scale. Mouth is large, oblique and is forming angle of 15–30° to horizontal axis of head and body. Posterior margin of maxilla is just short of vertical drawn through posterior margin of orbit. There are villiform teeth on upper and lower jaws but no canine teeth; and tooth band is narrowing posteriorly. Tooth bands of upper and lower jaws are subequal in width. 4–5 rows of small teeth are seen at front of vomer. There are 2–4 rows of small teeth on each palatine. Underside of lower jaw is without ridges.

Dorsal profile of snout is steep, forming angle of 50–60° to horizontal axis of head and body. Nasal spine is simple, and its tip is extending well beyond top of ascending process of premaxilla in lateral view. Occipital region is slightly concave (nearly flat or slightly convex in young and juveniles), without distinct transverse ridges at rear of occiput. Occiput is surrounded laterally by bases of tympanic and parietal spines. Origin of pelvic-fin spine is slightly posterior to vertical drawn through origin of first pectoral-fin ray. Posterior tip of depressed pelvic fin is just reaching to anal-fin origin. Origin of first anal-fin spine is anterior to vertical drawn through last dorsal-fin spine base. Body coloration is reddish, mottled with dark red blotches; and anterior part of caudal peduncle is chalky red. Underside of head is pale red or chalky red. Subopercle is with black blotch the size of which is subequal to pupil diameter. Spinous portion of dorsal fin is reddish, with white membranes distally. Soft-rayed portion of dorsal fin and anal fin are reddish, with broad white band obliquely across middle of each fin. Pelvic fin is red, with white margin. Pectoral fin is semitransparent red, with poorly defined black blotch (Motomura et al., 2014).

Fisheries/aquarium/food/pharmaceutical values: Not reported.

Sebastapistes strongia (Cuvier, 1829) (=*Sebastapistes oglinus*)

Photo credit: John E. Randall.

Common name(s): Barchin scorpionfish

Distribution: Indo-Pacific: Red Sea and East Africa; Society Islands, north to Taiwan, south to Queensland, and Australia.

Ecology: This reef-associated species is common in areas with mixed sand and rubble in reef flats, shallow lagoons and channels; solitary and cryptic; with camouflage ability, it hides in algae during the day; depth range 0–37 m.

Biology

Identification: This species has the following diagnostic features: dorsal fin is with 12 spines and 8–9 soft rays; anal fin is with 3 spines and 5 soft rays; 14–17 pectoral fin soft rays; ctenoid scales; presence of palatine teeth; low spines on top of head; absence of occipital pit; no coronal spines; prominent head pores; pored lateral line scales, 20–22; scales in lateral series 41–45; and depth of body is 2.6 in SL. Eyes are large with radiating stripes, sometimes with branched cirri above. Body coloration is brown to reddish brown, mottled with brown, white and black. Lower jaw is with dark and light bars. This species can grow to a maximum length of 9.5 cm (TL).

Food and feeding: This nocturnal species feeds on small fishes and crustaceans.

Major threats: There have been no confirmed population declines for this species. However, because of its affinity with coral reefs, it may be experiencing population declines due to habitat loss in parts of its range. However, it is also found on rocky reefs, and significant global population declines are not suspected. As of 2008, 15% of the world's coral reefs were considered under imminent threat of being "Effectively Lost" (with 90% of the corals lost and unlikely to recover soon), with regions in East Africa, South and South-east Asia, and the wider Caribbean being the most highly threatened. Of 704 zooxanthellate reef-building coral species which were assessed by using the International Union for Conservation of Nature Red List Criteria, 32.8% are in categories with elevated risk of extinction.

Fisheries/aquarium/food/pharmaceutical values: This species is of no commercial importance. However, it is often caught as a bycatch at shallow depths over coral reefs.

It is also an aquarium species, and its aquarium parameters are given below.

Min. aquarium size—20 US Gal. (76 L)

Foods and feeding—It is fed 2–4 times a week with its diet items which include marine fish and crustacean flesh, live grass shrimp, etc.

Reef compatibility—As it will eat on its associates, viz., ornamental shrimps, smaller fishes, etc. utmost care has to be taken in the maintenance of community aquarium.

Sebastapistes taeniophrys (Fowler, 1943)

Image not available

Common name(s): Eyebrow scorpionfish

Distribution: Indo-West Pacific: Madagascar and Philippines.

Ecology: This pelagic-neritic, demersal species is inhabiting shallow sandy bottoms. (All other species of *Sebastapistes* are known to live in rocky or coral reefs.)

Biology

Identification: This species has the following diagnostic features: dorsal fin is with 12 spines and 9 soft rays; anal fin is with 3 spines and 5 soft rays; 15 pectoral-fin rays; 31–33 scale rows in longitudinal series; 11–14 pored lateral-line scales; 3 predorsal scale rows; and 12 gill rakers. There are 3 suborbital spines and coronal spines are absent. Lower opercular spine is with a median ridge and is not covered with scales. Body is covered with ctenoid scales. Several dark transverse bands are seen on ventral surface of mandible; a distinct elongate black blotch is between second or third and seventh dorsal-fin spines; and there is no black blotch on the nape. It is the smallest member of this genus as its maximum length is only 2.5 cm (TL) (Motomura, 2009).

Major threats: Threats of this species are possibly from degradation of its shallow-water habitat due to coastal development.

Fisheries/aquarium/food/pharmaceutical values: Not reported.

Sebastapistes tinkhami (Fowler, 1946)

Photo credit: John E. Randall.

Common name(s): Darkspotted scorpionfish, Tinkham's scorpionfish

Distribution: Indo-Pacific.

Ecology: This reef-associated, benthic, solitary, and cryptic species inhabits deeper reefs, seaward reefs and mainly at oceanic islands; also in lagoon and coral platform, sand and coral gravel mixed terrain; depth range 1–30 m. With camouflage ability, it often hides its body.

Biology

Identification: This rare species has the following diagnostic features: dorsal fin is with 12 spines and 9 soft rays; anal fin is with 3 spines and 5 soft rays; and pectoral fin is with 15–16 soft rays. Body coloration is brown or yellow-brown; and body side and fins are all covered with black spots; head is speckled with small dark; and iris of his big eye is dark red. It grows to a maximum length of 8 cm (TL).

Food: It is a nocturnal species feeding on mainly other fishes and crustaceans.

Fisheries/aquarium/food/pharmaceutical values: Not reported.

Taenianotus triacanthus (Lacepède, 1802)

Photo courtesy: John E. Randall.

Common name(s): Yellow leaf scorpionfish, leaf scorpionfish, paperfish, three-spined scorpionfish, flat Scorpionfish

Distribution: Indo-Pacific: widespread from east African coast and the Red Sea to the tropical Indo-Pacific, north to the Galapagos.

Ecology: This reef-associated, benthic, nonmigratory species inhabits rocks, reef flats, outer reef slopes, current-swept channels in sublittoral and intertidal habitats with strong wave action; it can be sometimes found hiding in amongst corals and gorgonians; depth range 1–135 m. Small juveniles may drift down and settle in estuaries and harbors.

Biology

Identification: This species is monotypic genus (only species in the *Taenianotus* genus). It is a very pretty and colorful fish and is almost as flat resembling a leaf in many other ways. It has a very compressed body and a sail-like, large dorsal fin. It has a variably developed "beard" of fine appendages around the mouth and sometimes real algae and hydroids grow on its skin. Head and mouth are large. A dark line is seen through their eyes. Dorsal fin which is high and attached to the caudal fin starts just behind eyes and has 12 spines and 8–11 soft rays. Anal fin has 3 spines and 5–6 soft rays. The third or fourth spine is the longest; suborbital ridge without spines or with lump at head of ridge; and preopercle with 2 indistinct spines only. Skin often has blotches that enhance a camouflage effect. It has prickly papillae instead of scales. Its color varies from green, red, pink, brown, ocher, and yellowish to a ghostly white. It grows to a maximum length of 10.0 cm (TL). The maximum size is approximately 25 cm.

Behavior: It is observed singly or in a small group. When it is solitary, it is usually immobile among algae or seagrass but effects hip movements resembling that of a leaf falling down from a tree. To enhance this camouflage, it even makes gentle sideways movements in its pelvic area which make it resemble a drifting inert object. This species periodically sheds its skin as a nearly single piece (almost like a snake). It molts twice a month with the skin breaking off first in the head region. These fishes are regularly molt their skin to remove algae and invertebrates that grow on their sides and can change color after molting.

Food: It feeds on small crustaceans and fishes; also feeds on larvae. It is an ambush predator, waiting until suitable prey approaches. Then it slowly moves with its pectoral fins close to the victim. When the leaf scorpionfish is close enough, the prey is sucked in by a sudden opening of its mouth.

Major threats: There have been no confirmed population declines for this species. However, because of its affinity with coral reefs, it may be experiencing population declines due to habitat loss in parts of its range. However, it is also found on rocky reefs, and significant global population declines are not suspected. As of 2008, 15% of the world's coral reefs were considered under imminent threat of being "Effectively Lost" (with 90% of the corals lost and unlikely to recover soon), with regions in East Africa, South and South-east Asia, and the wider Caribbean being the most highly threatened. Of 704 zooxanthellate reef-building coral species which were assessed by using the International Union for Conservation of Nature Red List Criteria, 32.8% are in categories with elevated risk of extinction.

Fisheries/aquarium/food/pharmaceutical values: Not reported.

Aquarium parameters

It is an aquarium fish and is taken for the aquarium trade due to its unusual appearance and interesting behavior. It is a poor swimmer and is easily caught with hand nets. It is a reef safe fish that can be kept in reef tank with no harm to corals. But it is a threat to mobile inverts like shrimps and crabs, including much smaller fishes. It is best kept in peaceful community tank or a small reef tank that does not contains aggressive or picky fishes like triggers and butterflyfish. Good tank mates of these fishes are lionfish, squirrelfish, etc. In tanks, they do attain to the maximum size of 10 cm. in captivity (http://www.theaquariumwiki.com/wiki/Taenianotus_triacanthus).

Feeding: In the aquarium, these fish demonstrate similar behavior as in nature. They do not hide secretly in a dark place waiting for the prey to pass by and quickly dash toward the prey before sucking them into their stomach like other scorpions, but instead, they perch on a well-chosen position. It might be underneath the coral or some rockwork where the water current flows swiftly, even though in front of the tank or a bright area. They use their "unique" sitting action, and bend their head lower than tail, move themselves side to side as if imitating a leaf moving by the water current. By this way, the prey may think that they were only a leaf and when the prey came to a place close enough, they will gasp the prey and eat it immediately. Further, these fishes are not finicky eaters. They might require a living food such as mollies, guppies, or a small shrimp at first. But as they do acclimatize, they can be switched to a piece of cut seafood frozen or otherwise. In aquarium tanks, it is better to place the prey in front of them so that it could easily be seen and do not overfeed them.

Molting in tanks: In aquarium tanks, the soft water current will help these fishes to release their molt easier. The slimy molt will be appeared in transparent form, floating in the tank or sticking to something in the tank and should be taken out if seen.

Thysanichthys crossotus (Jordan and Starks, 1904) (=*Scorpaenodes crossotus*)

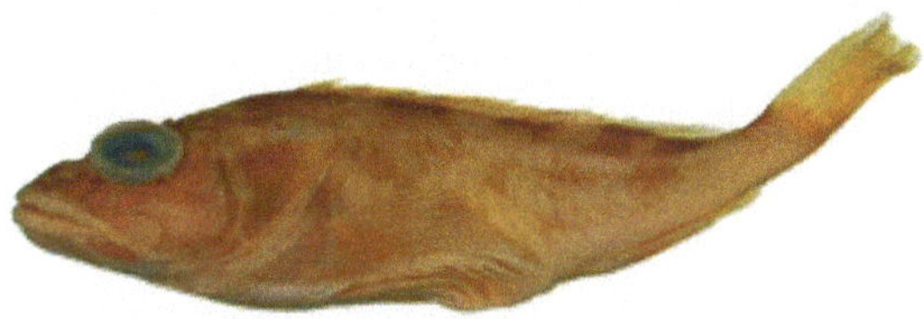

Photo courtesy: Shao, Kwang-Tsao.

National Museum of Marine Biology and Aquarium, Taiwan

Common name(s): Not designated

Distribution: Northwest Pacific: Japan, East China Sea, and Taiwan.

Ecology: This demersal species mainly inhabits the bottom of 120–130 m deep.

Biology

Identification: It is diagnosed by the following features: dorsal fin is with 13 spines and 11 soft rays; anal fin is with 3 spines and 5 rays; and lateral line scales, 45. There are many scalp flaps on the head and sides. Head, orbital bone and the former gill cover are large. Body coloration is brown, pale white belly. Spine of dorsal fin is with brown spots. It grows to a maximum length of 8.4 cm (SL).

Fisheries/aquarium/food/pharmaceutical values: It is a small fish. In addition to academic research, it has no economic value.

Trachyscorpia verai (Béarez and Motomura, 2009)

Image courtesy: Béarez and Motomura.

Common name(s): Ecuadorian deep-sea scorpionfish
Distribution: Pacific Ocean: TEP, Ecuador (only from off Puerto López, Manabí).
Ecology: It is demersal fish inhabiting in deeper waters from 200 m.

Biology

Identification: Body of this species is moderately compressed anteriorly and is progressively more compressed posteriorly. Nape and anterior body are not strongly arched. Body is not very deep as its depth is 1.5 in head length. There are no distinct small papillae on head. There are two filiform tentacles on dorsal margin of eye membrane, and one below anterior nostril. One filiform tentacle is present on posterior end of most head spines, including nasal, preocular, postocular, sphenotic, parietal, suborbital ridge, posterior lacrimal, upper preopercular, nuchal, and tympanic spines. There are no distinct tentacles on supraocular spine base. There are no tentacles associated with posterior nostril or on occiput, midinterorbital space, maxilla, lips, lower jaw, preopercle, opercle, fins, and lateral body surface. Pectoral-fin axil is without skin flaps.

Mouth is large and is slightly oblique. Jaws are subequal. Posterior margin of maxilla is reaching a vertical line through the posterior limit of the second third of orbit. Upper edge of midmaxilla is slightly swollen laterally. Lower jaw is with a blunt but conspicuous symphysial knob. Upper jaw is with a band of villiform teeth; tooth band is broadening posteriorly. Lower jaw is with a single row of villiform teeth anteriorly, becoming a narrow band (3 or 4 rows) posteriorly. Underside of lower jaw is smooth, without ridges. Dorsal profile of snout is steep. Nasal spine is simple, small and is directed anterodorsally. Most spines on head are well developed.

Origin of first dorsal-fin spine is reaching a vertical through the midpoint between posterior end of supracleithral spine and upper opercular spine. Posterior margin of opercular membrane extends to a vertical between third and fourth dorsal-fin spine bases. Posterior tip of pectoral fin is reaching a vertical through origin of eleventh dorsal-fin spine and not reaching anal-fin origin. Origin of pelvic-fin spine is slightly posterior to origin of pectoral fin. Pelvic fin is short and is not reaching anus. Origin of first anal-fin spine is slightly posterior to origin of first dorsal-fin ray. Coloration of body is uniformly bright red with some indefinite dusky spots or markings on head, and a dark blotch on the opercle. Ventrum is somewhat paler. It is a medium-sized scorpionfish, and it grows to a maximum length of 29.8 cm (SL) (Béarez and Motomura, 2009).

Fisheries/aquarium/food/pharmaceutical values: It is probably a deepwater species rarely caught by the longline fisheries.

Ursinoscorpaenopsis kitai (Nakabo and Yamada, 1996)

Common name(s): Not reported as it is a new genus and species
Distribution: Northwest Pacific: Japan.
Ecology: It is a demersal species inhabiting depths up to 138 m.

Biology

Identification: This species is diagnosed by its dorsal fin which has 12 spines and 10 soft rays; and anal fin is with 3 spines and 5 soft rays. Body of this species is deep, somewhat compressed, is bent upward at caudal peduncle, and is covered with cycloid scales. Head is large and snout is blunt. Mouth is large and oblique; and maxilla is ex tending just posterior to anterior margin of eye. Teeth on jaws are villiform, in narrow bands. Teeth on prevomer are villiform, in a small triangular band. Tongue is smooth, pointed at tip. Lachrymal is with 2 spines overlying maxillary; anterior spine is angled forward with a small cirrus, and posterior spine is angled downward with a large cirrus. A pair of very short cirri is present between supraocular and postocular spines. Occipital pit is triangular. Lateral line is curving downward and is ex tending to base of caudal fin. Swim bladder is present and ovate. Dorsal fin is with 12 spines and 19 soft rays (last ray is branched at base); anal fin is with 3 spines and 5 rays (last ray is branched at base); and pelvic fin is with 1 spine and 5 rays. Dorsal fin is originating at posterior tip of posttemporal spine; second to fifth dorsal fin spines are curved backward; and fourth spine is the longest. Anal fin is beginning at a vertical through first dorsal fin spine; and third spine is the longest. Pectoral fin is large, almost reaching anus. Pelvic fin is moderately pointed, not reaching to anus when depressed. Caudal fin is round. Head and body are orange-red. In formalin-preserved specimens, head and body, including dorsal, anal, and pectoral fins, are creamy white with indistinct dark marks. Pelvic fin is with a darker posterior margin. Caudal peduncle is with an indistinct transverse dark band. Caudal fin is creamy white with a transverse indistinct dark band. It grows to a maximum length of 24 cm (SL) (Nakabo and Yamada, 1996).

Fisheries/aquarium/food/pharmaceutical values: Not reported.

3.2 Rockfishes (Subfamily Sebastolobinae)

Helicolenus dactylopterus (Delaroche, 1809)

Photo courtesy: Østergaard, Thorke A.S.

Common name(s): Blackbelly rosefish, blue-mouth, blue-mouth redfish, rosefish

Distribution: Western Atlantic: Nova Scotia, Canada to Venezuela. Eastern Atlantic: Iceland and Norway to the Mediterranean and the Gulf of Guinea, including Madeira, the Azores, and the Canary Islands; also Walvis Bay, Namibia to Natal, South Africa.

Ecology: It is a bathydemersal species; adults are found in soft bottom areas of the continental shelf and upper slope; depth range 50–1100 m.

Biology

Identification: It is a robust fish with a large head. Profile of the nape is relatively steeply inclined. It has villiform teeth on both jaws and its large mouth is dark colored inside. Dorsal fin has 11–13 spines and 10–14 rays; anal fin has 3 spines and 5 rays; and pectoral fin has between 17 and 20 rays. These fishes have 55–80 vertical rows of ctenoid scales and their lateral line has tubular scales; chest, cheek and maxilla are usually scaled but the snout and ventral part of the head are naked. They usually have 25 vertebrae. Gill rakers are well developed: 7–9 on the upper arch, and 16–21 on the lower arch. Body coloration is variable; back and sides are red and the belly is pink, with 5–6 dark bands below anterior, middle, and posterior dorsal spines; and Y-shaped dark bar is seen between soft dorsal and anal fin. A dark blotch is present on the posterior part of the spinous dorsal fin. As with other species of scorpionfish, the spines of the blackbelly rosefish contain toxic venom and have reportedly caused injuries to humans. However, there has been little research on the venom produced by this species. Its maximum length is 50.0 cm (TL); weight is 1.6 kg; and reported age is 43 years.

Food and feeding: They feed on both benthic and pelagic organisms such as crustaceans, fishes, cephalopods, and echinoderms. The crustacean food items include isopods and brachyuran crabs. This Atlantic species is a typical sit-and-wait predator with a highly cryptic coloration. The proportions of these prey types in their diet vary according to the size of the fish and the depth distribution (Gastoni et al., 2011).

Lipid distribution: Blackbelly rosefish (*Helicolenus dactylopterus lahillei*) has a lipid content of 13%–28% (on a wet weight basis) in skull, spine, and liver and 1%–3% (on the same basis) in gonads and flesh. Wax esters represent 21.1% of the oil content in muscle and triacylglycerols 84.4% of the oil content in skull and 87.8% in spine. Flesh wax esters and poorly calcified bones exert buoyancy control; bone triacylglycerols play a role in energy reserves (Mendez et al., 1993).

Fisheries/aquarium/food/pharmaceutical values: Though this species is a commercial one in the Mediterranean and in the Catalan coast, it is not targeted commercially. It is caught by artisanal longline and gillnet fisheries near the Strait of Gibraltar and along the Portuguese continental coast and the Azores. In the western Mediterranean, blackbelly rosefishes are mostly caught as bycatch in bottom trawl fisheries targeted at deep-sea crustaceans. It is sold fresh.

Sebastiscus marmoratus (Cuvier, 1829) (=*Sebastes marmoratus*)

Photo courtesy: Joi Ito, Flickr, and Wikipedia.

Common name(s): False kelpfish, dusky stingfish, Japanese stingfish, marbled rockfish, marbled perch, marbled sea bass, marble sea ruff

Distribution: Western Pacific: southern Hokkaido, Japan to the Philippines.

Ecology: It is a reef associated, demersal species found near shore; on rock and gravel bottom and reef sediments; it is oceanodromous; and depth range 0–50 m.

Biology

Identification: This species is diagnosed by its dorsal fin which is with 12 spines and 11–13 soft rays; anal fin is with 3 spines and 5 soft rays; and pectoral fin is with 17–19 soft rays. Among the pectoral rays, 1 upper is unbranched, 8 are branched and 9 lower are unbranched and are often thick. In the caudal fin usually 32–33 rays. There are 22–23 gill on the first arc and 25 vertebrae. Spines on top of head are well developed. Coloration of body is dark olive or almost black, or from brownish to dark red with more light spots of irregular shape, which in some individuals can be rounded. It has a maximum TL of 41 cm. The species is viviparous.

Food and feeding: It is a carnivorous fish when it is small. It mainly feeds on fish; brachyuran crab, *Xantho pilipes*; and anomuran decapods.

Parasites: Parasitological examination revealed infections with the locally common generalist parasites *Derogenes varicus* (Digenea) and *Hysterothylacium aduncum* (Nematoda) that likely have been acquired through prey fish.

As a sentinel species to assess petroleum hydrocarbons pollution: This species has been reported as a potential sentinel organism to monitor marine pollutants in aquatic environments (Bo et al., 2017).

Sound production: Several rockfish species of the genus *Sebastes* have been reported to be soniferous during the nonspawning season. In this species, three distinct sounds were identified based on frequency features. The common characteristics among all sounds were low frequency (below 300 Hz) and rapidly dampened pulses consisting of 3–5 acoustic energy cycles. During free-swimming conditions in the canvas tank, the fish produced voluntary sounds with lower frequencies than the disturbance sounds produced by individuals during prodding. Two types of sounds were identified in the disturbance context: one type consists of single or double pulses with two peak frequencies, and another type consists of a series of pulses with a single peak frequency. These results suggest that specific-species and behavior-associated sounds are potentially useful in passive acoustical surveys to monitor rockfish populations and distributions remotely (Zhang et al., 2015).

Fisheries/aquarium/food/pharmaceutical values: It is a commercially cultured in Japan. It is a high-value economic species with higher yields in northern coast of Taiwan. Main fishing methods are longline and bottom trawl. It has sweet and flexible meat.

Sebastolobus alascanus (Bean, 1890)

Photo credit: Wikipedia.

Common name(s): Shortspine thornyhead, channel rockcod, idiotfish, idiot cod, rock cod, shortspine channel rockfish, spinycheek rockfish

Distribution: North Pacific: Baja California, Mexico, Japan.

Ecology: It is a demersal species found on soft bottoms; depth range 17–1600 m. Adults are commonly found on muddy bottoms and bottoms with mixtures of mud and cobble and mud and boulders where sea urchins are abundant. Highest abundance of adults has been reported between 200 and 400 m. Recently, settled individuals are more abundant at the deep end of their range than the shallow end; midsized individuals are more abundant at the shallow end; and adults are more abundant at deep locations.

Biology

Identification: This species is one of the longest-lived fishes in the world and is diagnosed by its spiny ridge across its cheek. Dorsal fin is with 15–17 spines and 8–9 soft rays; third dorsal spine is not much longer than second, fourth or fifth; and anal fin is with 3 spines and 4–5 soft rays. Caudal fin is slightly rounded. Head is large and body is elongated and tapering. Head spines are very strong; and nasal, preocular, supraocular, postocular, tympanic, parietal, and nuchal spines are present and coronal spines are absent. A strong spiny ridge is seen on head and a strong notch is on pectoral fin. These fishes are usually red or orange-red and often have white patches on their sides, back, cheeks, and spiny dorsal fin tips when seen underwater. Their fins can have black patches and there may be dark spots or speckles on their sides. Gill chamber is mostly pale. This species has a thick glandular tissue on posterior margin of fin spines which is believed to be venomous. It grows to a maximum length of 80.0 cm (SL); it has a published weight of 9.0 kg; and a maximum reported age of 115 years.

Abundance estimate: The abundance of this species in the Gulf of Alaska has been reported to range from 0 to 7.5/100 m^2, with a mean of 1.22/100 m^2, which was positively correlated with depth and amount of hard substrate (Else et al., 2002).

Migrations and movements: Shortspine thorny heads undergo ontogenetic migration from shallow into deep water. Larvae and young juveniles are pelagic for 14–15 months. These early life history stages are widely transported, primarily via water currents. Juveniles settle onto the continental shelf and then move into deeper water as they become adults. The ontogenetic migration transports particulate organic carbon from the bottom to the surface by eggs and particulate organic matter from the surface back down to the bottom as recruiting juveniles.

Food and feeding: Benthic individuals of this species are sit-and-wait predators that rest on the bottom and remain motionless for extended periods of time. It eats a variety of invertebrates such as shrimp, crabs, and amphipods, as well as fishes and worms. In the western Bering Sea, their primary dietary components were decapod copepods and gammarid amphipods, representing 83% and 15.7% by weight, respectively. Longspine thornyheads are commonly found in the stomachs of shortspine thornyhead. Cannibalism of newly settled juveniles is important in the life history of thornyheads because it decreases the number of juveniles but in turn provides more food for adults.

Conservation status: It is an endangered species.

Fisheries/aquarium/food/pharmaceutical values: These fishes are caught using otter-trawls and longline gear. Shortspine thornyheads have also been caught in traps and on setlines. The commercial fishery for this species has grown over the last several decades, especially along the Pacific coast due to the development of a North American export market to Japan. These fishes are occasionally caught in recreational fisheries also.

Sebastes auriculatus (Girard, 1854)

Photo credit: Wikipedia.

Common name(s): Brown rockfish

Distribution: Eastern Pacific: Prince William Sound, Alaska to central Baja California, Mexico.

Ecology: Adults and subadults are commonly found near the sea bottom over both high and low terrain, and sometimes among eelgrass or other vegetation. Pelagic juveniles are found at shallower depths; depth range 0–135 m.

Biology

Identification: This species is diagnosed by the following characters: 13 dorsal spines and 12–15 dorsal soft rays; 3 anal fin spines and 5–8 anal soft rays; and 26–27 vertebrae. Caudal fin rounded. Brown rockfishes are heavy bodied and colored various shades of brown, with dark brown to blackish mottling over the top. Red-brown, brown or orangish stripes radiate back from the eye and upper jaw. Fins are pinkish. One key characteristic for this species is a prominent dark brown blotch on the upper part of the gill cover, which tends to become faint in large individuals. Fin spines are sharp and mildly venomous and can cause annoying wounds. These fishes can grow to about 56 cm in length, with the females being potentially larger than the males. Both sexes have similar growth rates and life spans and are known to live to at least 34 years of age.

Predators: Birds, dolphins, seals, sharks, lingcod, cabezon, and salmon have been observed to feed on juvenile and adult brown rockfish.

Food and feeding: During the day this fish tends to lie on the seabed among seaweed or next to rocks or in other hidden locations. As juveniles they feed on small crustaceans, amphipods, and copepods but at approximately 13 cm, shift to crabs and small fish. An adult brown rockfish (over 30 cm) will feed on larger fish, shrimp, crabs and other crustaceans, and polychaetes. In general, predation most likely lessens as individuals grow.

Migrations and movements: Movements of greater than 3 km are rare for brown rockfish and they are said to have a strong homing tendency. Juveniles gradually move into deeper water as they mature.

Fisheries/aquarium/food/pharmaceutical values: Brown rockfishes are commonly taken by recreational fishers in Puget Sound, Washington, and off central to southern California. They are also caught from private boats, piers, and shore. Brown rockfishes are a valuable hook-and-line species for the commercial live-fish fishery along the central California coast. Small live specimens of this species make excellent bait for large lingcod. Flesh is firm and tasty but rarely found in markets.

Sebastes caurinus (Richardson, 1844)

Photo credit: Wikipedia.

Common name(s): Copper rockfish, chucklehead, white belly

Distribution: Eastern Pacific: Kenai Peninsula, Gulf of Alaska to central Baja California, Mexico.

Ecology: This species is particularly abundant in shallow protected bays and inlets, among rocks and kelp beds. This species is also seen around pilings and jetties or under floats; depth range 10–183 m. Adults are found either in schools or as single individuals. Larval and small juvenile copper rockfishes are pelagic for several months to a year and are frequently associated with surface waters containing surface-forming kelp such as *Macrocystis* sp., *Cystoseira* sp., and *Nereocystis* sp. After several months, at about 4 cm, the juveniles settle to the bottom on rocky reefs as well as sandy areas in shallow habitats, up to about 6 m (Love, 1996) and are referred to as benthic juveniles.

Biology

Identification: This species is diagnosed by the following characters. Dorsal fin has 13 spines and 11–14 dorsal soft rays; anal fin has 3 spins and 5–7 soft rays; and 25–26 vertebrae. Body of the copper rockfish is moderately deep and compressed. Head is large with a slightly curved upper profile; mouth is large; and the lower jaw projects slightly. Caudal fin is rounded. Copper rockfishes are variable in color from olive brown to copper to orange, with pink or yellow blotches. Rear two-thirds of the lateral line are light pink to white and the belly is white. Several copper orange, brown, or yellowish bars radiate back from the eyes. Dorsal fins are dark copper brown to black with some white. It also has a characteristic light-colored horizontal band along the lateral line, which is noticeable underwater. Dorsal and anal spines of this fish are mildly venomous. Males normally grow larger than females. It has a maximum length of 66.0 cm (TL) and maximum published weight of 2.7 kg. Its maximum reported age is 57 years.

Migrations and movements: Copper rockfish may move inshore to release their young. Once adults find a good reef, many do not seem to move about much. Tagging studies indicated that copper rockfish show little movement once they have settled to the bottom. Movement of up to 1.6 km has been noted, but the majority of tagged and recaptured copper rockfishes is from the locality where they were originally taken. In northern waters, copper rockfish probably withdraw in winter deep within crevices to avoid storms.

Food and feeding: Copper rockfishes are opportunistic carnivores. Crustaceans, followed by fish and mollusks, are the most important food groups of adult copper rockfish in terms of volume, number, and frequency of occurrence. Generally, copper rockfish rely less on reef-associated food organisms as their age (size) increases. They feed during the day and at night. Smaller copper rockfish (<45 mm SL) in the kelp canopy eat primarily calanoid copepods, with some harpacticoids and zoea. Fishes 110–115 mm eat small crustaceans such as amphipods, shrimp, caprellids, isopods, and pinnixid crabs. Copper rockfish 1–3 years old eat juvenile Dungeness crab and anchovies, with fish increasing and crustaceans decreasing as the fish grow.

Reproduction/Growth: Males are known to mature between 3 and 7 years, while females mature between 4 and 8 years. Generally, the larger a female is, the more young she will bear. These fishes are viviparous. Fertilization and development of the embryo take place in the body of the mother. They give birth to live young after a gestation period of around 10 months. Spawning takes place between February and May.

Predators: As juveniles and adults, copper rockfishes are preyed upon by a variety of fishes including other rockfishes, lingcod, cabezon, salmon, several species of birds, and marine mammals.

Aquaculture: Rockfish culture remains at a preliminary stage due to difficulties in obtaining larvae and establishing optimal culture conditions. As attempts to obtain larvae from hook-and-line captured wild rockfish have failed due to high larval mortalities, maintaining mature reproducing rockfish in tanks has been reported to be an alternative to obtaining viable larvae. Gravid females of this species were collected and held in isolation until they naturally released live larvae. Survival of larvae at 18°C was <15%, while survival was >40% at 10 and 14°C. An average growth rate of 0.13 mm/day was observed for all the females of different sizes, 2.95, 0.86, and 0.77 kg experimented. Survival and initial size of larvae were greater for larvae from the largest female. Results from these experiments are helpful in optimizing rockfish larval culture conditions for possible restoration of threatened rockfish species of the Eastern Pacific and for aquaculture (Gilbert, 2006).

Habitat associations: The copper and quillback rockfishes have been reported to occur primarily in areas with a high percent coverage (60%–100%) of understory kelp at shallow depths (1.5–4.5 m). They are found patchily distributed in these areas, and predator densities did not seem to be a factor in determining which areas were utilized. Preserving shallow-water understory kelp habitats is a critical part of the conservation and management of copper and quillback rockfishes in the San Juan Archipelago (Hayden-Spear and Gunderson, 2007).

Fisheries/aquarium/food/pharmaceutical values: Copper rockfishes are moderately important in the recreational catch from southern California northward to at least southeastern Alaska. Adults are commonly taken by party and private vessels and young are occasionally taken from piers, jetties, and rocky shores. These fishes are also part of the commercial catch off California and are caught primarily by hook and line, and gill nets. This fish is sold mainly as fresh fillets or live in Chinese restaurants and fish markets; excellent for fish chips or for pan frying as fillets.

It is also an ideal species for aquarium tanks.

Sebastes constellatus (Jordan and Gilbert, 1880)

Photo credit: Wikipedia.

Common name(s): Starry rockfish, spotted corsair, spotted rockfish, chinafish, red rock cod, spiked head

Distribution: Eastern Pacific: San Francisco in California, USA to southern Baja California, Mexico.

Ecology: This reef-associated, demersal species is found over hard bottoms, usually around large rocks, boulders, and occasionally over cobble or wrecks. They have also been observed inside vase sponges, and rarely over mud near rocks; depth range 24–275 m.

Biology

Identification: This species is diagnosed by the following characters. Dorsal fin is with 13–14 spines and 12–14 soft rays; anal fin is with 3 spines and 5–7 soft rays; 25–26 vertebrae; and 7 branchiostegal rays. Body of the starry rockfish is elongate, robust, and heavy forward tapering to the tail. Head is rather pointed in profile and mouth is large with the lower jaw projecting only slightly beyond the upper jaw when the mouth is closed. Body is red orange and profusely covered with small white spots. There are four or five large whitish blotches along the back. It is a very distinctive fish that is not easily confused with any other rockfish. It grows to a maximum length of 46.0 cm (TL); maximum published weight of 1.0 kg; and maximum reported age of 32 years.

Migrations and movements: Starry rockfishes are usually solitary but occasionally they form small aggregations. It is unlikely that they move from reef to reef.

Food and feeding: These fishes feed on small fishes, crabs, shrimp, and other small invertebrates.

Fisheries/aquarium/food/pharmaceutical values: Starry rockfishes are important to both sport and commercial fisheries. They are a minor part of the party- and private-vessel sport fishery in southern California and central California. These fishes are primarily taken by hook and line, and gill nets in the commercial fishery. They are occasionally caught by sport fishers.

Sebastes crameri (Jordan, 1897)

Photo courtesy: Gjernes, Michael, Archipelago Marine Research Ltd.

Alaska Fisheries Science Center

Common name(s): Darkblotched rockfish

Distribution: Northeast Pacific: southeast of Zhemchug Canyon in the Bering Sea to Santa Catalina Island, California, USA.

Ecology: Pelagic juveniles of this species are found over a depth range of 20–45 m. They settle into benthic habitat and may be associated with either soft muddy or rocky bottom habitat. As individuals increase in size and age, they migrate to deeper waters but remain as bottom dwellers, usually in areas of cobbles or boulders; adults occur in depths of 25–910 m and 95% are between 50 and 400 m. Immature fishes have low dispersal capability (<100 km) and adults appear to be highly sedentary.

Biology

Identification: This species is diagnosed by the following features. Dorsal fin has 13 spines and 12–14 dorsal soft rays; anal fin is with 3 spines and 5–7 soft rays; second anal fin spine shorter than third; presence of 7–8 head spines; and caudal fin moderately indented. It is a deep-bodied fish. Adult fishes range in color from white to pink or red and are characterized by four to five discrete dark brown or black blotches on their backs which extend across the lateral line. Juveniles are white with a dark patch on their gill cover, in addition to four to five brown to reddish-brown wide vertical bars, one of which is on the head, and the others extending from the dorsal fin almost to the belly. These fishes exhibit sexual dimorphism, with males achieving maximum length faster than females but with mature females being larger at any given age than males. The maximum age recorded for the species is 48 years old in Canada and 105 years old in the United States. The generation time (average age of parents in the population) is 23 years. Males, however, tend to have longer spines, fin rays and upper jaw length. These fishes have venom glands in their dorsal spines. It grows to a maximum length of 58.0 cm (TL).

Association: These fishes associate with several other groundfish, including Pacific Ocean Perch, Arrowtooth Flounder, and Yellowmouth Rockfish.

Migrations and movement: Darkblotched rockfish migrate to deeper waters with increasing size and age. They make limited migrations after they have recruited to the adult stock.

Food and feeding: Little is known about the feeding strategy of immature fishes. Adults feed mainly at midwater depths on amphipods, copepods, euphausiids, gammarids, salps, and occasionally on other fish and octopus. Euphausiids were found to be the dominant food for Darkblotched Rockfish from Vancouver Island to northern California.

Predation: Juveniles of this species are preyed upon by Albacore (*Thunnus alalunga*), Chinook Salmon (*Oncorhynchus tshawystscha*), and Pacific Hake (*Merluccius productus*). Juvenile rockfish species comprise a significant portion of seabird diets in the California Current System.

Fisheries/aquarium/food/pharmaceutical values: Like other rockfish, these fishes have closed swim bladders, which make them vulnerable to injury when captured from deep water. Consequently, bycatch mortality is assumed close to 100%. This species is a commercially important in the United States, representing the fourth most common species caught by the commercial trawl fishery in 2004. In Canada, however, no directed catch of Darkblotched Rockfish exists and instead the species is caught as a bycatch in the Pacific

Ocean Perch fishery. In the 2007–08 fishing season, the total Canadian catch of this species had a landed value of approximately $61,000.

Conservation and management: This species has been designated as overfished in the United States and is currently managed under a rebuilding plan that regulates where, when, and by how much it can be harvested.

Sebastes dalli (Eigenmann and Beeson, 1894)

Photo credit: Wikipedia.

Common name(s): Calico rockfish

Distribution: Eastern Central Pacific: San Francisco, California, USA to central Baja California, Mexico.

Ecology: It is a demersal species. At rest, calico rockfish seek crevices or are exposed on the bottom, rarely swimming more than 2 m above the bottom. Adults often live at sand-rock interfaces, especially rocky shelf areas where there is a mud-rock or a sand-mud interface. Adults are also associated with areas of high- and low-relief, including artificial reefs. Juvenile calico rockfishes are found in areas of soft sand-silt sediment, at sand-rock interfaces, and on artificial reefs over a wide depth range, including intertidally. Calico rockfishes are a warm-temperate species and are more abundant during warm water years (17°C or warmer). Its depth range is 18–256 m.

Biology

Identification: This species is diagnosed by the following characters. Dorsal fin has 12–14 spines and 12–14 soft rays; anal fin has 3 spines and 6–7 soft rays; 26–27 vertebrae; and 7 branchiostegal rays. Adults of this species are easily distinguishable from all other rockfish by their reddish brown bars slanting obliquely on the brown or yellowish green body; and brown streaks and spots on caudal fin. It is a viviparous species with planktonic larvae and pelagic juveniles. It grows to a maximum length of 25.0 cm (TL) and its maximum reported age is 12 years.

Food and feeding: Juvenile calico rockfish feed on zooplankton such as copepods, barnacle cypriots, and larval fish. Adults feed on larger crustaceans, such as euphausiids, copepods, and crabs; and on fishes, gammarid amphipods, bivalves, and cephalopods. Calico rockfish probably compete with other foraging rockfish species and other finfishes with similar food habits. They may also compete with other fish and with other calico rockfish for favorable habitat because they are a residential, nonschooling species

Predators: Larger rockfish species, lingcod, cabezon, and salmon prey upon adult calico rockfish. Sea birds and dolphins have also been known to feed on calico rockfish. As larvae, calico rockfishes are preyed upon by siphonophore and chaetognaths.

Fisheries/aquarium/food/pharmaceutical values: Because of the relatively small size of adult calico rockfish, they are not usually targeted by either sport or commercial fishermen but are caught incidentally when other finfish species are targeted. Calico rockfish frequently appear as a bycatch in prawn trawls in southern California and are caught by sport anglers on commercial passenger fishing vessels and private boats when they are fishing for other, larger benthic species.

Sebastes elongatus (Ayres, 1859)

Photo courtesy: Gjernes, Michael: Archipelago Marine Research Ltd.

Common name(s): Greenstriped rockfish, cucumber, watermelon, poinsettia, chili, garnet. Older names include reina, serena, strawberry rockcod, and striped rockfish

Distribution: Northeast Pacific: Chirikof Island in the Gulf of Alaska to Cedros Island, Baja California, Mexico.

Ecology: Greenstriped rockfish is a deep-water, sedentary species that can inhabit waters 52–828 m although it is commonly encountered inshore and offshore. About 95% of greenstriped rockfish in survey catches occurred in 100–250 m. As they age, greenstriped rockfish move to deeper water. They are solitary and are often found resting on the seafloor and living among cobble, rubble, or mud. These fishes are also common around oil platform supports and adjoining mussel mounds. In areas off California, they have been observed with other rockfishes over rock outcrops.

Biology

Identification: This species is diagnosed by the following features. Dorsal fin is with 13 spines and 12–14 soft rays; anal fin is with 3 spines and 6 soft rays. This species has a relatively slim body and is easily identified from other rockfishes by its distinct color pattern, which includes four green horizontal stripes. Background color for both adults and juveniles ranges from reddish to white when seen underwater. Areas of the body often turn red or orange after capture. Black patches can sometimes occur. Females are larger than males at any given age after maturity. Both sexes are roughly the same size at maturity. Small venom-secreting glands have been discovered in the dorsal, pelvic, and anal spines of this species. It grows to a maximum length of 43.0 cm (TL); weight of 630.00 g; and life span of 54 years.

Food and feeding: Juvenile and adult greenstriped rockfish prey upon planktonic prey such as euphausiids, copepods, and pelagic tunicates, as well as small fishes (e.g., hake, anchovies, and lanternfishes), shrimp, and squid.

Predator: Small greenstriped rockfishes are eaten by chinook salmon.

Reproduction: It is a viviparous species. About 11,000–295,000 eggs are produced by the females each season. They release one brood of larvae, except off southern California, where females release two. Larval release varies, occurring generally from January to July, depending on geographic location. When released, larvae are about 0.5 cm long.

Fisheries/aquarium/food/pharmaceutical values: From British Columbia, and sometimes off southeast Alaska, to at least southern California, these fishes are often taken in commercial fisheries, usually by trawl. Gillnets and hook and line are also used. Although they are popular in the southern California Asian markets, the greenstriped rockfishes are commonly discarded from the catch due to their small size. Only the largest individuals of this species are filleted for human consumption and smaller ones are processed into fishmeal or used as animal feed. The flesh has a delicious flavor and appealing texture. It is rarely caught by sports anglers but an excellent bait for large lingcod. This species has got rare appearances in public aquariums.

Sebastes eos (Eigenmann and Eigenmann, 1890)

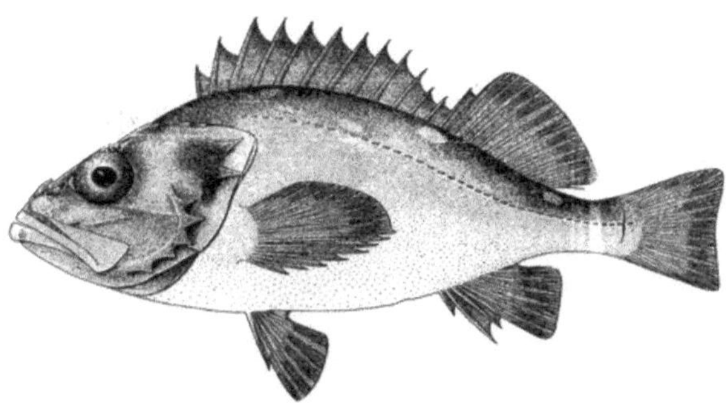

Photo credit: Fish Biosystem.

Common name(s): Pink rockfish

Distribution: Eastern Central Pacific: Monterey Bay in California, USA to central Baja California, Mexico.

Ecology: Pink rockfishes are common in waters, from 45 to 366 m. Adults have been observed in boulder fields, resting on soft bottom sediments; and they have also been reported near rocky bottoms on the shelf, slope, and in canyons; whereas, juveniles have been reported inhabiting primarily soft bottom sediments.

Biology

Identification: Pink rockfish is squat, stout and spiny. It is orange or almost white with backs that lack distinct spots or vermiculations. Dorsal area has a faint dusting of brown or green. Dorsal fin is white or light-pink and it has 3–5 whitish blotches on back. Soft dorsal, caudal, and anal fins are trimmed in white. There are 17–18 pectoral rays; spines are seen on the lower edge of gill covers; and 30 or more long and smooth gill rakers are present on first gill arch. It has a maximum length of 56.0 cm (TL) and weight of 2.4 kg.

Migrations and movements: No information.

Food and feeding: No information.

Reproduction: This is a viviparous species and almost nothing is known regarding the life history of this species.

Fisheries/aquarium/food/pharmaceutical values: Pink rockfishes are taken in commercial fisheries and occasionally in sport fisheries off California (McCain et al., 2005).

Sebastes fasciatus (Storer, 1854)

Photo credit: Wikipedia.

Photo: J.D. Dutil (MPO/DFO Canada)

Common name(s): Acadian redfish, redfish, ocean perch, sharp-beaked redfish

Distribution: Northwest Atlantic: Gulf of St. Lawrence to shelf waters of Nova Scotia in Canada. Northeast Atlantic: off Iceland and western Greenland.

Ecology: It is a deep water fish inhabiting rocky or clay-silt bottom; found in shallower waters during the colder winter months; depth range 70–592 m; and is gregarious during all life. Late juveniles (11–20 cm TL) are primarily associated with boulder reefs that have deep interstices amongst the boulders; as well as in exposed boulders that do not have crevices along their lower margins but are surrounded by dense patches of cerianthid anemones, *Cerianthus borealis*. The use of both boulder and cerianthid habitats are either on an encounter basis, regardless of habitat saturation or predation pressure, or because boulder reefs serve as recruitment habitats, and cerianthid habitats serve as a conduit for redfish moving away from saturated boulder reef sites.

Biology

Identification: This species is distinguished by the following characters. Dorsal fin is with 14–16 spines and 12–17 soft rays; anal fin is with 3 spines and 6–9 soft rays; pectoral fin is with 17–20 rays; 31–38 scales on lateral line; and 29–32 vertebrae. Redfish which have perchlike in appearance have a flattened body that is three times longer than it is deep. Their mouth is large and is lined with many small teeth. One continuous dorsal fin runs from the nape of their neck to their caudal peduncle. Their tail fin is noticeably small. Coloration of body is bright or grayish-red with a dusky zone on upper and posterior part of opercle. It is a slow growing species and has low fecundity. It grows to a maximum length of 51 cm and can live up to 50 years or more.

Food and feeding: Their diet includes euphausiids (*Meganyctiphanes norvegica*, Thysanopoda), decapods, mysids, small molluscs, and fishes. They are known to hit almost any bait

Predators: Acadian redfish is preyed on by the halibut, the Atlantic cod, swordfishes and harbor seals.

Reproduction: It spawns in the fall to late winter. It is ovoviviparous bearing live young. After the insemination of the females, the ripe oocytes (15,000–20,000) are fertilized and the larvae are released.

Conservation: Due to its slow growth rate, low fecundity, and being considered a great food fish, the redfish has been classified as Endangered by the IUCN since 1996. However, due to conservation efforts, the redfish population has rebounded, and in 2012 this species was described as fully rebuilt, sustainably managed and responsibly harvested under US regulations.

Fisheries/aquarium/food/pharmaceutical values: It is a popular display fish in public aquaria and an important fishery resource. Although they are considered to be an excellent food fish, they offer the angler little in the way of a struggle when caught.

Sebastes goodei (Eigenmann and Eigenmann, 1890)

Photo credit: Wikipedia.

Common name(s): Chilipepper, red snapper

Distribution: Eastern Pacific: Queen Charlotte Sound, British Columbia, Canada to Magdalena Bay, Baja California Sur, Mexico.

Ecology: This demersal species inhabits deep rocky reefs as well as sand and mud bottoms and is also thriving in brackish waters. Chilipepper is considered to be a parademersal species as it is occurring in the middle-shelf mesobenthal to outer-shelf. Though it has been collected from a depth of 425 m, it is common between 50 and 250 m. Larvae and small juveniles which are associated with kelp Canopies are generally found near the surface. Chilipeppers are found in water with salinities of 32–34 ppt, temperatures of 5–25°C, and at dissolved oxygen levels of 1.0–6.8 ppm.

Biology

Identification: This species has the following distinguishing characters. Dorsal fin is with 13 spines and 13–16 softrays; anal fin has 3 spines and 7–9 soft rays; 26 vertebrae; absence of coronal and nuchal spines; and moderately indented caudal fin. Body of the chilipepper is slender and rather elongate. Head is elongate, pointed and with no spines; it has a smaller mouth with an upper jaw that extends only to about the center of the eye, not past it; and lower jaw is projecting. Coloration of body is generally pinkish becoming whitish below. Though body fins are pink, caudal fin dusky. Its maximum length, weight and reported age are 56 cm, 1.5 kg and 35 years, respectively. Females are growing substantially larger than males.

Movement and migration: Movements and migrations of chilipepper rockfishes are poorly understood. However, movements of up to 2.4 km/day have been recorded. Further, they may also move as far as 45 m off the bottom during the day to feed. Chilipepper may also school by sex just prior to spawning.

Food and feeding: Food and feeding: Larval and juvenile chilipepper eat all life stages of copepods and euphausiids and are considered to be somewhat opportunistic feeders. Adults prey on large euphausiids, squid, and small fishes such as anchovy, lanternfishes, and young hake.

Predators: Its predators include marine birds and mammals.

Fisheries/aquarium/food/pharmaceutical values: t is considered to be an excellent food fish. Chilipepper rockfish have been one of the most important commercial target species in California waters since the 1880s, as well as an important recreational target in Southern California waters historically, and an important recreational target in central and northern California more recently (following the movement of recreational fishing effort to deeper waters in the 1970s and 1980s). Catches were estimated to have begun in 1892 and are estimated to have ranged from several hundred to nearly 1000 t throughout the first half of the 20th century. Gear types of this species are grouped into four general categories; trawl, hook and line, setnet, and recreational. Since World War II a majority has been taken with trawl gear, although hook and line, setnet, and recreational gear have accounted for between 20% and 40% of landings for most of the last three decades (Field, https://www.pcouncil.org/wp-content/uploads/ChilipepperJan2009final.pdf). It is marketed fresh and frozen and is eaten fried, boiled, microwaved, and baked.

Sebastes helvomaculatus (Ayres, 1859)

Photo credit: Wikipedia.

Common name(s): Rosethorn rockfish

Distribution: Eastern Pacific: Kodiak Island, Alaska to Guadalupe Island, Baja California, Mexico; south of San Francisco, California.

Ecology: These fishes are found demersal in benthic environments over mixed rock and mud substrate at depths between 25 and 549 m with water temperatures as low as 4.1°C (39°F). Most (96%) occur from 100 to 350 m. They are common resting solitary in crevices or caves on vertical wall habitat or in mud protected by boulders or cobble. They are typically solitary or found in small aggregations. Occasionally they are found in in association with sea lilies, demosponges, and brittlestars (*Ophiacantha* spp.).

Biology

Identification: This species is diagnosed by the following characters. Dorsal fin is with 13 spines and 12–14 soft rays. Anal fin is with 3 spines and 6–7 soft rays; pectoral fins have 15–18 rays; presence of 27–33 gill rakers; very strong to strong head spines; and slightly indented caudal fin. These fishes have small and elongated bodies with a depth that is 28%–32% of SL. They vary in color from crimson, orange, yellow, or yellowish-green arranged in a series of diffuse stripes. Their anal and dorsal fins are usually striped in pink or green and the top of their head may be green with white stripes. They have four or five white blotches on their back. Their anal spines are shorter than, or equal in length to, the anal rays. Their head is relatively narrow and spine-laden with a short pointed snout, large eyes, and a small terminal mouth. They reach a maximum length of 43 cm, weight of 1.5 kg, and reported age of 87 years.

Migrations and movements: No information.

Food and feeding: These fishes feed on a wide variety of invertebrates such as euphausiids and other crustaceans; and small fish.

Predators: These fishes are preyed upon by lingcod and yelloweye rockfish.

Fisheries/aquarium/food/pharmaceutical values: It is not important as food fish. It is of minor importance to commercial fisheries and is somewhat uncommon in sport catches. It is caught as a by-catch of trawl and longline fisheries as well as by deep water recreational anglers with good frequency.

Sebastes levis (Eigenmann and Eigenmann, 1889)

Photo credit: Wikipedia.

Common name(s): Cowcod, cow rockfish, cowfish, red snapper

Distribution: Eastern Central Pacific: Usal, Mendocino County in central California, USA to central Baja California and Guadalupe Island (off north central Baja California) in Mexico.

Ecology: This is a deeper water (bathydemersal) species occurring at depths from 20 to 500 m. Adults are commonly found at depths of 152–244 m and juveniles are most often found in 20–100 m of water. They are generally solitary but occasionally aggregate. Juveniles recruit to soft bottom habitats and to habitats consisting of low-relief rocks and sedimentary outcrops bounded by mud and sand, commonly at depths of 40–224 m. Solitary subadult cowcod have been found in association with large white sea anemones on outfall pipes as well as submarine canyons under ledges and in crevices of isolated rock outcrops surrounded by mud. Adults may excavate semiconsolidated mudstone to create desirable habitats. Juveniles occur over sandy and clay bottoms and near oil platforms. Solitary juveniles have been observed resting within a few centimeters of soft-bottom areas where gravel or other low relief was found.

Biology

Identification: This species is diagnosed by the following features. Dorsal fin is with 13–14 spines and 12–13 soft rays; anal fin is with 3 spines and 6–7 soft rays; 26 vertebrae; 7 branchiostegal rays; heads are large and spined; dorsal fins are deeply notched; and there is an unusually wide space between the eye and the upper jaw. Body and head of the cowcod are somewhat compressed. Head is very large. Mouth is large with a projecting lower jaw. Adults are uniform pale pink to orange in color. Young fish have four dark vertical bands on their sides which gradually fade into dusky blotches as they increase in size. Maximum length, weight and reported age are 1 m, 13 kg and 55 years respectively.

Migrations and movements: Although the cowcod is considered as generally not migratory; it may move to some extent to follow food.

Food and feeding: Food and feeding: It is an ambush predator. The diet of the cowcod includes mainly fishes, octopus and squid. Juvenile cowcod eat small shrimp and crabs

Conservation: The main threats to cowcod are overharvesting and bycatch in other fisheries. The species was estimated by the US National Marine Fisheries Service to be at 4%–7% of its unfished biomass in 2000. They were declared overfished in 2000. At current population levels, it will take on the order of 75–100 + years to recover the species (with a modest 2–4 metric t of harvest/bycatch per year).

Fisheries/aquarium/food/pharmaceutical values: Cowcod have some commercial importance and are prized by sport fishers. Cowcod are overfished in the southern California Bight; biomass during the 1998 season was about 7% of the virgin level and recent catches have been near 20 metric t/year. Projections based on recent recruitment levels indicate that biomass will decline at catch levels >5 t/year. Trend data indicate that recruitment will be poor in the near future. Recreational fishing effort in deep water has increased and has become more effective for catching these fishes. Areas with relatively high catch rates for this species are fewer and are farther offshore. Cowcod die after capture and cannot be released alive. Two areas recently closed to bottom fishing will help rebuild the cowcod stock (Butler et al., 2003).

Sebastes maliger (Jordan and Gilbert, 1880)

Photo credit: Wikipedia.

Common name(s): Quillback rockfish

Distribution: Eastern Pacific: Prince William Sound, Gulf of Alaska to Point Sur, central California, USA. This species is present throughout the coastal waters of BC, Canada.

Ecology: Quillback rockfishes are a common, shallow-water benthic species. They are taken from subtidal depths to 275 m, but they occur mainly from 9 to 147 m. Young quillback rockfish occur along the shores at depths less than 60 m and adults usually in deeper waters to 140 m. These fishes are solitary reef-dwellers, living close to or on the bottom. Occasionally they will rise up 9–12 m in the water column. In Puget Sound, they occupy a wide variety of habitats, having the highest densities on shallow (<30 m) reefs. Quillback rockfish live among rocks or sometimes on coarse sand or pebbles next to reefs, particularly in areas with a lot of flat-bladed kelp. They are either found perched on rock or kelp or wedged into crevices and holes and are rarely seen out in the open or unstructured areas of reefs. Adults tend to associate with high-relief substrata, and young-of-the-year tends to associate with low-relief substrata. In British Columbia, young fishes are frequently associated with sponge beds. Young-of-the-year tends to be on the most complex areas of low-relief reefs and use eelgrass/sand habitat as temporary habitat. Young settle at 18–25 mm TL to shallow, vegetated habitats such as beds of kelp and eelgrass. Densities on low-relief reefs and sand/eelgrass increased during the summer coincident with peak plant cover. These benthic juveniles (18–25 mm TL) have also been reported to gradually settle in shallow waters along the shores and are associated with a variety of habitats, including drifting aggregates of benthic macrophytes, established bull kelp (*Nereocysis luetkeana*) beds, natural rock configurations, and artificial reefs.

Biology

Identification: This species is diagnosed by the following characters. Dorsal fin has 13 spines and 12–14 soft rays; anal fin has 3 spines and 6–7 soft rays; strong head spines; spinous dorsal with deeply incised membranes; and rounded caudal fin. Quillback rockfishes are mottled orange-brown with yellow spotting around the head, gills, and front of the dorsal spines. Their coloring allows them to blend well with rocky bottom reefs and boulder barriers where they usually live. Juvenile is finely spotted and barely recognizable. Maximum length, weight and reported age of this species were 61.0 cm TL, 3.3 kg and 95 years respectively. Though it has a long life span, it had only three strong reproductive year classes in the last century in BC waters. Quillbacks obtain their name from the sharp, venomous quills or spines on the dorsal fin. At the base of the spines are venomous glands, which excrete poison into the spines. The stinging spines protect the quillback from predators. They are not extremely toxic to humans but can still cause pain and infection. Rockfish have a gland that produces and absorbs gas as needed, to expand or deflate the swim bladder, which enables the fish to move through different water depths. When a rockfish is caught at depths of 60 feet or more and pulled to the surface, the swim bladder cannot adjust quickly to the change in water pressure and overfills with air—pushing the fish's stomach out of its mouth and causing damage to other organs. This does not affect the eating quality of the fish; however, significant numbers of rockfish inadvertently caught with other species are wasted in this manner.

Migrations and movements: On high-relief rocky reefs in Puget Sound, quillback rockfish maintain small home ranges (within 30 m^2). Off-reef movement occurs during the summer. During the fall and winter, they remain on artificial reefs. On low-relief rocky reefs, they maintain considerably larger home ranges (400–1500 m^2). These fishes only inhabit low-relief reefs during the summer and only return from displacements in the summer coincident with peak algal cover. They move from artificial reefs to low-relief reefs during the summer and return to artificial reefs in the fall when kelp disappears on low-relief reefs. Returns to original reefs when artificially displaced indicate site fidelity. Quillbacks can return to their home sites when experimentally displaced up to 6.4 km. They are not territorial of their home range. They may use navigation or olfactory cues to relocate home sites. They maintain small home ranges during the day, night, and high currents. Female quillback rockfish probably move to other habitat to release larvae because no pregnant individuals were observed in these studies. Tagging studies in central California and Washington have shown quillback to be residential (no movement other than diurnal) or to show movement of less than 9.6 km. They have also demonstrated homing ability and specific diurnal movement patterns (Mc Cain et al., 2005).

Food and feeding: Quillback rockfish consume a wide range of prey taxa but are more dietary generalists than other rockfish species. Off British Columbia, quillback rockfish feed on herring and demersal and pelagic crustaceans. They feed primarily during midday and are inactive, sheltering in holes and crevices, during the night. As planktonic larvae, quillback rockfishes are known to consume nauplii, invertebrate eggs, and copepods. After they settle in the shallow, nearshore areas, they remain zooplanktivorous and feed on crustaceans such as barnacle cypriots, shrimp, and calanoid copepods. As adults their habit is more benthic, and they are known to feed on a variety of prey such as crustaceans, small fish including rockfishes and flatfishes, bivalves, polychaetes, brachyuran crabs, gammarid amphipods, euphausiids, and calanoid copepods. And fish eggs such as from lingcod.

Predators: Quillback rockfish larvae are subject to predation by siphonophore and chaetognaths. As juveniles, they are preyed upon by fishes, including larger rockfishes (such as yellow eye), lingcod, cabezon, and salmon. Various marine birds and pinnipeds take juvenile quillback as well. Adults are also subject to predation by larger fishes including some sharks, as well as pinnipeds, and possibly river otters.

Conservation: It is a threatened species.

Aquaculture: Rockfish culture remains at a preliminary stage due to difficulties in obtaining larvae and establishing optimal culture conditions. Attempts to obtain larvae from hook-and-line captured wild rockfish have failed due to high larval mortalities; however, maintaining mature reproducing rockfish in tanks is an alternative for procuring viable larvae. Visibly gravid females were collected monthly from the Oregon Coast Aquarium in Newport and held in isolation until they naturally released live larvae. *Nannochloropsis oculata* (Nanno) and *Isochrysis galbana,* Tahitian strain (T-Iso) were tested in a green-water and enrichment study with larvae of this species. T-Iso strain was found to be better than Nanno for green-water conditioner and as a food for rotifers and either newly hatched or enriched *Artemia.* An average growth rate of 0.13 mm/day was observed. Survival and initial size of larvae were greater for larvae from the largest female. Results from these experiments have been found to be helpful in optimizing rockfish larval culture conditions for possible restoration of threatened rockfish species of the Eastern Pacific and for aquaculture (Gilbert, 2006).

Fisheries/aquarium/food/pharmaceutical values: Quillback Rockfish is a prized food fish that is targeted in commercial, recreational and Aboriginal fisheries coastwide. Quillback Rockfish is a preferred species for the local live-fish markets and maintains a premium price over the fresh (dead) product. Given the accessibility of this species to harvest, they have likely always been an important component of Aboriginal fisheries. Quillback rockfish have white, flaky meat with a delicate flavor. It is marketed as fillets. From Oregon to southeast Alaska quillback rockfishes are an important part of the inshore sport commercial fishery and are also taken by party and private vessels and divers. They are often caught in deep, cold water by bottom fishing with herring or jigging. According to the Oregon Department of Fisheries and Wildlife, the sport harvest of nearshore rockfish (such as quillback, copper, and china rockfish) off the Oregon coast has been 6–12 MT annually between 2004 and 2009. This is within the sport harvest limits set to maintain the resource.

Sebastes melanops (Girard, 1856)

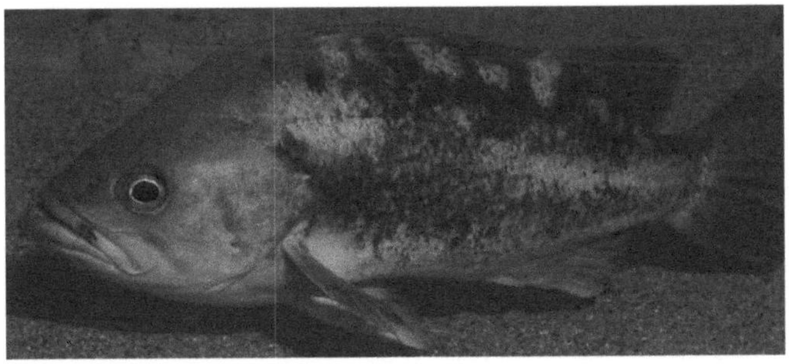

Photo credit: Wikipedia.

Common name(s): Black rockfish, black snapper, black bass, gray rockfish, red snapper

Distribution: Eastern Pacific: Amchitka Island, Aleutian Islands, Alaska to Paradise Cove, Baja California, Mexico.

Ecology: Black rockfish occur from the surface to 366 m but are most common at depths shallower than 55 m. Juveniles occur well up in the water column, usually near or in such shelter as kelp or pilings, though they may live in deeper waters in the winter. Adults are also usually observed well up in the water column. In the water column, the frequency of black rockfish occurring closer (<0.5 m) to vertical *Macrocystis* fronds is significantly greater than that expected had they been randomly distributed. Off California, black rockfish make up the kelp-rockfish assemblage along with the blue, olive, kelp, black-and-yellow, and gopher rockfishes. Adults inhabit the midwater and surface areas over high-relief rocky reefs. They are found in and around kelp beds, boulder fields, pinnacles, and artificial reefs. These fishes often form schools of thousands of individuals, often in association with reefs and with other species including yellowtail and dusky, silvergray and blue rockfishes. The abundance of black rockfish in shallow water declines in the winter and increases in the summer. Densities of black rockfish decrease with depth during both the upwelling and nonupwelling seasons.

Biology

Identification: This species is diagnosed by the following characters. Dorsal fin is with 13–14 spines and 13–16 soft rays; anal fin is with 3 spines and 7–9 soft rays; 26 vertebrae; 8 weak to very weak head spines; moderately large eyes; 7 branchiostegal rays; anal fin profile is rounded or with greater portion slanted posteriorly; and posterior margin of caudal fin is indented. Body of the black rockfish is oval or egg-shaped and compressed. Head has a steep upper profile which is almost straight; and mouth is large and the lower jaw projects slightly. Body coloration is brown to black on the back, paler on the sides, and dirty white below. There are black spots on the dorsal fin. Dorsal and anal spines of this fish are mildly venomous. Maximum length, weight and reported age are 64 cm, 4.8 kg, and 50 years, respectively.

Movement and migration: Tagging studies off Washington, Oregon, and Southeast Alaska made in this species have shown that although for the most part they spend most of their life in a small area, some long-distance travel does occur. Black rockfish may travel up to 600 km. In the summer, schools of feeding black rockfish are commonly seen at the surface along the kelp-lined shores of the western Juan de Fuca Strait, particularly near Duncan Rock (about 2.8 km north-northwest of Cape Flattery). Schools of black rockfish occur above shallow water British Columbia reefs only from June to September. Although black rockfish may occur with blue rockfish, particularly in central and northern California, they are not considered to be competitors because their diets share little in common. Black rockfishes are commonly associated with other nearshore fish species, particularly other rockfishes such as gopher, China, and brown rockfishes.

Movement patterns: Studies relating to the movement patterns of black rockfish have been made in Oregon coastal waters. The results showed that the fish were temporarily absent from the monitored area for short periods especially during the reproductive season (November, January, and February), but both sexes had the longest absences during April through July. These data indicate that black rockfish in open coastal waters live in a very restricted area for long periods as adults but may relocate periodically. A small home range could make them susceptible to local depletion from targeted fishing but also make them good candidates for protection using marine reserves (Parker et al., 2007).

Food and feeding: Adult black rockfish primarily prey on pelagic nekton (anchovies and smelt) and zooplankton such as salps, mysids, and crab megalops. Juveniles feed on copepods, zoea, other crustaceans (such as carangids and mysids), barnacle cypriots, fish larvae, and juvenile polychaetes. During upwelling periods the adults have been reported to prey on small fishes (including juvenile blue and other rockfishes), euphausiids, and amphipods; during nonupwelling periods they primarily consume invertebrates, such as crustaceans, polychaetes, cephalopods, chaetognaths, and jellyfish. During the summer months when recruitment of juvenile is relatively high, juvenile rockfishes are the primary prey of adult black rockfish. Most feeding probably occurs during the day or at twilight. They feed almost exclusively in the water column.

Predators: Adults are subject to predation by large rockfish, lingcod, sharks, salmon, dolphin, pinnipeds, marine birds, and possibly river otters. Larval black rockfishes are subject to predation by siphonophore and chaetognaths. Juveniles fall prey to other rockfishes, lingcod, cabezon, salmon, marine birds, and porpoise.

Fisheries/aquarium/food/pharmaceutical values: This species is maintained in aquariums. Black rockfish is also an excellent food fish and a popular sport fish in Alaska, where it is known as "black bass." These fishes are harvested in Oregon, California, Washington, British Columbia, Alaska, and the Pacific. They are a fairly common catch along the West Coast of North America, where populations are stable. Schools of adults often aggregate over rocky bottoms or at the sea surface, habits that make them susceptible to targeted fishing. These fishes are commonly caught from commercial passenger fishing vessels and when trolling for salmon. The black rockfish is also an important recreational and commercial species in the nearshore rockfish group, particularly in areas north of San Francisco. Black rockfishes are an important commercial nearshore species in California, specifically in the Crescent City port complex; 74 percent, by weight, of all black rockfish were landed there over the past decade.

Sebastes miniatus (Jordan and Gilbert, 1880)

Photo credit: Wikipedia.

Common name(s): Vermillion rockfish, red snapper, red rock cod

Distribution: Eastern Pacific: Queen Charlotte Island, British Columbia, Canada to central Baja California, Mexico.

Ecology: Vermilion rockfish occur in shallow water when young and in deeper water as larger adults. Adults inhabit shallow to deep rocky reefs. They occur at depths up to 436 m, and commonly occur at depths of 50–150 m. Newly released larvae are pelagic and found near the surface for 3–4 months. They then settle to the bottom in waters between 5 and 30 m deep. Young vermilion rockfish recruit to sand, to sand/low-rock substrata without algae or kelp, and to other structures, such as worm tubes, eelgrass, and pilings, which are surrounded by sand. In general, young are recruited to the bottom on soft or hard substrata with low relief (<1 m). Juveniles are secretive and often take refuge in dense algae and kelp beds.

Biology

Identification: This species is diagnosed by the following characters. Dorsal fin has 13 spines and 13–15 soft rays; anal fin is with 3 spines and 7 soft rays; moderate to weak head spines; lower jaw is rough and scaly; caudal peduncle is deep; and caudal fin is slightly indented. Body of the vermilion rockfish is moderately deep and compressed. Upper profile of the head is somewhat curved; mouth is large, with the lower jaw slightly projecting. It has scales on the bottom of the lower jaw which make it rough to the touch. Vermilion rockfish range in color from bright red to orange to red. Sides are mottled with gray; mouth and fins are red; fins are usually edged with black; and 3 obscure stripes are radiating from each eye. Maximum length, weight and reported age are 91.0 cm TL, 6.8 kg and 60 years respectively.

Migrations and movements: Vermilion rockfishes are usually found aggregating near or slightly above the bottom, often over high relief or artificial structures such as wastewater discharge pipes and oil drilling platforms. They probably move from reef to reef, particularly in deep water, but it is unknown how far they move. This species has strong site fidelity and moves very little from its primary habitat type. Movements of vermilion rockfish off reefs may be associated with following schools of prey, such as squid.

Food and feeding: Vermilion rockfish prey on other fishes (anchovies, lanternfishes, small rockfishes), octopi, squids, and krill. Pelagic young feed primarily upon crustaceans.

Fisheries/aquarium/food/pharmaceutical values: Vermilion rockfishes are popular in both sport and commercial fisheries. They are highly prized by party- and private-vessel anglers throughout California with the majority of catches occurring from Monterey Bay south. Divers on the central California coast occasionally take large solitary individuals. Juveniles are caught from piers from about Santa Barbara northward. Adults are taken primarily by gill net and hook and line, and make up a substantial part of the rockfish commercial catch off California. It is an excellent food fish but does not keep well in the freezer. It is sold as fillets together with other rockfish. Its flesh is firm, lean, and mild-flavored with a nutty accent. Its flavor profile lends itself to a wide range of rubs, marinades, and sauces. Further, it is a good source of niacin, Vitamin B6, Vitamin B12, phosphorus, potassium, and selenium.

This species is also a game fish and is a popular aquarium fish.

Sebastes nebulosus (Ayres, 1854)

Photo credit: Wikipedia.

Common name(s): China rockfish, yellowstripe rockfish, yellowspotted rockfish

Distribution: Eastern Pacific: southeast Alaska to Redondo Beach and San Miguel Island, southern California, USA.

Ecology: China rockfish occur both inshore and along the open coast from 3 to 128 m. They are most commonly found in waters between 18 and 92 m. They inhabit shallow subtidal waters during summer. The juveniles are pelagic, but the adults are sedentary, associated with rocky reefs or cobble. They are residential, and generally are found resting on the bottom or hiding in crevices and kelp beds. These fishes are solitary, sedentary, and probably territorial also. They occupy progressively deeper waters in the southern portion of their range.

Biology

Identification: This species is diagnosed with its dorsal fin which has 13 spines and 13–14 soft rays; and anal fin is with 3 spines and 6–8 soft rays. Head spines are thick and strong and the parietal ridges are very high and thick. Coronal and nuchal spines are absent. Parietal ridges are very high and thick. Mouth is small. Caudal fin is rounded. Body coloration of this species is blue or black, mottled with yellow. Some individuals are white ventrally. A broad yellow stripe from about the third dorsal spine runs into or along the lateral line. Pelvic, anal and caudal fins are dark. This species has mildly venomous spines. Maximum length, weight and reported age are 45 cm, 1.7 kg and 79 years, respectively.

Migrations and Movements: They remain close to home crevices and take shelter when disturbed. Tagging studies indicate that they probably move little as adults.

Food and feeding: China rockfish larvae are planktivores. They eat invertebrate eggs and nauplii, and copepods as their primary prey. Juveniles eat crustaceans, such as barnacle cypriots. Prey taxa of the China rockfish off southeastern Alaska include brittle stars, rock crabs, decorator crabs, brachyuran crab larvae, caridean shrimp, hermit crabs, and fish. China rockfish forage on sedentary invertebrates comprising the living turf. Off Oregon they consume crustaceans, particularly decorator or spider crabs, and rock crabs. Off California, China rockfish consume crustaceans (primarily brachyuran crabs), octopi, abalones, chitons, fishes, and brittle stars. China rockfishes are dietary specialists and have the greatest dietary similarity with quillback rockfish.

Predators: Predators of larvae of this species include siphonophore and chaetognaths. Juveniles are prey of birds, porpoises, and fishes, including rockfishes, lingcod, cabezon, and salmon. Predators of adult China rockfish include sharks, dolphins, seals, lingcod, and possibly river otters. China rockfishes are likely to compete with other demersal species such as kelp greenling, cabezon, and lingcod; and other rockfishes such as grass, quillback, copper, black and yellow gopher, and vermilion snapper, all of which also inhabit rocky areas.

Reproduction: In this species, fertilization is internal. No further information is available on its reproductive biology.

Fisheries/aquarium/food/pharmaceutical values: China rockfishes are valuable to the commercial rockfish fishery with most of the catch by hook-and-line gear. This species is overexploited by commercial fishing in many locations. China rockfishes are moderately important in the sport catch. They are taken by party and private vessels from central California to southeastern Alaska and are occasionally speared by divers. It is commonly caught by recreational harvesters off the northern Washington coast. Recreational harvest within Puget Sound has been closed.

It is also a popular aquarium fish and maintained in aquariums.

Sebastes nigrocinctus (Ayres, 1859)

Photo credit: Wikipedia.

Common name(s): Tiger rockfish

Distribution: Eastern Pacific: Cape Resurrection, Kenai Peninsula, Alaska to Point Buchon, central California, USA.

Ecology: It is a reef-associated, semidemersal to demersal species commonly found in caves along undersea cliffs or on the sea floor, generally in high relief areas with strong currents; often associated with "wall" habitat; solitary and territorial, defending a home crevice among rocks; depth range 10–305 m; juveniles of the species are pelagic, around rocky reefs, as shallow as 9 m.

Biology

Identification: It is diagnosed by the following characters. Dorsal fin has 13 spines and 13–15 soft rays; anal fin is with 3 spines and 6–7 soft rays; strong head spines; nasal, preocular, supraocular, postocular, tympanic, coronal, parietal and nuchal spines are present; strongly concave between eyes; parietal ridges are very thick and coarse; and caudal fin is rounded. This species is striped like a tiger, as it has shades of pink, gray or rose, with five black or red bars radiating backwards from the eyes. It quickly changes color when disturbed or when there is a change in lighting. In younger individuals, tips of the ventral and anal fins are darkened. It has a maximum length of 61.0 cm (TL); weight of 2.2 kg; and reported age of 116 years.

Migrations and movements: As these fishes are territorial, they are reported to make short storm-related movements.

Food and feeding: Tiger rockfish exit their caves in the evening to feed. They are known to prey upon caridean shrimp, crabs (particularly rock crabs), amphipods and small fishes like herring and juvenile rockfish. This species is a generalized feeder that depends on currents bringing food items near its home territory. Larvae are planktonic and prey on plankton such as copepods, siphonophores and chaetognaths.

Reproduction: It is a viviparous species. In Puget Sound, the spawning season peaks are in May and June. No other information.

Fisheries/aquarium/food/pharmaceutical values: In the northeastern Strait of Georgia, tiger rockfishes are generally captured in 21–140 m of water. Its flesh is firm and tasty.

Sebastes norvegicus (Ascanius, 1772) (=*Sebastes marinus*)

Image not available

Common name(s): Golden redfish, ocean perch

Distribution: Eastern Atlantic: Kattegat and North Sea, northward to Spitsbergen, southern part of Barents Sea eastward to Kanin Banks and Novaya Zemlya, White Sea, Iceland and eastern Greenland. Western Atlantic: Greenland and southeastern Labrador in Canada to New Jersey in USA.

Ecology: This benthic, pelagic-oceanic species is found at depths over 100 m. Juveniles may be found in shallower, inshore waters and in fjords. This oceanodromous species has a depth range of 100–1000 m.

Biology

Identification: This species is diagnosed by the following characters. Dorsal fin is with 14–16 strong spines and 13–16 soft rays; and anal fin is with 3 spines and 7–10 soft rays. Like similar species, it has relatively large and tall fins including an almost truncate tail fin, rounded anal, pectoral and pelvic fins, and a continuous dorsal fin. Membrane of the pectoral fins extends almost to the tips of the rays, which number 18–21 rays. Spines on the preoperculum are flattened, and ventral spines are downward pointing. Lateral line is with 31–39 scales and there are 30–32 vertebrae. While the young fishes are of brownish color, the adults are bright red with a dusky area behind the opercle. Adults are slow-moving and gregarious throughout life. This species appeared on a 15-pfennig stamp of West Germany in 1964. Maximum length, weight and reported age are 100.0 cm (TL), 15.0 kg, and 75 years, respectively.

Food and feeding: It feeds mostly on euphausiids in summer; herrings in autumn and winter; capelins, herrings, euphausiids and ctenophores in spring.

Reproduction: This ovoviviparous/viviparous species enter reproduction rather late. Gonads of male and female do not mature at the same time. The spermatozoa are kept in the ovary of the female after copulation until such time that the eggs ripen paving the way for fertilization. Copulation takes place in late summer or early autumn. In winter, females give birth to 50,000–350,000 pelagic larvae of 8 mm length.

Ciguatera poisoning: This species has also been reported to be responsible for Ciguatera poisoning in certain areas.

Fisheries/aquarium/food/pharmaceutical values: The fisheries of this species are highly commercial and recreational as it is a gamefish. These fishes are caught in large numbers by vessels from northern Europe, and to a lesser extent from North America. The total catch reported for this species to FAO for 1999 was 7264 t. One of the main fishing areas of the rose fish is the Irminger Sea between Iceland and southeastern Greenland. While annual catches during the 1980s and 1990s were less than 20 kt, this has increased considerably since 1999, to between 40 and 60 kt. In 2000, almost 80 kt were caught. Since then, annual catch has declined back to between 40 and 60 kt. The countries with the largest catches were Faeroe Islands (7129 t) and France (128 t). Since the mid-2000s, populations have been considered severely overfished. According to Greenpeace, some populations are no longer reproducing sufficiently, and their chances of recovery are slim. In 2010, Greenpeace International added the rose fish to its seafood red list. These fishes are utilized fresh and frozen and are eaten fried, broiled, microwaved, and baked.

Sebastes oculatus (Valenciennes, 1833)

Common name(s): Patagonian redfish, blackbelly rosefish

Distribution: Southeast Pacific and Southwest Atlantic: Chile, Argentina, and the Falkland Islands.

Ecology: It is a demersal species inhabiting rocky reefs; depth range 15–280 m.

Biology

Identification: This species is diagnosed by the following characters. Dorsal fin has 12–13 spines and 12–14 soft rays; anal fin is with 3 spines and 6–7 soft rays; and presence of wide mouth and large eyes. Body of the fish is gray to red colored with up to 5 white blotches on back and sometimes on sides. Its spines may be traumatogenic for the victims. It grows to a maximum length of 40.8 cm (TL).

Parasite: A new species of *Clavellotis* (Copepoda, Lernaeopodidae) has been reported to be parasitic on this species collected from southern coast of Argentina (Castro and Gonzalez, 2005). For another parasitic copepod *Caligus quadratus*, this species was the host.

Reproduction: It is a viviparous species. No other information is available on the reproduction of this species.

Fisheries/aquarium/food/pharmaceutical values: A commercial fisheries exists for this species.

Sebastes paucispinis (Ayres, 1854)

Photo credit: Wikipedia.

Common name(s): Bocaccio rockfish, salmon grouper, red snapper, Pacific red snapper

Distribution: Eastern Pacific: Stepovak Bay, Alaskan Peninsula to Punta Blanca, Baja California, Mexico.

Ecology: It is a middle shelf-mesobenthal species and is found as deep as 475 m. However, these fishes were found to be most common between 50 and 250 m. According to some reports (McCain et al., 2005) it is a both a nearshore and offshore species. Larvae and small juveniles are pelagic; large juveniles and adults are semidemersal. Juveniles frequently settle out over rocky areas associated with algae or on to sandy areas with eelgrass or drift algae. Larvae and small juveniles are commonly found in the upper 100 m of the water column, often far from shore. Juvenile and subadults were found to be more common in shallower water than adults. Adults have two primary habitat preferences: some are semipelagic, forming loose schools above rocky areas; and some are nonschooling, solitary benthic individuals. Young and adults also occur around artificial structures, such as piers and oil platforms. Benthic juveniles and adults are usually found around vertical relief; over sand-mud bottoms with little relief and in areas with mixtures of rocks and boulders, rock ridges, and rocks and boulders among mud. Solitary adults have been found in association with large sea anemones, as well as under ledges and in crevices of isolated rock outcrops. Larval stages of this species were found in euhaline waters, congregating in local areas with salinities of 31–34 ppt, temperatures of 6–15.5°C and dissolved oxygen concentrations of 1.0–7.0 ppm.

Biology

Identification: This species is diagnosed by the following features. Dorsal fin has 13–15 spines and 13–16 soft rays; anal fin is with 3 spines and 8–10 soft rays; 26 vertebrae; 7 branchiostegal rays; and slightly indented caudal fin. Body of the fish is elongate and compressed. Head is pointed; mouth is large; and lower jaw is greatly protruding. Color of the fish varies from shades of brown to reddish and extends down over the belly. Young fishes are generally light bronze with speckling over the sides and back. As they mature, their color generally becomes darker and the speckling gradually disappears. Maximum length, weight and reported age are 91.0 cm (TL), 9.6 kg, and 50 years, respectively.

Migrations and movements: Based on a limited amount of data, it was reported to be a nonmoving species. But elaborate studies made on this species confirmed the movement of these fishes to large distances. The tagged adults made frequent small vertical movements, indicating that they were at the top or just above rock habitats during the day, and down in the lower reaches of these habitats at night. Young-of-the-year Boccaccio recruits into shallow water during their first year of life, then move into deeper water with increased size and age. Subadults probably wander from one rocky area to another.

Food and feeding: Larvae of this species often eat diatoms, dinoflagellates, tintinnids, and cladocerans. Copepods and euphausiids of all life stages (adults, nauplii, and egg masses) are common prey for juveniles. Adult fishes feed on mainly fishes such as surfperch, jack mackerel, sablefish, anchovies, sardines, Pacific mackerel, deepsea lanternfish, other rockfishes, and sand dabs. Squid, octopus, and crab also are eaten. These fishes probably locate prey by sight and feed mostly at night.

Predators: These fishes are eaten by sharks, salmon, other rockfishes, lingcod, and albacore, as well as sea lions, porpoises, and whales. They also directly compete with chilipepper, widow, yellowtail, and shortbelly rockfishes for both food and habitat resources.

Reproduction: In this ovoviviparous/viviparous species, females start maturing when they are 43 cm long. In Southern California, they spawn their larvae in 2 or more batches and spawning occurs in almost every year. In Central and Northern California they spawn from January to May, while further north spawning is restricted to January to March. As with all rockfish, fertilization is internal and development of the embryos takes place within the ovaries of the female until they are ready to hatch. A 70-cm female was estimated to contain nearly 1.5 million eggs per season. The main hatching period runs from December through April. The newly hatched young, about 0.6 cm long, does not completely absorb the yolk from the egg stage for a period of 8–12 days.

Environmental effects: Certain effects of strong and weak upwelling affect the food sources of this fish and the survival of its larvae. Larval rockfishes are abundant in or near front upwelling fronts. When the water is cold the upwelling is strong with more productivity and warmer water produces a weaker upwelling with a low amount of resources. Also, a weak upwelling may affect reproduction in egg size, egg amount, and egg quality.

Conservation status: It is a critically endangered species as per IUCN.

Fisheries/aquarium/food/pharmaceutical values: In the commercial fishery, these fishes are caught primarily in bottom trawls. These fishes are also caught by rigs. A typical rockfish rig consists of one or two hooks with enough weight to get to the bottom on a fairly straight course. Bait is usually strips of squid, live anchovies or sardines or lead head scampi; and sometimes jigs are used. It is also a famous sport fish throughout its range. Flesh of this species is of excellent quality when kept chilled and is sold with other rockfish species.

Recreational and commercial fisheries off the coast of California rely heavily on this species. They are caught by trawling, gillnetting and hook and line. Overfishing has occurred over the past decade. Commercial fishermen tend to target these fishes due to their abundance and longer shelf life. The California Department of Fish and Game has set a regulation limit of 2 fishes per day at a minimum length of 25 cm. Also, the depths of fishing have decreased now as older and larger fish tend to stay deeper because the deepest fishermen can fish at is around 73 m.

This fish is also used in public aquaria.

Sebastes pinniger (Gill, 1864)

Photo credit: Wikipedia.

Common name(s): Canary rockfish, red snapper, fantail, canary, orange rockfish

Distribution: From Alaska to Baja California, Mexico.

Ecology: Canary rockfishes are considered a middle shelf-mesobenthal species as they occur at a depth range from the surface (juveniles) to 425 m but primarily inhabit waters 50–250 m deep. Larvae and juveniles are pelagic. Larvae can be captured over a wide area, from 13 to 306 km offshore, and pelagic juveniles occur mostly beyond the continental shelf. They inhabit shallow water when they are young, and deep water as adults. Adults have two primary habitat preferences: some are semipelagic, forming loose schools above rocky areas; and some are nonschooling, solitary benthic individuals. Adult canary rockfishes are associated with pinnacles and sharp drop-offs. They are also found near, but usually not on the bottom, often associating with yellowtail, widow, and silvergray rockfish. Further, they are most abundant above hard bottoms, and they have been observed among mixtures of mud and boulders. In the southern part of its range, they appeared to be a reef-associated species. On Heceta Bank, near Oregon, they were commonly found in boulder and cobble fields in association with rosethorn, sharpchin, yelloweye, and pygmy rockfish. Off Southeast Alaska canary rockfish primarily associated with complex bottoms composed of rocks and boulders, and a few individuals were seen near soft sediments.

Biology

Identification: This species is distinguished by its dorsal fin which is with 13 spines and 14–15 soft rays; anal fin has 3 spines and 7 soft rays; strong to moderate head spines; absence of coronal and nuchal spines; strong anterior slant in anal fin; and embedded scales on lower jaw. Body of the canary rockfish is elongate, moderately deep and compressed. Head is large with an upper profile that is somewhat curved. Pectoral, pelvic, and anal fins are moderately pointed and are large. Color of the fish is yellow orange with gray mottling on the back and paler, near white, below. Fins are also yellow orange. Middle of the sides is in a clear, gray zone. There is often a black spot near the back of the first dorsal fin especially in fish shorter than 35 cm. Although the canary rockfish resembles the vermilion rock fish superficially, both are easily separated; underside of the lower jaw of the canary rockfish has no scales and feels smooth to the touch when rubbed from back to front; and in vermilion rockfish scales are seen on the underside of its lower jaw so that it feels rough when rubbed forward. This species reaches a maximum length of 76 cm and age of 75 years.

Migrations and movements: Canary rockfishes are densely aggregating fish. Juveniles descend into deeper water as they mature. They move into deeper water with age and also are capable of major latitudinal movements (up to 380 nautical miles). Juveniles have been reported to be associated with rocky sandy areas during the day, and with sand flats during the night.

Food and feeding: Canary rockfish juveniles and adults primarily prey on crustaceans, primarily planktonic euphausiids and mysids, and occasionally on fish like anchovies, sand dabs, and other small fishes. Their feeding rate increases during the spring–summer upwelling period when euphausiids are abundant.

Conservation: In this species, minimum population doubling time 4.5–14 years. Various state restrictions on fishing have been put in place over the years, including banning retention of canary rockfish in Washington in 2003. Because this species is slow-growing, late to mature, and long-lived, recovery from these threats will take many years, even if the threats are no longer affecting the species.

Fisheries/aquarium/food/pharmaceutical values: Canary rockfishes are usually caught at depths of 50–300 ft., although juveniles have been taken at the surface and adults have been taken from depths as great as 300 m. They are found around reefs and over soft bottoms. This species is a major constituent of the commercial trawl fishery off Oregon and Washington. Off California, this fish is caught both in the sport fishery and commercial longline fishery. It is moderately important in the party and private vessel sport fishery, from central California northward. The typical rockfish rig and bait are used for catching this species.

Sebastes rosaceus (Girard, 1854)

Photo credit: Wikipedia.

Common name(s): Rosy rockfish

Distribution: Eastern Pacific: Puget Sound, Washington, USA to central Baja California, Mexico; but doubtful north of California.

Ecology: This demersal species inhabits from 7 to 262 m, however adults are common between 30 and 46 m. Juveniles are found from 30 to 61 m, and recruit to rocky areas. Adults are solitary, bottom-dwelling rockfish, found over hard, high relief, and low relief among rocks and sand. Both juveniles and adults are sometimes associated with oil platforms. These fishes are solitary and hide under dark ledges during the day.

Biology

Identification: This species is diagnosed by its dorsal fin which is with 13–14 spines and 11–14 soft rays; anal fin is with 3 spines and 5–7 soft rays; pectoral fin is with 16–18 rays; 27–34 gill rakers; squarish caudal fin; and body covered with scales. It is a small fish with wide body that had a depth that is 33%–37% of SL. Head is short with a short snout; eyes are large and bulging; and mouth is small and terminal, heavily covered with spines. It ranges in color from various shades of yellow, to red to a pale pink and striated with purple on the head and back. They have four to six white "squares" just below the dorsal fin, with pink and yellow fin membranes. These fishes normally change their colors quickly when collected to a deep red uniform coloration. They are found mixed in with Blue, Pygmy, Speckled, Squarespot, and Starry Rockfish. They reach a maximum length of 36 cm.

Food and feeding: Rosy rockfish primarily eat small, bottom-dwelling animals, such as shrimp and crabs.

Recovery of visual performance: Rapid ascent during fishing capture can cause exophthalmia ("pop eye") in physoclistic fishes, resulting in stretching of the optic nerves and extraocular muscles, but it is not known whether exophthalmia affects vision temporarily or permanently. Experiments were conducted to assess changes in visual performance of rosy rockfish (*Sebastes rosaceus*) that had experienced exophthalmia. Vision was functional 4 days after recompression and was improved after 1 month of recovery evidenced by individuals being able to track both smaller and faster-moving gratings. Further, the results suggested that, after recompression, rosy rockfish recover from exophthalmia fairly rapidly and perhaps fast enough to minimize significant adverse impacts on survival. This measured recovery from exophthalmia, in addition to evidence of high short-term, postrelease survivorship, shows that recompression of unwanted rosy rockfish may be a viable management technique, and may be appropriate for other rockfish species, some of which are at low population densities due to high fishing pressure (Rogers et al., 2011).

Fisheries/aquarium/food/pharmaceutical values: Rosy rockfishes are commonly caught aboard sport fishing party and private vessels in southern and central California (to about Morro Bay). Though they are caught with regularity by recreational fishermen, they are routinely discarded due to their small stature. They are also occasionally taken in the commercial gill net and hook-and-line fishery for rockfish. The majority of the catch is sold to Asian markets and restaurants.

Sebastes saxicola (Gilbert, 1890)

Photo credit: Milton Love.

Common name(s): Stripetail rockfish

Distribution: Eastern Pacific: Yakutat Bay, Alaska to Rompiente Point, Baja California, Mexico.

Ecology: Stripetail rockfish occur from 10 to 547 m, but 97% of survey catches occurred at depths between 100 and 350 m. They inhabit the outer shelf upper slope. They are the dominant soft-bottom fishes off southern California, along with Dover sole, slender sole, Pacific sand dab, plainfin midshipman, yellowchin sculpin, and speckled sanddab. Pelagic juveniles of this species are found over a relatively narrow depth range, 50–60 m. Juveniles recruit to soft bottom habitats and to habitats consisting of low-relief rocks and sedimentary outcrops bounded by mud and sand, at depths of 60–100 m. Some juveniles have also been reported in waters as deep as 224 m. Most adults are demersal, associated with mud bottoms and bottoms containing mud and scattered small rocks, although some adults are parademersal. Juveniles have also been reported to recruit to soft substrata (sandy bottoms) in association with macrophytes, to cobble, and in some cases to oil platforms.

Biology

Identification: This species is diagnosed by its dorsal fin which is with 13 spines and 12–14 soft rays; and anal fin is with 3 spines and 5–8 soft rays. Head spines are strong to moderate; nasal, preocular, postocular, tympanic and parietal spines are present; nuchals are usually absent; and supraocular and coronal spines are absent. Eye is large; 2 strong, sharp, hooked spines are seen on lower margin of suborbital bone; and second anal fin spine is longer than third in depressed fin. Caudal fin is moderately indented. Coloration of body is pinkish red to yellowish red; silvery ventrally; dusky saddles on back; and green streaks on caudal fin. Anterolateral glandular grooves of dorsal fin are with venom gland. It reaches a maximum length of 38.0 cm (TL) and age of 38 years.

Migrations and movements: Juvenile stripetail rockfish may be diurnally active and adults are probably nocturnally active. Once recruited to shallower depths, they gradually move to depths commonly used by adults.

Food and feeding: Adult stripetail rockfish feed on pelagic prey such as euphausiids and juveniles feed primarily on calanoid copepods.

Fisheries/aquarium/food/pharmaceutical values: It is commonly caught with trawls but usually discarded because of its small size. However, it is an important bycatch species. Few are caught north of Northern California. Further, it is rarely taken on baited hook in deep water. It is also a gamefish.

Sebastes semicinctus (Gilbert, 1897)

Photo credit: Milton Love.

Common name(s): Halfbanded rockfish

Distribution: Eastern Central Pacific: Point Pinos, Monterey County in central California, USA to central Baja California, Mexico.

Ecology: This demersal, offshore species is seen on flat bottom; and depth range is 58–402 m. While adults are found in deep water reefs, young of year are found in slightly shallower waters (more than 30 m).

Biology

Identification: This smallest species is diagnosed with the following characters. Dorsal fin is with 13 spines and 12–14 soft rays; and anal fin is with 16–18 soft rays. Shape between eyes is flat. Body color is dusky-pink above and silvery-pink on sides; two conspicuous dark-red blotches are seen on posterior half of body; brown on back; brown streaks are on membranes of caudal fin; and fins are light pink, light yellow-green. Left pelvic spine which serves as a sting has deep glandular grooves with poorly developed unilateral venom glands. It reaches a maximum length of 25.0 cm (TL) and reported age of 15 years.

Reproduction: It is a viviparous species and further detailed information is not available.

Fisheries/aquarium/food/pharmaceutical values: Not reported.

Sebastes serriceps (Jordan and Gilbert, 1880)

Photo credit: Wikipedia.

Common name(s): Treefish

Distribution: Eastern Central Pacific: San Francisco in California, USA to central Baja California, Mexico.

Ecology: Treefishes are found from 5 to 97 m depths but are most common at depths less than 60 m. It is a residential species with a limited home range; and it does not exhibit migrational activity. Adults are normally found in nearshore, shallow rocky reefs; caves and crevices; and on artificial reefs. These fishes normally shelter during the day in holes along rocky reefs at Catalina Island. Pelagic juveniles are often found in drifting kelp mats, which have broken free from beds and are traveling with the currents. They recruit to hard substrata with high relief (>1 m) at shallow, subtidal depths to 30 m. Juvenile habitat is artificial reefs.

Biology

Identification: This species is identified with the following characters. Dorsal fin has 13 spines and 13–15 soft rays; anal fin is with 3 spines and 5–7 soft rays; pectoral fin has 17–18 rays; and 27–29 gill rakers. It is one of the most striking, unusually marked rockfishes, with 5–6 black bars over a yellowish to olive body and red lips and chin. It has compact body with large head spines which are venomous to humans. Its maximum length and age are 41.0 cm (TL) and 25 years respectively.

Migrations and movements: Treefishes are solitary and are highly territorial, defending their shelter against intruders. Translocated adults normally return to the same shelter.

Food and feeding: Treefish normally feed on bottom invertebrates (such as shrimp, mollusks, and crabs) and small fishes. It is most probably either crepuscular (feeding at dawn and dusk) or nighttime ambush predator.

Competition: Treefishes are solitary and highly territorial. They may compete with other treefish and nearshore rockfish species such as gopher, grass, and black-and-yellow rockfishes for food and shelter habitat.

Predators: The young are fed upon by lingcod, cabezon, birds, rockfish, salmon, porpoises, and terns (sea birds), while adults are preyed upon by sharks, dolphins, and seals.

Reproduction: No data are available for size at maturity for this species. Treefishes are thought to spawn once annually in late winter. It is a viviparous species and its reproductive biology is as same as other *Sebastes* sp.

Fisheries/aquarium/food/pharmaceutical values: *Sebastes serriceps* is an important species in both the nearshore recreational fishery and in the commercial live fish fishery. Treefishes are occasionally taken by party- and private-vessel anglers and by divers, mainly from Santa Barbara southward. Of late, they have become an important component of the live-fish fishery.

Sebastes zacentrus (Gilbert, 1890)

Photo courtesy: Gjernes, Michael, Archipelago Marine Research Ltd.

Common name(s): Sharpchin rockfish

Distribution: Eastern Pacific: Semisopochnoi Island in the Aleutian chain to San Diego, California, USA.

Ecology: Sharpchin rockfish is an outer shelf-mesobenthal species. They occur from 25 to 495 m deep but about 96% occur from 100 to 350 m. Adult rockfish can occur over soft bottoms but they apparently prefer mud and cobble, and mud and boulder substrata, and are associated with boulder and cobble fields; and demosponges, and brittlestars (Ophiacantha). They also occur in dense patches on and within 2 m of the bottom, often mixed with pygmy rockfish. Larval sharpchin rockfish have been reported to occur at depths of 270–2800 m. Small sharpchin rockfishes are found over rocky banks off Oregon associated with vase sponges and fields of crinoids. These fishes are sometimes found in small schools and are solitary normally.

Biology

Identification: This species is diagnosed by the following characters. Dorsal fin has 13 spines and 14–15 soft rays; anal fin is with 3 spines and 7–8 soft rays; strong head spines; absence of supraocular, coronal and nuchal spines; second anal fin spine is longer than third; and caudal fin is slightly indented. Body of the sharpchin rockfish is elongate with a distinct knob at the tip of the lower jaw. Underwater they are white, pink, or red and display 4–5 distinct brown or orange saddle marks. One of the most distinct marks is the forked bar ("<") radiating back from the eyes. Once captured the saddle marks become less distinct although the forked bar is still recognizable. It reaches a maximum length of 45.0 cm (TL) and maximum reported age of 58 years.

Food and feeding: The diet of *Sebastes zacentrus* includes euphausiids, shrimp, amphipods, copepods, and small fishes. Off Oregon, euphausiids made up 85% of stomach contents (by weight). In the Gulf of Alaska, this species utilized a variety of prey taxa; however, 84% of stomach contents were amphipods, euphausiids, and copepods

Reproduction: It is an ovoviviparous/viviparous species. Parturition (giving birth to young) occurs from March through July off Oregon and from May through June off northern and central California. Sharpchin rockfish transform from larvae to pelagic juveniles when they are between 13.5 and 20 mm in length. Transition from pelagic to benthic habitat takes place at lengths somewhere between 35 and 65 mm.

Fisheries/aquarium/food/pharmaceutical values: Sharpchin rockfishes are taken in commercial fisheries along the West Coast. It is rarely caught by commercial harvesters off the Washington coast. Recreational harvest within Puget Sound has been closed. It is also a gamefish for recreational fisheries.

3.3 Lionfishes (Turkeyfishes) (Subfamily Pteroinae)

Brachypterois curvispina (Matsunuma et al., 2013)

Photo courtesy: Muddula Krishna, Naranji.

Common name(s): Australian sawcheek scorpionfish

Distribution: Western Pacific: Australia (Queensland); India.

Ecology: It is a reef associated, pelagic-neritic species; depth range 32–130 m.

Biology

Identification: This species is diagnosed by its dorsal fin which has 12 spines and 10 soft rays; anal fin is with 3 spines and 5 soft rays; long pectoral fin has 14–16 soft rays; and 24 vertebrae. Body of this species is oblong and is moderately compressed. Head is large and its length is greater than body depth. Interorbital spaces are shallow and are about one-seventh of orbit diameter extending above dorsal profile of head. Interorbital ridges are moderately developed and separated by a shallow channel. Mouth is moderately large, and oblique; and upper edge of maxilla is extending slightly beyond a vertical through midorbit. Four to five rows of tooth are present at front of upper jaw; and about 4 tooth rows are at front of lower jaw. Vomer is with a V-shaped patch of two to four rows of small teeth. Tongue is smooth and rounded apically. Nostrils are close to each other; anterior one is round and tubular; and posterior nostril is entire and elliptical. Dorsal profile of snout is steep. Gill openings are wide; gill membrane is free from isthmus; pseudobranchiae are present; and there is a slit below fourth gill arch. Gill rakers in middle of arch are short and are reduced to spiny knobs at either end. Air bladder is thin and spherical in shape. Dorsal fin is continuous, with notch after penultimate spine; and dorsal origin is above middle of opercle; origin of first dorsal-fin spine is above cleithral spine; bases of first and second dorsal-fin spines are closer than those of succeeding adjacent spines; length of the longest spine is about half of the maximum body depth; penultimate spine is shortest; and intespinous membrane is distinctly notched. All dorsal fin soft rays are branched; and fourth soft ray is the longest. Origin of first anal spine is below base of penultimate dorsal fin spine; and third spine is the longest. Six to seven rays of pectoral fin are divided and the longest ray reaches a vertical through base of last anal soft ray. Pelvics are shorter than pectorals, not reaching vent. Caudal fin is with rounded edge. It grows to a maximum length of 9.0 cm (SL) (Naranji and Kandula, 2017).

Fisheries/aquarium/food/pharmaceutical values: Not reported.

Brachypterois serrulata (Richardson, 1846) (=*Sebastes serrulatus*)

Photo credit: Wikipedia.

Common name(s): Sawcheek scorpionfish; shortfin firefish, pygmy lionfish

Distribution: Northwestern Pacific: from South China Sea north to southern Japan, including the Natuna Islands, Borneo, the Gulf of Thailand, Ng Tran Bay, Vietnam and Taiwan.

Ecology: This marine and brackish water demersal species lives in muddy habitat that is very sheltered; deep or very sheltered estuaries.

Biology

Identification: This species has the following combination of characters: dorsal fin is with 13 spines and 10–11 soft rays; anal fin is with 3 spines and 5–6 soft rays; pectoral-fin rays 14–16; scale rows in longitudinal series 44–51; scales above lateral line 4–6; scale rows between last dorsal-fin spine base and lateral line 4–7; relatively short head, its length 38–43% SL; relatively small orbit, its diameter 11%–15% SL; relatively short caudal fin, its length 30%–45% SL; posterior lacrimal spine is usually directed posteroventrally, with no upward distal curve; and spine(s) at corner of outer angular ridge have developed only in adults.

Head and body are pinkish to reddish-brown, darker dorsally, and paler ventrally; and a poorly defined large black blotch (subequal to orbit diameter) is seen on opercle. Five dark red to black bands are seen on head; anteriormost band from anterior margin of eye is running over snout to tip of lower jaw; second is extending from anteroventral margin of eye onto lacrimal and both jaws; third is from just below eye to posterodorsal corner of maxilla; fourth is from posteroventral margin of eye to posterior portion of opercle; fifth is from posterior margin of eye to opercle, more or less continuous with blotch on opercle. Five poorly defined dark or black vertical bands are seen on body; anteriormost band is extending from between first and third dorsal-fin spine bases; second is from between fourth and sixth spine bases; third is from between eighth and tenth spine bases; and fourth is Y-shaped; and fifth is on caudal peduncle just in front of caudal-fin base. Ground color of fins is similar to that of head and body. Dorsal fin is with a large dark red to black blotch on spinous portion between approximately sixth and ninth spines; and about 4–5 more or less dotted dark red lines are on soft-rayed portion. Anal fin is with about 6–8 somewhat dotted dark red lines. Pectoral fin is darker distally, with about 6–10 oblique narrow dark red bands. Pelvic fin is with about 6–7 narrow dark red to black bands. Caudal fin membrane is translucent with dark red band marginally; and about 60–110 small dark red spots are present on rays. Eye is vivid yellow and iris is dark blue. This species grows to a maximum length of 10 cm (TL) (Matsunuma et al., 2013).

Food: It feeds on small fishes and crustaceans.

Fisheries/aquarium/food/pharmaceutical values: Sometimes caught by trawl net.

Brachypterois serrulifer (Fowler, 1938)

Photo courtesy: Muddula Krishna, Naranji.

Common name(s): Sawcheek scorpionfish, sawmaxilla scorpionfish

Distribution: Indo-West Pacific: Madagascar, Red Sea and Persian Gulf, east to the Timor Sea and the Philippines, South and East China seas, and north to Tosa Bay, Kochi, southern Japan.

Ecology: This pelagic-neritic species lives in soft bottoms; depth range 6–82 m.

Biology

Identification: This species is diagnosed by the following characters: dorsal fin is with 13 spines and 10 soft rays; anal fin is with 3 spines and 5 soft rays; and pelvic fin is with 1 spine and 5 soft rays. Gill rakers are simple, short and robust. Swim bladder is present. Body of this species is oblong, moderately compressed anteriorly and is strongly compressed posteriorly. Head is large and its length greater than body depth. Low membranous tube is associated with anterior and posterior nostrils; and posterior edge of tube on anterior nostril is slightly expanded. Numerous small papillae are sometimes present on frontal surface of bulge of snout and chin. Ctenoid scales are covering on head, including cheek, lacrimal, suborbital pit, interopercle, subopercle, and opercle. Scales are absent on both lips and bulge of snout. Maxilla is with weakly ctenoid scales; and posterior portion of maxilla is almost entirely scaled or partly scaled as one-third maxilla depth. Mandible is usually with exposed weakly ctenoid scales. Ctenoid scales are covering on body and extending onto base of caudal fin; scales on ventral body surface are relatively weak; pectoral-fin base is covered with weakly ctenoid scales; and a row of elongated cycloid scales is present on bases of soft-rayed portions of dorsal and anal fin. Lateral line is complete. Three sensory pores are present on underside of each dentary; and a small pore is seen on each side of symphysial knob. Mouth is moderately large, oblique and is forming an angle of c. 40–50° to horizontal axis of head and body; and upper edge of maxilla is swollen laterally forming a low ridge. Upper jaw is with a band of small slender conical teeth; about 4–6 tooth rows are at front of upper jaw; and about 4 tooth rows are present at front of lower jaw. Vomer is with a V-shaped patch of 1–5 rows of small teeth. Dorsal profile of snout is steep, forming an angle of c. 40° to horizontal axis of head and body. Interorbital ridges are moderately developed, with a row of

0−19 spines. Interorbital space is shallow, about one-seventh of orbit diameter extending above dorsal profile of head. Dorsal-fin soft rays are all branched; length of the longest ray is subequal to or slightly longer than that of the longest fin spine; and posteriormost ray is not joined by membrane to caudal peduncle. Anal-fin soft rays are all branched. Pectoral fin is long; and usually fifth (branched ray) and tenth (unbranched ray) rays are the longest in relatively small and large specimens, respectively. Pelvic-fin soft rays are all branched and third ray is the longest, much longer than upper jaw. Caudal fin is rounded. It grows to a maximum length of 95 cm (SL) (Matsunuma, 2013).

Fisheries/aquarium/food/pharmaceutical values: Not reported.

Dendrochirus barberi (Steindachner, 1900)

Photo credit: Wikipedia.

Common name(s): Hawaiian lionfish, green lionfish

Distribution: Eastern Central Pacific: Hawaiian Islands; Johnston Islands.

Ecology: It is a reef-associated, nonmigratory, and benthic species and is found under ledges in lagoons and seaward reefs; normally found in either crevices or caves during the day and it may become benthopelagic during the nighttime; depth range 1−50 m during day and 1−134 m at night.

Biology

Identification: This species is diagnosed by the following characters: dorsal fin is with 13 spines and 9 soft rays; and anal fin has 3 spines and 5 soft rays. It has a large pectoral fin with its upper rays branched distally; and dorsal fin spines are longer than 1/2 body depth. Scales are ctenoid with about 50−55 vertical scale rows. Coronal spines are present. Suborbital ridge is of a single row of spines and it is not as broad patch of tiny spinules. Supraocular tentacle when present is short. It grows to a size of 16.5 cm in TL.

Fisheries/aquarium/food/pharmaceutical values: This species occasionally makes its way into the aquarium trade and its aquarium parameters are given below.

Minimum tank size: 50 gal Care level: Moderate temperament: Semiaggressive reef compatible: With caution max. size: 17.5 cm.

Dendrochirus bellus (Jordan and Hubbs, 1925)

Photo courtesy: Shao, Kwang-Tsao.

Common name(s): Bricked firefish

Distribution: Northwest Pacific: Japan and Taiwan; New Caledonia; India.

Ecology: This demersal species inhabits rocky littoral and sublittoral zones and occurs over sandy mud or gravel bottoms; depth range 10–200 m.

Biology

Identification: This species is diagnosed by the following characters. Dorsal fin spines 13 and dorsal fin rays 10; anal fin spines 3 and anal fin rays 5; pectoral fin rays 17; pelvic fin spine 1 and pelvic fin rays 5; total gill rakers, 20; and vertical scale rows 36. Body is deep (depth 38.77%–40.85% of SL) and is moderately slender. Head is long and is robust with short snout (8%–10% of SL). Mouth is large, and its gape is slightly oblique. Jaw teeth are slender and conical; vomerine tooth patch is V shaped. Anterior nostril is with short tentacle. Eyes are large. Nasal spine is two-tipped; preocular spine is with four tips; supraocular spine is with three points and postocular spine is with nine tips. Scales are large ctenoid. Lateral line is complete and is sloping downwards. Caudal peduncle is short and deep. It has mottled brown body which has broad red vertical bars on a pale background. It grows to a maximum length 15.0 cm (SL) (Padate et al., 2017).

Food: It feeds on small fishes and crustaceans.

Fisheries/aquarium/food/pharmaceutical values: This species is rarely taken as a food fish. However, this species can be maintained in aquaria. Its aquarium parameters are given below:

Aquarium size: Approximately 800 L (approx.)

Temperature: 22–24°

Nutrition: Small fish and live mysids; sometimes frozen foods

Lighting: Prefers somewhat dimmed lighting.

Dendrochirus biocellatus (Fowler, 1938)

Photo courtesy: John E. Randall.

Common name(s): Twospot Turkeyfish, double-ocellated scorpionfish, ocellated lionfish, twinspot lionfish, two-eyed lionfish

Distribution: Indo-Pacific: Mauritius, Reunion, Maldives and Sri Lanka; Society Islands; north to southern Japan; south to Scott Reef.

Ecology: It is a reef associated, demersal species commonly found at depths of 1–40 m. This species is very secretive, and has been found in clear waters with a high concentration of corals. It spends its days hiding in caves and sponges, and only comes out at night to feed.

Biology

Identification: This species is diagnosed by its dorsal fin which has 13 spines and 9 soft rays; and anal fin has 3 spines and 5 soft rays; has a massive bod which is, compressed laterally. It is orange-red with darker areas forming more or less diffuse vertical bands, one at the level of the head, two on the body and one at the caudal peduncle. In front of it an almost white area is seen. The scales are large and conspicuous. Head has 2 globular, prominent and high-set eyes and muzzle is well developed. There is a white area between the eyes and another on the muzzle, in front of the eyes. On each side of the snout, at the level of the white zone, is a long horizontal cutaneous flap, curved toward the rear at its extremity. Mouth is very wide. Its long spines of dorsal fin are more or less free. They are striped with white and brown. Soft part of this fin (9 soft rays) is white with 2 large black ocelli circled with yellow (rarely a single or 3). Rays of pectoral fins are connected by a membrane to their extremity. This fin is broad, fan-shaped and has alternating white concentric zones and brown areas on its inner face. Anal fin (3 spines and 5 soft rays) also has these alternations of white and brown areas or may have some large brown dots. Caudal fin is transparent and only its rays are zebra of translucent and brown zones. It grows up to 13 cm (TL). Transparent caudal fin with only visible rays.

Food: This fish feeds at night, mainly fish but also crabs and shrimps. Its jaw is protractile and it opens and sucks its prey.

Reproduction: It is a gonochoric species (i.e., sexes are separated). Breeding begins with a courtship display, during which male and female move quickly up to the surface, to descend immediately. Then, during their last ascent to the surface, the pair stays close to the surface and the gametes are released into the water column where fertilization takes place. Each female expels two mucous tubes containing the eggs. These tubes then gradually fill with seawater and after a few minutes they took a form of rugby ball and drift in the water column. The emerging larvae are pelagic.

Fisheries/aquarium/food/pharmaceutical values: It is an aquarium fish. In captivity, it is somewhat shy. Furthermore, it may hide in crevices during daylight. This fish has been reported to breed in captivity. Other suitable aquarium parameters of this species are given below:

Temperature: 22–28°

Nutrition: Living small fish, shrimp, and mysids

Aquarium size: From 100 L

This fish does better in smaller aquaria without any other fish, because they are easier to feed. Large aquarium, however, may be required when these fishes are kept with other fishes.

Dendrochirus brachypterus (Cuvier in Cuvier and Valenciennes, 1829) (=*Brachirus brachypterus*)

Photo credit: John E. Randall.

Common name(s): Fuzzy dwarf lion, shortfin turkeyfish, dwarf lionfish

Distribution: Throughout the tropical waters of the Indo-West Pacific; Red Sea and East Africa to Samoa and Tonga, north to southern Japan, south to Lord Howe Island; Mariana Islands in Micronesia; the Arafura Sea and Australia.

Ecology: It is a reef-associated demersal species. It is common in reef flats and shallow lagoons; areas with weed-covered rocks on sandy substrates; and also in estuaries. Adults are often found on sponges and juveniles are sometimes found in small aggregations on remote bommies (outcrop of coral reef) with 10 or so individuals. Its depth range is 2–80 m.

Biology

Identification: This species is diagnosed by its dorsal fin which has 13 spines and 9–10 soft rays; and anal fin is with 3 spines and 5 soft rays; has large pectoral fins with distinct spotted bands. There is usually a short tentacle above the eye and variably developed leafy appendages on the head and lateral line. Body is reddish with vague broad bars; paired fins are with bars; and median fins are with small dark spots. Middorsal spines are shorter than body depth. All 13 dorsal spines are about the same length or slightly shorter than the greatest body depth. Pectoral-fin rays lack filaments. Body coloration is variable from red to brown. This species grows to a maximum length of 17 cm (TL).

Food and feeding: It is nocturnal feeding on small crustaceans.

Reproduction: During breeding, a distinct pairing is seen in this species. Females possess specialized ovarian structures including stalk-like ovigerous lamellae and secretory epithelia. Detailed information on reproduction of this species is wanted.

Fisheries/aquarium/food/pharmaceutical values: It is an aquarium fish and it occasionally makes its way into the aquarium trade. Its aquarium parameters are given below.

Minimum volume: 100 L

Hardiness: Hardy

Suitable for aquarium: Suitable with care

Reef safe: Reef safe with caution

Aggressiveness: Mostly peaceful but might be aggressive toward similar species

Food: Larger crustaceans

Dendrochirus tuamotuensis (Motsunuma and Motomura, 2013)

Photo Credit: Matsunuma and Motomura.

Common name(s): Dwarf lionfish

Distribution: Makemo Atoll, Tuamotu Archipelago, French Polynesia.

Ecology: It is a demersal and reef associated species; collected from a depth of 120 m.

Biology: This little-known species was only scientifically described in 2013, and, for some reason, its remarkable discovery was mostly ignored by aquarists. Just a single specimen has ever been found, having been recovered from a trap set at 120 m in the Tuamotu Islands of French Polynesia.

Identification: This species is diagnosed by its dorsal fin which has 13 spines and 9 soft rays; anal fin has 3 spines and 5 soft rays; and 24 vertebrae. It is uniquely characterized by having bilobed pectoral fins. Body of this species is oblong, moderately compressed anteriorly and is more strongly compressed posteriorly; depth is moderate, greater than length of the longest dorsal-fin spine. One simple tentacle is seen on posterior edge of low membranous tube associated with anterior nostril, its length distinctly greater than diameter of anterior nostril, its tip extending beyond posterior margin of posterior nostril when depressed posteriorly. One short, simple tentacle is on supraocular. There is no skin flap on orbit; small skin flap is on uppermost preopercular spine base; and 2 small skin flaps are on posterior margin of preopercular. One short tentacle is on anterior portion of ventral margin of lacrimal. Large skin flap is on tip of posterior lacrimal spine and its tip is not extending beyond profile of lower jaw when depressed ventrally (lateral view). Lateral line is complete and is weakly sloping downward. Sensory pores of cephalic lateralis are opening at tips of spine-like tubes. Three sensory pores are on underside of each dentary. Mouth is moderately large and is slightly oblique, forming angle of about 30° to horizontal axis of head and body; upper edge of posterior maxilla is swollen laterally, forming low ridge and displaying poorly developed median lateral ridge; and posterior margin of maxilla is just reaching midorbit level. Symphyseal gap is separating premaxillary teeth bands distinctly narrower than width of each band. Upper jaw is with band of small, slender, conical teeth; about 5 tooth rows are at front of upper jaw; about 4 tooth rows are at front of lower jaw; small teeth in 5 rows are forming blunt V-shaped patch on vomer; and there is no palatine teeth.

Dorsal profile of snout is relatively steep, forming angle of about 40° to horizontal axis of head and body. Nasal spine is with single small, upwardly directed spinous point distally; and its length subequal to anterior nostril diameter. Origin of first dorsal-fin spine is above lower posttemporal spine; bases of first and second dorsal-fin spines are closer to each other than those of subsequent adjacent spines; length of first spine is 0.8 times that of second spine; sixth spine is the longest; lengths of fifth to seventh spines are subequal; 12th spine is shortest; and membrane of spinous portion of dorsal fin is strongly incised. Dorsal fin soft rays are all branched; third ray is the longest; and posterior most ray is branched, with extremely low membrane extending to caudal peduncle. Origin of first anal-fin spine is just below base of last dorsal-fin spine; third spine is the longest; and first spine is 0.6 times longer than second and third spines. Anal-fin soft rays are all branched; and second ray is the longest. Pectoral fin is long; fifth ray is the longest; lengths of rays are decreasing from fifth to ninth; middle 8 rays are branched into two; and lower unbranched rays are weakly thickened. Pelvic-fin soft rays are all branched; third ray is the longest; last ray is joined by membrane to abdomen for less than one-fourth of its length; and tip of the longest ray is just reaching anterior margin of anus when depressed. Caudal peduncle is moderately deep and its depth is 0.6 times caudal-peduncle length. In preserved specimens, head and body are brownish-white without markings. Skin flap on uppermost preopercular spine base is with brown pigmentation; and other skin flaps and tentacles on head are pale brownish-white. Eye is blackish-blue, without radial markings. Dorsal fin membrane is translucent; each fin spine is with a small brown blotch on middle to upper portion; and soft-rayed portion of dorsal fin is with 33 small brown spots on rays. Anal fin membrane is translucent, and soft-rayed portion is with 17 small brown spots on rays, expressed as short lines basally. Pectoral fin membrane is transparent, with 6 transverse broad, brown bands. Pelvic fin membrane is translucent, with 5 broad, brown bands. Caudal fin membrane is transparent, with 34 brown spots on rays. It grows to a maximum length of at 5.7 cm (SL) (Matsunuma and Motomura, 2013).

Fisheries/aquarium/food/pharmaceutical values: It can be maintained in aquaria. Its aquarium parameters are given below:

Aquarium size: 500 L (approx.)

Temperature: 22–28°

Nutrition: Small fish, shrimp, and shellfish; substitute feed

Compatibility: It is better not to keep other fish with this fish that fit into its mouth.

Dendrochirus zebra (Cuvier, 1829)

Photo credit: John E. Randall.

Common name(s): Zebra lionfish, butterfly scorpionfish, dwarf lionfish, zebra turkeyfish

Distribution: Indo-West Pacific: Red Sea and East Africa to Samoa, north to southern Japan and the Ogasawara Islands, south to Australia and Lord Howe Island.

Ecology: It is a bottom-dwelling, reef-associated inshore species and is found on coral, pebble, and rock bottoms on reef flats, outer reefs and lagoons and also in caves, sometimes in small groups; found in inshore waters at depths of 0–98 m.

Biology

Identification: The zebra turkeyfish is an unusual looking fish. Dorsal fin has 13 tall, quill-like spines and 10–11 soft rays. Middorsal spines longer than body depth. Anal fin has 3 spines and about 10 soft rays. Large, banded fan-like pectoral fins that flare out on either side as the fish lies on the seabed. It has six dark brown to reddish-brown bars on its body (alternating with narrow bars in larger individuals), a dark brown band through the eye, a large black spot on the lower gill cover, and pale and darker bands on the dorsal-fin spines. Soft rays of dorsal fin, anal fin, and rounded caudal fin are transversely banded in black and white. This species has a maximum length of 20 cm (SL).

Food: The zebra lionfish feeds on small crustaceans and is in turn preyed upon by groupers.

Reproduction: It spawns in pairs. Courtship and spawning occur at night. Males are aggressive; and females are smaller developing almost white face when in courtship. Spawning occurs at the apex of a short and rapid paired ascent resulting in a gelatinous mass of 2000–15,000 eggs. Hatching occurs 36 hours later and the pelagic larvae settle out in a few weeks at a size of 10–12 mm. During pelagic stages, they have been reported to travel great distances and expatriate to subtropical zones.

Fisheries/aquarium/food/pharmaceutical values: This species is of minor use to commercial fisheries. It is however an aquarium fish and is taken for the aquarium trade. Its aquarium parameters are given below.

Min. tank size: 30 gal
Disposition: Semiaggressive
Mature size: 25 cm
Diet: Carnivore
Reef compatibility: With Caution
Care level: Intermediate

Ebosia bleekeri (Döderlein, 1884)

Photo credit; Gloerfelt-Tarp, Thomas.

Common name(s): Bleeker's lionfish, cockscomb firefish venomous species, Indonesia

Distribution: West Pacific: southern Japan to Hong Kong; Australia, Taiwan, China and Korea; Indonesia.

Ecology: It is a demersal species found in the sublittoral zone and in areas of cool upwelling on fine sand and mud bottom; depth range 53–220 m.

Biology

Identification: This species has the following diagnostic features. Dorsal fin has 8 spines and 10 soft rays and an anal fin is with 3 spines and 8 soft rays. It is a sexually dimorphic lionfish. Males have pair-thin elongated curved bony crests on the top rear of the head, and yellowish to orange median fins. Females have a low spiny ridge on the head and reddish median fins. It grows to a maximum length of 22.0 cm (SL).

Fisheries/aquarium/food/pharmaceutical values: It is an aquarium fish.

Ebosia falcata (Eschmeyer and Rama-Rao, 1978)

Photo Credit: Sujatha and N.M. Krishna.

Common name(s): Cocks comb fire fish

Distribution: Indian Ocean: off Somalia, Pakistan; west coast of India; off the Andaman Sea coast of Thailand.

Ecology: It is a demersal species; depth range 47–243 m.

Biology

Identification: This species is diagnosed by its dorsal fin which has 12–13 spines and 10 soft rays; anal fin is with 3 spines and 8–10 soft rays; and 24 vertebrae. In this species, crest of adult males is very elongate and low in females. Body of this species is oblong, compressed; dorsal profile is more convex than ventral; and maximum depth is at dorsal origin. Predorsal profile is with deep slope from nape to snout; and snout is with a concavity before eyes. Interorbital is wide and is deeply concave with coronal ridges, ending on parietal crests; and eyes are large with elevated supraorbital ridges. Mouth is medium, and oblique; and maxilla reaches to below front or middle of eye. Preorbital is deep. Jaws are equal; and tip of lower jaw with less developed symphysial knob, which fits in notch on upper jaw. Villiform teeth are seen on jaws and vomers; patches of teeth are on roof and floor of pharynx; and there are no palatine teeth. Tongue is small, smooth, and pointed at free end. Nostrils are close to each other; and anterior one is round, tubular with lanceolate flap on posterior edge. Gill opening is wide; gill membrane is free from isthmus; pseudobranchiae and slit are seen below fourth gill arch; and gill rakers are in middle of arch, moderately long and are gradually reduced to spiny knobs at either end. Preopercle is slightly rounded; opercle flap is pointed; and flap ends in origin of the pectoral base. Dorsal fin has 12–13 spines and 10 soft rays; ad anal fin is with 3 spines and 8–10 soft rays. Dorsal fin is continuous, with a notch; its origin is above middle of opercle; and fifth spine is the longest. Pectorals are long, with rounded edge and are reaching caudal peduncle. Third anal spine is the longest. Caudal fin is elongate, with rounded edge. Supraorbital tentacle is small. Scales are ctenoid and are often shed during capture. Cheeks and postorbital area are scaled; and there are no scales on interorbital and snout. Spines on head are well developed. Lachrymal is with three to four serrated ridges; two to six small spines are seen over maxilla; suborbital ridge is with 10–16 small spines of variable shape; and preopercle is with four spines (third one is broad and fourth spine is blunt). Head and body are red, with brown bars; and lower flanks are pale, with yellow tinge. On head, one faint bar is on each side of snout; and two dark bars radiate ventrally from eye. Five broad vertical bands are seen on body. A black blotch is seen on body above pectoral fin base. Soft dorsal, anal, and caudal fin are with small gray spots; and caudal base is yellow. It grows to a maximum length of 9.3 cm (SL) (Krishna and Sujatha, 2016).

Fisheries/aquarium/food/pharmaceutical values: Not reported.

Ebosia saya (Matsunuma and Motomura, 2014)

Photo credit: Mizuki Matsunuma and Hiroyuki Motomura.

Common name(s): Not designated as it is a new species

Distribution: Saya de Malha Bank (western Indian Ocean).

Ecology: It is a demersal species; depth of 95–126 m.

Biology

Identification: This species is diagnosed by its dorsal fin which has 13 spines and 9–10 soft rays; and anal fin is with 3 spines and 8 soft rays. Body of this species is oblong, moderately compressed anteriorly and is extensively compressed posteriorly. Its depth is moderate, and maximum body depth is slightly less than the longest dorsal-fin spine length. Head is large, and its length is greater than body depth. A short flap with a median ridge is seen. Supraocular is with a small flap; and its length is about half of iris diameter. Two extremely small flaps are seen on preopercle margin below third preopercular spine base. An extremely small skin flap is seen anterodorsally on orbit surface; and its length is less than one-third of posterior nasal pore diameter. Snout tip is with two extremely short barbels.

Mouth is set relatively low on head, is protruding moderately downward, and is moderately large and slightly oblique. Anterior region of maxilla is with a poorly developed median lateral ridge. Dorsal profile of snout is steep, forming an angle of c. 60 to horizontal axis of head and body. Nasal bone is without spines. Doral-, anal-, and pelvic-fin spines are with deep grooves. Origin of first dorsal-fin spine is above pterotic spine base.

Dorsal-fin soft rays are all branched; fourth (or fifth) ray is the longest and its length distinctly less than that of the longest dorsal-fin spine; and posterior most ray is free from caudal peduncle. Origin of first anal-fin spine is below 12th dorsal-fin spine base; and third spine is the longest. Anal-fin soft rays are all branched; fourth ray is the longest; and posteriormost ray is free from caudal peduncle. Pectoral fin is long and its sixth, branched ray is the longest.

All pelvic-fin soft rays are branched; and its second soft ray is the longest. Caudal fin is long and its length is 1.2 times head length; and profile of posterior margin of this fin is rounded. Caudal peduncle is relatively short and low and its depth is 0.7 times caudal-peduncle length. Ground color of head and body is cream-white to pale pink, darker dorsally, and whitish ventrally. Three relatively narrow dark red bands with narrow brown margin are seen on sides of head; anteriormost band is below eye to preopercular region; middle band is from posteroventral margin of orbit reaching obliquely to ventral margin of subopercle; and posterior most band is reaching mid posterior margin of opercle. Five moderately broad pale brown bands are seen on sides of body. A relatively small brown to black blotch is just above pectoral-fin base (slightly smaller than iris). Eyes are yellow; and iris is black, without radial markings (Matsunuma and Motomura, 2015).

Fisheries/aquarium/food/pharmaceutical values: Not reported.

Ebosia vespertina (Matsunuma and Motomura, 2016)

Photo Credits: Mizuki Matsunuma and Hiroyuki Motomura.

Common name(s): Western falcate lionfish

Distribution: Western Indian Ocean: off Mozambique and the east coast of South Africa.

Ecology: It is a demersal species; depth range 54–770 m.

Biology

Identification: This species is diagnosed by the following characters: dorsal fin is with 13 spines and 9 soft rays; anal fin has 3 spines and 7–8 soft rays; and pectoral fin is with 17–18 fin rays. Body of this species is oblong, moderately compressed anteriorly and is extensively compressed posteriorly. Depth is moderate, and maximum body depth is less than the longest dorsal-fin spine length. Head is large and its length is greater than body depth. A short simple, pointed flap with a median ridge is seen on posterior edge of low membranous tube associated with anterior nostril; and its tip is extending slightly beyond posterior margin of posterior nostril. Supraocular is with a small flap. Two small flaps are seen on preopercle margin below third preopercular spine base. An extremely small skin flap is seen on orbit surface. Snout tip is with two extremely short barbels.

Mouth is set relatively low on head; protruding moderately downward; moderately large and is slightly oblique. Anterior region of maxilla is with a poorly developed median lateral ridge; upper edge of posterior maxilla is swollen laterally, forming a low ridge; and posterior margin of maxilla is extending slightly beyond midorbit level. Lips are thickened; and lower lips are broadly extending onto mandible. Lower jaw is with a small symphysial knob.

Dorsal profile of snout is steep. Doral, anal and pelvic fin spines are with deep grooves. Origin of first dorsal fin spine is above pterotic spine base. Ground color of head and body is pale red, and whitish ventrally. Three relatively narrow, dark red bands are with narrow whitish margins on sides of head; anterior most band is from ventral margin of orbit, running ventrally; middle band is from posteroventral margin of orbit, reaching obliquely ventral margin of subopercle; and posterior most band is, reaching postero-entral margin of opercle at level of upper origin of pectoral-fin base. Both jaws are yellowish; and mandible and branchiostegal membrane are bright yellow. Eyes are orange and iris is black. Flaps on posterior margin of preopercular are yellow. Five vertical, moderately broad dark red bands are saddling body. Spinous dorsal fin coloration is similar to ground color of body; and soft-rayed portion of dorsal fin pale is yellow and somewhat reddish basally. Anal fin pale is yellow and is somewhat reddish basally. Pectoral fin is dark yellow and is somewhat reddish basally, with numerous relatively large black blotches (larger than iris). Pelvic fin is red and broadly blackish. Caudal fin is broadly bright yellow and is paler basally, with 6 small black spots which are scattered basally and dorsally. It grows to a maximum size of 11.0 cm (SL) (Matsunuma and Motomura, 2016).

Fisheries/aquarium/food/pharmaceutical values: Not reported.

Erosa erosa (Cuvier, 1829)

Photo credit: Gloerfelt-Tarp, Thomas.

Common name(s): Pacific monkeyfish, Daruma stinger, pitted scorpionfish, pitted stonefish

Distribution: Western Pacific: from southern Japan to northern Australia.

Ecology: It is a reef-associated, benthic species inhabits trawling grounds, and sandy and muddy bottoms in tropical inshore waters; depth range 0−1750 m.

Biology

Identification: This species is diagnosed by the following characters; dorsal fin has 14 spines and 9 soft rays; and anal spin is with 2 spines and 5−6 soft rays. It is a variably-colored, stocky fish ranging from reddish to orange or brown with white mottling, and dark crossbands and/or rows of irregular white spots on the pectoral, pelvic, anal and caudal fins. It has a large rounded head, with a bony hump over the eyes followed by a deep pit on the back of the head, and a very short snout that is almost vertical in profile. It grows to a maximum length of 15.0 cm (TL).

Fisheries/aquarium/food/pharmaceutical values: This species is not utilized. This species is taken as bycatch in trawls.

Parapterois heterura (Bleeker, 1856)

Photo credit: Wikipedia.

Common name(s): Blackfoot firefish, blue fin lionfish

Distribution: Indo-West Pacific: Natal, South Africa to southern Japan; Indonesia, New Guinea and Australia.

Ecology: It is a reef-associated, demersal fish inhabiting sheltered coastal bays and fine sand or muddy habitats; depth range 40–300 m is largely nocturnal and may partially bury itself in the substrate during the day making it difficult to spot.

Biology

Identification: Dorsal fin of this fish has 13 spines with long filaments on their tips, and 9 soft rays. Anal fin has 2 spines and 7–8 soft rays. Outer rays of caudal fin also have filamentous extensions. Upper margin of the eye is below the base of the first dorsal fin spine. Scales are present in a pit between the posterior nostrils. It is a lionfish with irregular dark bars on the head and body, the inner surface of the pectoral fin dark brown with vivid blue bands, and dark spots forming bands on the soft dorsal and caudal fins. It grows to a maximum length of 38 cm (TL).

Food and feeding: When disturbed, the brightly colored pectoral fins of this species probably serve to startle predators. The large fins serve primarily to corner prey when hunting. This fish feeds on small fish and bottom-dwelling invertebrates.

Fisheries/aquarium/food/pharmaceutical values: Trawled on soft bottom in coastal waters. This species can be found in the aquarium trade.

Parapterois macrura (Alcock, 1896)

Photo credit: Raju Saravanan.

Common name(s): Not designated

Distribution: Indian Ocean: west coast of India; Oman and Somalia.

Ecology: It is a benthopelagic species; depth range 64–430 m.

Biology

Identification: This species is distinguished by the following characters: dorsal fin has 13 spines and 9 soft rays; anal fin is with 2 spines and 7 soft rays; 18–20 pectoral fin rays; 5–6 upper gill rakers; 11–12 lower gill rakers; and 43–49 longitudinal scale series. There are no scales on midline of interorbital space throughout life; longitudinal scale series 43–49; head length 42.4% of SL; snout length 15.8% of SL; head depth (23.0% of SL); tip of tentacle on posterior spine of lacrimal is not extending beyond level of posterior margin of maxilla; and upper margin of orbit is higher than or level with base of first dorsal spine in adults. Body coloration is red and white, and its eir pectoral fins are expanded. This species generally grows to a length of 12.4 cm (SL).

Fisheries/aquarium/food/pharmaceutical values: It is an aquarium fish and is occasionally seen in aquarium trade.

Pterois andover (Allen and Erdmann, 2008)

Photo credit: Gloerfelt-Tarp, Thomas.

Common name(s): Andover lionfish

Distribution: Western Pacific: Indonesia and Papua New Guinea; Sabah, Malaysia, and the Philippines.

Ecology: It is a benthopelagic species, found at depths from 3 to 70 m; occurs on or near turbid inshore reefs surrounded by soft sand–mud bottoms, often in the proximity of freshwater discharge.

Biology

Identification: All the species of Pterois are well known for their ornate beauty, venomous spines, and unique tentacles. Pterois andover is distinguished by the following characters: dorsal fin is with 13 spines and 11 soft rays; anal fin is with 3 spines and 7 soft rays; pectoral rays 12–14; vertical scale rows in longitudinal scale series 62–72; horizontal scale rows above lateral line 9–10; below lateral line 13–15; dorsal spines are with narrow, inconspicuous membrane on posterior edge with exaggerated pennant-like flap distally; relatively few dark spots on median fins, 5–28 on dorsal fin, 10–19 on anal fin, and 17–41 on caudal fin;

and first dorsal spine is relatively short, with an average length of 15% of SL. It grows to a maximum length of 23.0 cm (TL) (Allen and Erdmann, 2008).

Fisheries/aquarium/food/pharmaceutical values: Not reported.

Pterois antennata (Bloch, 1787)

Photo credit: Wikipedia.

Common name(s): Spotfin lionfish, broadbarred firefish, broadbarred lionfish, ragged-finned firefish, ragged-finned scorpionfish

Distribution: Indo-Pacific: East Africa to Marquesan and Mangaréva islands, north to southern Japan, south to Queensland, Australia and Kermadec and Austral Islands.

Ecology: This reef-associated species inhabiting lagoon and seaward reefs; hides in crevices under rocks and coral formations during the day; found singly or in small group; depth range 1–180 m.

Biology

Identification: This species is diagnosed by its dorsal fin which has 13 spines and 11–12 soft rays; and anal fin is with 3 spines and 6 soft rays. It is a reddish to reddish-brown lionfish with many darker narrow to broad bars on the body that are narrow and oblique on the caudal peduncle, scattered dark spots on the median fins; and three dark brown bars are seen on the head including an oblique bar through the eye ending in a large spot on the lower gill cover. It has a long banded tentacle above each eye, and large adults have bluish-black blotches near the base of the pectoral fin. It grows to a maximum length of 20 cm (TL).

Food: It feeds on shrimps and crabs.

Major threats: Due to its affinity with coral reefs, this species may be experiencing population declines due to habitat loss in parts of its range. However, this species also occupies rocky reefs, and therefore significant global population declines are not suspected.

Fatty acid content: The testis and ovary of this species have shown maximum level of SFA and the values SFA were found to be 45.4%, 42.0%, and 37.9% in the liver, abdominal muscle, and ovarian tissues, respectively (Gautam et al., 2014).

Fisheries/aquarium/food/pharmaceutical values: This species is found in subsistence fisheries but does not support large-scale fisheries due to its small size and dangerous spines. It is also commonly found in the aquarium trade. Its aquarium parameters are given below.

Minimum tank size: 50 gal

Care level: Moderate

Temperament: Semiaggressive

Reef compatible: With caution

Water conditions: 72–78°F, dKH 8–12, pH 8.1–8.4, sg 1.020–1.025

Diet: Carnivore

It should be provided with numerous hiding places that are suitable. It will hide while acclimating to its new environment. When first introduced into the aquarium, live saltwater feeder shrimp should be used to entice this fish to eat. After acclimating, the diet should consist of meaty foods such as live shrimp (including ornamental shrimp), live fish, and sometimes, crustacean flesh.

Pterois brevipectoralis (Mandrytsa, 2002)

Photo credit: Mizuki Matsunuma and Hiroyuki Motomura.

Common name(s): Not designated

Distribution: Western Indian Ocean.

Ecology: It is a demersal species; depth range 70–80 m.

Biology

Identification: The species is characterized by the following characters: dorsal fin has 13 spines and 10 soft rays; 16 pectoral-fin rays; 6–10 dark blotches on the pectoral-fin membrane; short pectoral fin; ctenoid scales are covering the head and dorsolateral body; and the posterior portion of the pectoral-fin rays (not associated with fin membrane) is with several bands. Further, extremely large (greater than orbit diameter) fan-like skin flaps are seen on the tip of the posterior lacrimal spine. Ground color of head and body is white and is somewhat pinkish dorsally. Jaws and skin flap on posterior lacrimal spine tip are pinkish. Reddish band is crossing eye, reaching posterodorsal corner of preopercle. Large blackish blotch is present on ventral corner of opercle. Dorsal-fin membrane is translucent; 2–3 reddish bands are on each dorsal-fin spine; and 25 small black spots are seen on soft rays. Anal-fin membrane is translucent; and 11 small black spots are present on soft rays. Pectoral fin is dark red, with several relatively large black blotches on membrane and narrow black band on distal margin; and free portion of rays is creamywhite with 1–3 reddish bands. Pelvic fin is dark red and is strongly blackish marginally. Caudal-fin membrane is translucent with about 50 small black spots on rays. Eye is yellowish and iris is black. It grows to a size of 9.1 cm (SL) (Matsunuma and Motomura, 2013).

Fisheries/aquarium/food/pharmaceutical values: Not reported.

Pterois lunulata (Temminck and Schlegel, 1843)

Photo credit: Shao, Kwang-Tsao.

Common name(s): Luna lion fish, dragon's beard fish, Japanese lionfish

Distribution: Indo-Pacific: Mauritius, Japan, East Asia, Indonesia, NE Australia, New Caledonia.

Ecology: This reef-associated species is common in sheltered bays over soft bottom near rocks; depth range 27–172 m.

Biology

Identification: This species is diagnosed by its dorsal fin which has 13 spines and 11–12 soft rays; anal fin is with 3 spines and 7–8 soft rays; and 13 pectoral fin rays. Medium sized individuals of this species are almost clear but become spotted in large adults. It is best identified by its 13 pectoral rays. Its pale-centered scales on the principal dark bands on the body are forming a net-like pattern and V-like markings across the pectoral fin rays. It grows to a maximum length of 35.0 cm (TL).

Food and feeding: It feeds on fish and crustaceans at dawn and dusk.

Reproduction: This species may also be bred in captivity.

Threats: Though this species is caught for food and aquarium trade, neither of these threats are considered to have a significant impact on the population size of this species, as these are localized threats only.

Fisheries/aquarium/food/pharmaceutical values: It is caught from the wild as a food source and for display in public aquarium and for aquarium trade. Its quality parameters are given below:

Minimum tank size: 55 gal or larger aquarium

Tank conditions: 72 and 78°F; pH of 8.1 or 8.4; sg 1.020–1.025 (when kept with invertebrates, and 1.020 and 1.023 when kept alone)

Feeding and diet: Carnivorous; meaty foods such as live shrimps and crustacean flesh.

Pterois miles (Bennett, 1828)

Photo credit: Wikipedia.

Common name(s): Devil firefish, common lionfish

Distribution: It is native to the western Indo-Pacific region. Indian Ocean: Persian Gulf and Red Sea south to Port Alfred, South Africa and east to Sumatra, Indonesia; eastern Mediterranean. This invasive Indo-Pacific lionfish species is now established throughout the Western North Atlantic.

Ecology: This reef-associated, benthic species lives in coastal waters in muddy habitats; depth range 25–85 m. This species is mainly nocturnal and may hide in crevices during the daytime.

Biology

Identification: Dorsal fin of this species has 13 long, strong spines and 9–11 soft rays; and anal fin has 3 long spines and 6–7 soft rays. Dorsal fin appears feathery and the pectoral fins are wing-like with separate broad, smooth rays. Coloration of this fish is from reddish to tan or gray; presence of numerous thin, dark, vertical bars on their head and body; tentacle above eye may be faintly banded; and adults have a band of small spines along the cheek and small spots in the median fins. Head is less angular. It has cycloid (rounded) scales (Belmore, 2013), and the head is large, up to one third its body length. Males and females look exactly alike except during times of mating. It grows to a maximum length of 43 cm and weight of 1.1 kg.

Behavior: Adults of this species are solitary and juveniles typically live alone or in small groups. When a predator approaches an adult, it expands its pectoral fins and there is no flight response. The predator moves back as the devil firefish approaches even when its pectoral fins are not fully on display. The devil firefish's prey avoids it as well even though it may not be a known predator species to that fish in nonnative areas. This is probably due to its striking color.

Age, growth and population structure: Studies relating to age, growth and population structure of this invasive lionfish made in northeast Florida showed the following findings: (1) rapid growth with seasonal variation in growth rates; (2) distinct cohorts were clearly identifiable in the length-frequency data, suggesting that lionfishes are recruiting during a relatively short period in summer; and (3) the majority of lionfish were less than 2 years old with no lionfish older than 3 years of age, which may be the result of culling efforts as well as ontogenetic habitat shifts to deeper water (Johnson and Swenarton, 2016).

Salinity tolerance: This migratory species has been reported to tolerate low-salinity conditions. When this species was exposed to four salinities [5, 10, 20 and 34 ppt (control)], no mortality was observed at salinities of 34, 20, or 10 ppt, but all fish died at salinity = 5 ppt within 12 days. It survived for at least a month at a salinity of 10 ppt and an average of about a week at 5 ppt. Fish started the experiment at an average mass of 127.9 g, which increased at a rate of 0.55 g/day, while they were alive, regardless of salinity treatment. These findings indicated that this lionfish can survive salinities down to 5 for short periods, and thus, it may penetrate and persist in a variety of estuarine habitats (Schofield et al., 2015).

Food and feeding: It feeds on other fishes and small crustaceans. In the Red Sea, this species has been reported to feed on assorted taxa of benthic fishes including damselfish, cardinal fish, and anthias. However, in the Pacific Ocean, it was observed to feed primarily on invertebrates including penaeid and mysid shrimps. Assessments of invasive lionfish feeding suggest that these fishes are largely piscivorous but also feed on a number of crustaceans.

Predators: It has few predators, probably because of its venomous spines but larger lionfish do prey on smaller ones. The bluespotted cornetfish (*Fistularia commersonii*) has been shown to feed on it, as also do groupers

Parasites: Simmons et al. (2014) reported on the occurrence of parasites (as listed below) of this species from the Western North Atlantic (Gulf of Mexico, and Caribbean Sea).

Digenea: *Lecithochirum floridense, Tergestia* sp.

Cestoda: *Nybelinia* sp., *Tentacularia* sp.

Acanthocephala: *Serracentis* sp., *Dollfusentis* sp.

Hirudinea: *Trachelobdella lubrica*

Isopoda: *Rocinela stignata*

Nematoda: *Raphidascaris* sp., *Contracaecum* sp., *Paracuaria adunca,* and *Hysterothylaceum* sp.

Reproduction: This species attains sexual maturity when it is 2 years old. It is a gonochoristic species; males and females exhibit minor sexual dimorphism only during reproduction. Courtship occurs right before dark and leads on late into the night, where the male and female circle around each other, then ascend near the surface of the water. The female releases her eggs and the male fertilizes it and the egg clumps float. Embryos then form in adhesive mucus and within a few days they become free floating larvae. The size of larvae on hatching is about 1.5 mm. During one mating session, a female can lay up to 30,000 eggs during spawning. Once warm water is present, it can reproduce easily throughout the year because it is an asynchronous batch spawner. In the Red Sea, a kind of parental care takes place where the adults isolate themselves from the juveniles, as a strategy to reduce predation on the juveniles. The seasonality of lionfish reproduction throughout their native range is unknown. Invasive lionfish collected off North Carolina and in the Bahamas suggests that this species is reproducing during all seasons of the year.

Reproductive biology: These lionfish have been reported to spawn buoyant eggs that are encased in a hollow mass of mucus produced by specialized secretory cells of the ovarian wall complex. Oocytes develop on highly vascularized peduncles with all oocyte stages present in the ovary of spawning females and the most mature oocytes placed terminally, near the ovarian lumen. Further, these lionfishes are asynchronous, indeterminate batch spawners and are thus capable of sustained reproduction throughout the year when conditions are suitable. This mode of reproduction could have contributed to the recent and rapid establishment of these lionfish in the northwestern Atlantic and Caribbean (Morris et al., 2011).

Fisheries/aquarium/food/pharmaceutical values

Pharmaceutical values: The venom of this species has been reported to possess anticancer and cytotoxic properties. The crude venomous spine homogenate showed an anticancer effect on the HepG2 cell lines at very low concentrations. Further, the fractions containing the venom protein band demonstrated cytotoxic effects on HepG2 cancer cell lines. These findings suggest that further investigations into the anticancer efficacy of the venom protein of this species have great potential for biotechnological advances in tumor research (Khalil, http://dar.aucegypt.edu/handle/10526/3344).

Aquarium values: It is an aquarium fish. Fully grown fishes should be housed in a 120-gal or larger aquarium with numerous hiding places. It will hide while acclimating to its new home but then is generally found out in the open. It may eat smaller fish, shrimp, and other crustaceans in the tank. When it is first introduced into the aquarium, live saltwater feeder shrimp should be used to entice this fish to eat. Its diet includes meaty foods such as live shrimp, live fish, and occasionally crustacean flesh. Once fully settled into the aquarium, with some diligence and patience, this fish can convert over to consuming a prepared diet consisting of chunks of fresh, uncooked table shrimp, frozen silversides, and pieces of frozen squid.

Pterois mombasae (Smith, 1957)

Photo credit: Dawn Goebbels.

Common name(s): African lionfish, deepwater firefish, or frillfin turkeyfish

Distribution: Tropical Indo-West Pacific: Durban in South Africa, Sri Lanka, India, New Guinea and Solomon Islands; northeastern Indian Ocean: Indonesia and Australia.

Ecology: This reef-associated species is found in soft-bottomed areas of continental shelf or muddy substrates with rich rubble ridges amongst rich invertebrate growth, especially sponges; rare inhabitant of rocky bottoms of deep offshore reefs; solitary or in small groups; depth range 10–70 m.

Biology

Identification: This species is diagnosed by its dorsal-fin with 13 spines and 10 soft rays; anal fin is with 3 spines and 6 soft rays; 18 pectoral-fin rays on each side of body; 50 scale rows in longitudinal series; more than 24 pored lateral-line scales; 8 scales above lateral line; 13 scales below lateral line; 7 scale rows between sixth dorsal-fin spine base and lateral line; 8 scale rows between last dorsal-fin spine and lateral line; 7 predorsal-fin scale rows; and 15 total gill rakers. Body is moderately compressed. Body depth is slightly less than the longest dorsal-fin spine length. Ctenoid scales are seen on lateral surface of body. Exposed cycloid scales are covering pectoral-fin base and ventral surface of body. There is no scale on snout and both jaws. Posterior margin of maxilla is reaching a vertical through middle of eye. About 5 rows of small conical teeth are forming a V-shaped patch on vomer. There are no palatine teeth. Three short barbells are seen on tip of snout. A tentacle is present on posterior edge of low membranous tube associated with anterior nostril. Supraocular tentacle is long and its length is 1.4 times of orbit diameter. A small skin flap is seen on upper portion of orbital surface. Two rounded skin flap is on posterior margin of preopercle. A short barbell is on anterior part of ventral margin of lacrimal. A rounded skin flap is seen on tip of posterior lacrimal spine. Nasal spine is simple. Median interorbital ridge is absent. Interorbital ridges are indistinct. Interorbital space is deep. Preocular is with two spinous points. Supraocular spine is simple and is covered with skin. Postocular spine is fan-shaped, with 7 (5 on right side) spinous points. Coronal and tympanic spines are simple. Occiput is nearly flat without surrounding ridges. Posterior margin of opercular membrane is reaching a vertical through fourth dorsal-fin spine base. Posterior tip of pectoral fin is extending beyond caudal-fin base but is not reaching to posterior tip of caudal fin. Origin of pelvic-fin spine is just below origin of pectoral fin. Posterior tip of depressed pelvic fin is just reaching to third anal-fin spine base. Origin of first anal-fin spine is just below origin of last dorsal-fin spine. Head and body are reddish white dorsally and whitish ventrally. Four reddish brown bars are on head; anteriormost bar is on snout from anteroventral margin of orbit to ventral margin of lacrimal; second bar is from ventral margin of orbit to upper edge of maxilla; third bar is from supraocular tentacle base to subopercle through eye; and fourth bar is saddled on nape. There are 16 bars on lateral surface of body, extending onto membrane of soft-rayed portions of dorsal and anal fins. Supraocular tentacle is reddish orange, with 4 reddish brown bands; skin flap on orbit surface is dark red; and other skin flaps on head are pale reddish white. Two to 5 reddish brown bars are present on dorsal-fin spines; and soft-rayed portion of the fin is semitranslucent, with 68 small black spots on rays. Soft-rayed portion of anal fin is semitranslucent, with 45 small black spots on rays. Caudal fin is semitranslucent, with 86 small black spots on rays. Pectoral fin membrane is reddish orange and is brownish distally, with 17 brownish red blotches. Its life span has been estimated as 8–15 years. It grows to a maximum length of 20.0 cm (SL) (Motomura et al., 2011a,b).

Major threats: Due to its affinity with coral reefs, this species may be experiencing population declines due to habitat loss in parts of its range. However, this species also occupies rocky reefs, and therefore significant global population declines are not suspected.

Fisheries/aquarium/food/pharmaceutical values: This species is of moderate commercial value. It is rarely sold fresh in small quantities in markets. It is also an aquarium fish and its aquarium parameters are given below.

Difficulty: Moderate
Min. tank size: 208 L (55 US G.)
Fish size for tank: 15.2–17.8 cm
Stocking ratio: 1:1 M:F
Temperature: 22–28°C (71.6–82.4°F)
sg: 1.020–1.025 pH: 8.2–8.4
Water hardness: 8–12 days
Diet: Carnivore; Live Foods

Pterois radiata (Cuvier, 1829)

Photo credit: Wikipedia.

Common name(s): Clearfin lionfish, tailbar lionfish, radiata lionfish, fireworks fish, or radial firefish

Distribution: Indo-Pacific: Red Sea to Sodwana Bay, South Africa and to the Society Islands, north to the Ryukyu Islands, south to New Caledonia; Mascarene Islands.

Ecology: This reef-associated benthic species inhabits lagoon and seaward coral reefs; prefers rocky reefs and seems to avoid stony corals; coastal, sometimes in surge zones where in narrow crevices, or along rock-walls in small caves; found under ledges during the day; depth range 1–30 m.

Biology

Identification: It is a generally uncommon species and only lionfish species which has spines without any markings. Dorsal fin which is tall, quill-like has 12 or 13 long, venomous spines and 10–12 soft rays. Anal fin has 3 spines and 5–6 soft rays. Large, fan-like pectoral fins flare out to the side and are clear and unbanded. Other fins are also colorless. Head and body color is reddish-brown with about six vertical dark bands of different color on the body separated by thin white lines. Two white horizontal lines occur on the caudal peduncle which distinguishes this species from other similar lionfish species. It can reach a maximum length of 24.0 cm (TL).

Food and feeding: This species is mainly nocturnal. It spends the day hiding in rock crevices, in small caves, or under overhangs. It emerges at night to feed exclusively on small crabs and shrimps

Fisheries/aquarium/food/pharmaceutical values: This species is very common in the commercial aquarium trade. It can also be found in some subsistence fisheries. It is also an aquarium fish. A 70-gal or larger aquarium with numerous hiding places is suitable. It will hide while acclimating to its new environment. When first introduced into the aquarium, live saltwater feeder shrimp should be used to entice this fish to eat. The diet of this species in aquarium consists of meaty foods such as live shrimp (including ornamental shrimp), live fish, and sometimes, crustacean flesh.

Pterois russelii (Bennett, 1831)

Photo credit: John E. Randall.

Common name(s): Plaintail turkeyfish, zebrafish

Distribution: Indo-Pacific: Persian Gulf and East Africa to New Guinea, south to Western Australia.

Ecology: It is a marine and brackish water species. This solitary, benthic species is found on muddy substrates, from well-protected shallow estuaries to quiet, deep offshore coral reefs; depth range 5–104 m.

Biology

Identification: It is a lionfish without rows of small dark spots on the soft dorsal, anal and caudal fins, and relatively short dorsal-fin spines. This species is distinguished by its 13 dorsal spines and, 10–12 dorsal soft rays; and 3 anal spines and 7–8 anal soft rays. It can reach a maximum length of 30 cm (SL).

Fisheries/aquarium/food/pharmaceutical values: It is an aquarium fish and is sometimes taken for the aquarium trade.

Pterois sphex (Jordan and Evermann, 1903)

Photo credit: John E. Randall.

Common name(s): Hawaiian lionfish, Hawaiian turkeyfish

Distribution: Eastern Central Pacific: known only from Hawaii.

Ecology: This reef-associated, nonmigratory and benthic species is found in seaward reefs and lagoons at depths from 3 to 122 m; generally hidden beneath ledges or caves during the day.

Biology

Identification: This species is distinguished by its dorsal fin which has 13 spines and 11–12 soft rays; and anal fin is with 3 spines and 6–8 soft rays. Pectoral rays are all unbranched, very long and free from membrane distally; and dorsal spines are very long, some are about as long as body depth. There are 50–55 vertical ctenoid scale rows in body; coronal spines are present; and most head spines become multiple with growth. Lachrymal and suborbital bones are densely covered with spines especially in individuals of 10 cm (SL) size. Supraocular tentacles are banded with black, and frequently tentacles are absent in large specimens. Body coloration is reddish-brown with numerous white bars and banded fins; eyes are silvery, crossed by a dark diagonal bar. Small specimens are with fewer bars on pectoral and pelvic fins. This species can reach a maximum length of 22 cm.

Food and feeding: It feeds on crustaceans during the night.

Fisheries/aquarium/food/pharmaceutical values: Not reported.

Pterois volitans (Linnaeus, 1758)

Photo credit: John E. Randall.

Common name(s): Red Lionfish, common lionfish

Distribution: It is mostly found in the warm Indo-Pacific waters. This invasive Indo-Pacific lionfish species is now established throughout the Western North Atlantic.

Pacific Ocean: Cocos-Keeling Islands and Western Australia; Eastern Indian Ocean: Marquesas and Oeno; north to southern Japan and southern Korea; south to Lord Howe Island; northern New Zealand; and the Austral Islands. This Pacific species has also invaded the Atlantic Ocean via aquarium trade.

Ecology: Found in marine and brackish waters; estuaries (Florida) and mangroves (Bahamas); inhabits lagoon and seaward reefs from turbid inshore areas; often solitary, they hide in unexposed places at daytime often with head down and practically immobile; pelagic juveniles expatriate over great distances; depth range 1–175 m.

Biology

Identification: This species is diagnosed by its dorsal fin which has 13 spines and 10–11 soft rays; anal fin is with 3 spines and 7–8 soft rays; and 14–16 pectoral fin rays. It is a striking, boldly-patterned fish with distinctive bands of red and white stripes and conspicuous, elongated, fan-like pectoral fins. Red and white stripes are intermingled with varying reddish, golden brown and black-brown bars or stripes, alternating against a pale yellowish-white background. A row of white spots is usually present along the lateral line. Long pectoral and dorsal fins are transparent and covered in rows of dark spots. Pelvic fins are black with numerous white spots. There are 90–120 lateral scale rows in body. Characteristic fleshy tabs around the face and numerous projecting spines on the head vary in their shape and size but are typically long on juveniles and somewhat leaf-like on adult. Scales are cycloid and are variable in color, usually in relation to habitat. Coastal species generally darker, sometimes almost black in estuaries. The larvae of the common lionfish have a large head with a relatively long, triangular snout, as well as long, serrated head spines. Further, they have very little distinctive coloring and are appearing mainly translucent, with only slight red-white coloration to the pectoral fins. Average maximum size of this species ranges from 28 to 43 cm with a maximum weight of 1.2 kg; and maximum reported age is 10 years.

Age, growth, and population structure: Studies relating to age, growth and population structure of this invasive lionfish made in northeast Florida showed the following findings: (1) rapid growth with seasonal variation in growth rates; (2) distinct cohorts were clearly identifiable in the length-frequency data, suggesting that lionfishes are recruiting during a relatively short period in summer; and (3) the majority of lionfish were less than 2 years old with no lionfish older than 3 years of age, which may be the result of culling efforts as well as ontogenetic habitat shifts to deeper water (Johnson and Swenarton, 2016).

Salinity tolerance: This migratory species has been reported to tolerate low-salinity conditions. When this species was exposed to four salinities [5, 10, 20, and 34 ppt (control)], no mortality was observed at salinities of 34, 20, or 10 ppt, but all fish died at salinity = 5 ppt within 12 days. It survived for at least a month at a salinity of 10 ppt and an average of about a week at 5 ppt. Fish started the experiment at an average mass of 127.9 g, which increased at a rate of 0.55 g/day, while they were alive, regardless of salinity treatment. These findings indicated that this lionfish can survive salinities down to 5 for short periods and thus it may penetrate and persist in a variety of estuarine habitats (Schofield et al., 2015).

Food and feeding: It is a voracious active predator. It hunts small fishes, shrimps, and crabs at night, using its widespread pectorals trapping prey into a corner, stunning it and then swallowing it in one sweep. In this species, an increase in piscivory with age was apparent. Further, there was an increase in the proportion of fish in its diet and evidence of diurnal feeding (Higgs, http://www.nickhiggs.com/uploads/1/5/7/5/15756104/higgs_lionfish_09.pdf).

Parasites: This species was found to be infected by the peridinian symbiotic parasite, *A. ocellatum* (Brown and Hovasse, 1946). Ramos-Ascherl et al. (2015) reported on the occurrence of 27 parasite taxa (as listed below) in this species collected from the coastal waters of Puerto Rico. Most of these parasitic species were found to be generalists and/or species that infect carnivorous fishes.

Acanthocephala: *Paracavisoma chromitidis*

Cestoda: *Otobothrium dipsacum, Scolex pleuronectis*

Digenea: *Bivesicula caribbensis, H. nimia, Lecithochirium floridense, Neotorticaecum* sp. *(metacercaria), Pachycreadium crassigulum, Stephanostomum* sp.

Hirudinea: *T. lubrica*

Isopoda: *Aegiochus tenuipes, Carpias serricaudus, Eurydice convexa, Excorallana quadricornis, Gnathia* spp., *Rocinela signata*

Monogenea: Gastrocotylinean postoncomiracidia, *Pseudempleurosoma carangis, Udonella caligorum*

Nematoda: *Anisakis simplex, Capillaria* sp., *Goezia* sp., *Raphidascaris* sp., *Spirocamallanus* sp., *Spirocamallanus partitus, Spirocamallanus spinicaudatus*

Simmons et al. (2014) reported on the occurrence of parasites (as listed below) of this species from the Western North Atlantic (Gulf of Mexico, and Caribbean Sea).

Digenea: *L. floridense, Tergestia* sp.

Cestoda: *Nybelinia* sp., *Tentacularia* sp.

Acanthocephala: *Serracentis* sp., *Dollfusentis* sp.

Hirudinea: *T. lubrica*

Isopoda: *R. stignata*

Nematoda: *Raphidascaris* sp., *Contracaecum* sp., *P. adunca*, and *Hysterothylaceum* sp.

Fogg et al. (2016) reported on the occurrence of parasites (as listed below) from this species collected from the northern Gulf of Mexico.

Parasite: Host site

Annelida: *T. lubrica* Gill chamber, operculum

Isopoda: *R. signata* Mouth, gills, pectoral fin

Copepoda: *Caligus lobodes* External surface

Lernaeenicus cf. *polyceraus* Tongue, operculum, mandible muscles

Nematoda: *Raphidascaris* sp Intestine

Trematoda/Digenea: *L. floridense* Stomach

Fatty acid profile: The level of SFA of this species was found to be 42.35%, 40.98%, and 35.04% in the liver, abdominal muscle and ovarian tissues of this species. Compared to *P. antennata*, a higher percentage of 37% MUFA was observed in the liver (37%) and testis (26.07%) of this species (Gautam et al., 2014).

Mercury content: Studies have been made on the mercury concentration of the lionfish species, *P. volitans* and *P. miles*. In the Florida Keys, total mercury content in dorsal muscle tissue from 107 lionfish ranged from 0.03 to 0.48 ppm, with all concentrations being less than the regulatory threshold for limited consumption. Further, the concentrations did not vary consistently with SLs or wet weights of lionfish. In 2010, lionfish from the upper Keys had mean concentrations that were 0.03–0.04 ppm higher than lionfish from the middle Keys, but mean concentrations did not differ consistently among years and locations. Overall, total mercury concentrations in lionfish were lower than those in several predatory fishes that support commercial and recreational fisheries in Florida (Huge et al., 2014).

Fisheries/aquarium/food/pharmaceutical values: *Pterois volitans* is a popular table fish. It is also collected for the aquarium trade. Its aquarium parameters are given below.

Minimum tank length/Size: A minimum 75-gal aquarium is recommended.

Maintenance difficulty: It is easy to keep and maintain this fish in aquarium tanks as it is one among the hardiest of all marine fish.

Maintenance: Live fish is fed in the beginning, by gradually enticing them to eat frozen of fresh foods such as silversides and lancefish. Other crustaceans and seafoods can also be tried. They should be encouraged to eat nonlive marine foods as soon as possible.

Social behavior: It is sociable and peaceful and it can be considered a community fish as long as the tankmates are not small enough to eat.

Sexual differences: Unknown.

Light: No special requirements.

Temperature: No special requirements; normal temperatures, 74–79°F.

Water movement: Weak, moderate, and strong; no special requirements.

Water region: Top, middle, and bottom; will swim anywhere their prey takes them.

Measures for the reversal of the lionfish invasion.

Consumption of lionfish in restaurants: Although consumption of lionfish is already present in some higher class restaurants, an increase in consumption is another short-term action in the reversal of the lionfish invasion. At present many chain restaurants of the world are unaware of the lionfish invasion and did not serve the lionfish on their menu. Several possible ways to increase lion-fish consumption are to encourage fishermen to catch and provide lionfish to seafood markets and increase the number of restaurants willing to prepare and serve lionfish as a delicacy. Also, convincing the general public that only the spine is poisonous, not the meat, and increasing the number of lionfish derbies conducted by seafood restaurants and popular seafood chains could also help reduce the invasion effects. Overall, these small but potentially effective measures will help raise awareness about the invasion and help decrease the thriving lionfish population (Brown, 2012).

Teaching sharks to consume lionfish: It has been over 10 years since the first lionfish was spotted in the US south Atlantic sea coast and the Caribbean Sea. As time passes more native species learn to consume the lionfish, creating a cycle of biological control for this booming population. Species that have been found to consume lionfish include native sharks, large grouper, and the blue spotted cornet fish. Studies already made showed that in locations with high numbers of large and medium-sized groupers had low numbers of lionfish. However, as grouper are over-fished in most areas, the invaded lionfish population is on the increase. It is reported that a more promising method of natural predation on the lionfish is through native shark species. The sharks will not be able to eliminate the lionfish, but through time, they should be able to regulate the numbers and keep the population in proportion to other native species' populations. Sharks will not naturally prey on the lionfish because they are not a native species, so some groups have tried to spear the fish and feed them to sharks to "teach" them that they are a food source. Once the lionfishes are speared they attract the sharks with the blood and the sharks will become curious and eventually try and eat the lionfish. They learn how to eat the lionfish without coming into contact with the poisonous spines by eating them tail first. The sharks will eventually begin to prey on the lionfish live and in some cases other sharks will learn the lionfishes are a food source by observation from other sharks or eating other shark's scraps. There have been at least four different species of sharks recorded consuming the lionfish. This could be an important measure in controlling the lionfish population in US waters. This method is something that is being practiced now and something that should continue to be encouraged as a long-term means in the control of invaded lionfish populations (Brown, 2012).

4 Venomology of Scorpionfishes

The family Scorpaenidae is second only to stingrays in marine fish envenomations. However, this family enjoys the dubious distinction of containing more venomous species than any other family of bony fishes. A total of 57 species of 23 genera viz. *Apistus, Brachirus, Centropogon, Choridactylus, Erosa, Gymnapistes, Helicolenus, Hypodytes, Inimicus, Leptosynanceja, Minous, Notesthes, Pterois, Ruboralga, Scorpaena, Scorpaenodes, Scorpaenopsis, Sebastapistes, Sebastes, Sebastodes, Sebastolobus, Snyderina,* and *Synanceja* have been reported to possess venomous properties (Halstead, 1970; Roche and Halstead, 1972). The severity of Scorpaenidae envenomations is progressively worse from stonefish to rockfish. That is, the stonefishes are the most dangerous followed by scorpionfish, lionfish, and rockfish in that order (Brown, 2012). There are estimated 40,000–50,000 marine fish envenomations that occur annually throughout the world and these are primarily due to stingrays followed by scorpaenid fishes. Although the majority of cases occur in their native habitat, viz., tropical Indo-Pacific region, multiple reports have also been reported from nonindigenous areas. In such envenomations, venom is released from the bony spines along the fins after puncturing the skin of the victim. The venom toxicity is due to antigenic, heat-labile proteins of high molecular weight. Typical symptoms include immediate pain followed by erythema and edema. Severe envenomation is associated with local necrosis developing over several days (Schult et al., 2017).

4.1 Venom Apparatus

In the scorpaenid fishes, the venomous portion of fins includes 12–13 dorsal spines, 2 pelvic spines, and 3 anal spines. Each spine is associated with a pair of venom glands. A loose integumentary sheath covers each spine. The pectoral fins, which are often ornate and plume like in these fishes, are innocuous.

4.1.1 Venom Apparatus of Scorpionfishes (*Scorpaena* spp.)

The most common type of venom apparatus in these fishes is the fin spine with grooves on both sides. A venom gland is located in each groove, and these glands range from poorly to well evolved. The spine and glands are surrounded by an enveloping sheath, known as the integument. These venomous spines, which are particularly efficient, are mostly used passively, in that they are erected when the fish feels threatened. If the spine penetrates the flesh of the victim, then the integument is pushed downward, and under the resulting pressure, the venom glands discharge their contents along the grooves and into the wound. While the scorpionfishes possess shorter and thicker spines with larger venom glands and a more

Biology and Ecology of Venomous Marine Scorpionfishes. DOI: https://doi.org/10.1016/B978-0-12-815475-5.00004-2

potent sting, the lionfishes have long, slender spines with small venom glands and a less potent sting (Gallagher, http://emedicine.medscape.com/article/770764-overview).

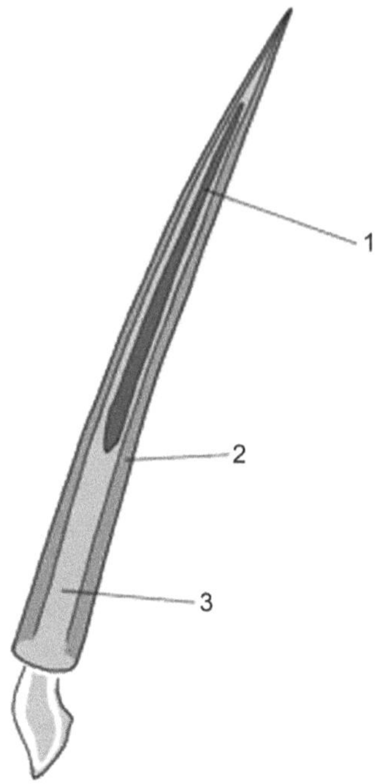

Venomous spine of a typical scorpionfish, (1) venom gland; (2) integument (enveloping sheath); (3) bony spine. Source: *Author.*

Scorpaena guttata: Like other species of *Scarpaena*, California scorpionfishes produce a toxin in their 12 dorsal, 3 anal, and 2 pelvic spines (Mariscal, 1974; Love et al., http://fishbull.noaa.gov/851/love.pdf). Each spine has two longitudinal grooves which are filled with epithelial tissue that contains venom-secreting glands. The spine is sheathed in a thin membrane. As the spine enters the victim, the membrane sheath is pushed back, helping to inject the venom into the victim. Fishermen are often affected when trying to remove the fish from the hook; when pulled from the water, the fish flares the spines on its fins and gill covers, making it difficult to handle. The spines themselves are sharp enough to inflict wounds, but the venom can cause a number of additional symptoms. If there are multiple punctures, the wound can induce shock, respiratory distress, or abnormal heart action and sometimes leads to hospitalization of the victim. It is one of the most noxious marine animals in California waters and this fish has been responsible for many human injuries. The effects of the sting are said to feel much like those of a rattlesnake bite. A spine prick to the finger was reported to cause a severe, throbbing pain and cyanosis of the digit, which became swollen and hard, then hot and red, then numb. Pain traveled up the arm to the axilla, which developed painful masses. Systemic symptoms included nausea and a feeling of faintness, and the skin was cool, clammy, and pale. In 2 weeks, the victim had recovered. The venom of this fish is also cardiotoxic, in laboratory studies, producing fluctuations in blood pressure and ECG changes such as ventricular tachycardia and bundle branch block.

Scorpaena plumieri: Like other members of the family Scorpaenidae, the spotted scorpionfish has venom-producing tissue associated with its fin spines (Hde et al., 2016).

Scorpaena brasiliensis: In this species, there are elongated venom glands in the grooves of the anterior portion of the spines of the dorsal, pelvic, and anal fins.

Scorpaena maderensis: Its anterolateral glandular groove of dorsal fin is with venom gland (Bayhan and Kaya, 2015).

Scorpaena notata: As other species of scorpionfish, *S. notata* has a row of spines on its back, linked to poison glands for defense (Bayhan and Kaya, 2015).

Scorpaena scrofa: In this venomous species, the poison glands, one per ray, are protected by a sheath which breaks at the impact, in the upper part of each spine. The quantity of venom produced by this species is very small compared to analogous tropical species.

4.1.2 Venom apparatus of rockfishes (Sebastes spp.) (Roche and Halstead, 1972)

Sebastes dalli: In this species, the fourth spine (29 mm, length) of the dorsal fin is venomous. Two glandular grooves are present in this spine. The grooves contain well-developed venom glands which are present for a distance of 7.5 mm (2–9.5 mm range). The venom tissue is discontinuous on one side, with gaps of 0.5 mm or more occurring several times. In these gaps, the normal connective tissue of the dermis replaces the venom tissue. Several colloidal areas are seen in the venom gland, and the integumentary sheath epidermis contains many large unicellular mucus glands (Anon., http://content.cdlib.org/view?docId=kt8z09n9rg;NAAN=13030&doc.view=frames&chunk.id=d0e1220&toc.depth=1&toc.id=d0e847&brand=cali).

4.1.2.1 Sebastes maliger

Dorsal stings: All dorsal spines contain well-developed bilateral venom glands except 10 and 12 spines in which the gland is very small and unilateral, and second spine in which there is only 1 shallow anterolateral groove with venom tissue. This groove eventually twists around to become anterior. Dorsal spines 1, 2, 4, and 6–9 contain colloidal areas in addition to intact venom cells.

Anal stings: All three anal spines contain well-developed bilateral venom glands. In addition, third spine contains colloidal areas.

Pelvic Stings: The left pelvic spine contains large well-developed bilateral venom glands with some scattered vacuolar areas.

4.1.2.2 Sebastes eos

Dorsal stings: All dorsal spines except spine 3 contain venom glands. In spines 2, 4, 5, and 13, these glands are unilateral and poorly to moderately developed. In spines 9 and 10, there are three glandular grooves with well-developed venom tissue. All the remaining dorsal spines contain bilateral venom glands moderately to well developed. No colloidal areas are present.

4.1.2.3 *Sebastes paucispinis*

Dorsal stings: No glandular grooves or venom glands are present in spines 1, 2, 12, and 13 but spines 7–11 have poorly developed unilateral venom glands. The remaining dorsal spines have moderately to well-developed bilateral glands with the exception of spine 4 which has a short section with three glandular grooves containing glandular tissue. No colloidal areas are present.

Anal stings: All three anal spines lack both glandular grooves and venom tissue.

Pelvic sting: The left pelvic spine contains a short, poorly developed venom gland, unilateral for most of its length. No colloidal areas are present.

4.1.2.4 *Sebastes elongatus*

Dorsal stings: Spines 2, 9, 12, and 13 have very shallow glandular grooves and no venom glands. Spines 6, 10, and 11 have poorly developed unilateral glands. The remaining dorsal spines have moderately to well-developed bilateral glands with the exception of spine 1 which also has a short section with three grooves and glands. No colloidal areas were found.

Anal stings: All three anal spines have bilateral, well-developed venom glands. No colloidal areas are present.

4.1.2.5 *Sebastes semicinctus*

Dorsal stings: Spines 1–7, 9, and 12 contain shallow glandular grooves, but no venom glands are present. Venom glands are unilateral and poorly developed in spines 8, 11, and 13. In spine 10, the glands are bilateral with colloidal areas on one side.

Anal stings: Anal spine 2 has no venom gland. Anal spine 1 has a unilateral, poorly developed gland and anal spine 3 has well-developed bilateral glands. No colloidal areas are present.

Pelvic sting: The left pelvic spine has deep glandular grooves with poorly developed unilateral venom glands. No colloidal areas were found.

4.1.2.6 *Sebastes nebulosus*

Dorsal sting: Only the spine 4 was examined in this species. Only one glandular groove is present in the distal portion of the spine. This groove contains a moderately developed venom gland which is present for a distance of about 2 mm. The pigmented layer forming the outer limits of the dermis is extremely well developed.

4.1.2.7 *Sebastes constellatus*

Dorsal sting: Only the spine 4 was examined in this species. Two glandular grooves are present in this spine and contain moderately developed venom glands which are present for a distance of about 6 mm. The glandular grooves appear wider than in most of the other species of *Sebastes* examined.

4.1.2.8 Sebastes serriceps

Dorsal sting: Only the spine 4 was examined in this species. Two glandular grooves are present, containing moderately to well-developed venom glands which are present for a distance of about 3 mm. The dermal pigment layer in this fish is well developed.

4.1.2.9 Sebastes goodei

Dorsal sting: Only the spine 4 was examined in this species. The spine is extremely slim and delicate with two glandular grooves containing well-developed venom glands which are present for a distance of about 3 mm.

4.1.2.10 Sebastes dalli

Dorsal sting: Only the spine 4 was examined in this species. The grooves contain well-developed venom glands which are present for a distance of 7.5 mm. The venom tissue is discontinuous on one side. Several colloidal areas are seen in the venom gland, and the integumentary sheath epidermis contains many large unicellular mucus glands.

4.1.2.11 Sebastes miniatus

Dorsal sting: Only the spine 4 was examined in this species. The spine contains two glandular grooves, one much deeper and wider than the other. Venom tissue appears at 4 mm and is present until 11.5 mm for a total distance of 7.5 mm. Venom tissue is present only in a miniscule amount in the shallower groove. Some colloidal areas are present.

4.1.2.12 Sebastes helvomaculatus

Dorsal sting: Only the spine 4 was examined in this species. It contains two glandular grooves. Poorly developed venom glands are present bilaterally for 2.5 mm. Dorsal spine 4 has two deep glandular grooves but contains a very small amount of venom tissue, only 0.5 mm.

4.1.2.13 Sebastes levis

Dorsal sting: Only the spine 4 was examined in this species. It contains two glandular grooves with well-developed venom glands beginning at 3.5 mm. By 4 mm, a third glandular groove with venom tissue appears. Venom glands are present in all grooves until 8.5 mm, for a total of 5 mm. In several places, a bridge of cells joins the epidermis to the venom gland.

4.1.3 Venom Apparatus in Lionfishes (Pterois spp.)

Lionfishes are venomous with their 13 dorsal spines, 3 anal spines, and 2 pelvic spines. The venom glands are located within two grooves of the spine. The spines are encased in an integumentary sheath or skin and contain two grooves of glandular epithelium that comprises the venom producing tissue. Spine glandular tissue extends approximately three quarters the distance from the base of the spine toward the tip. When the spines enter the skin of prey or a potential predator, the sheath around the spines is depressed and the tissue releases potent venom into the puncture wound (Anon., https://oceanservice.noaa.gov/facts/lionfish-facts.html).

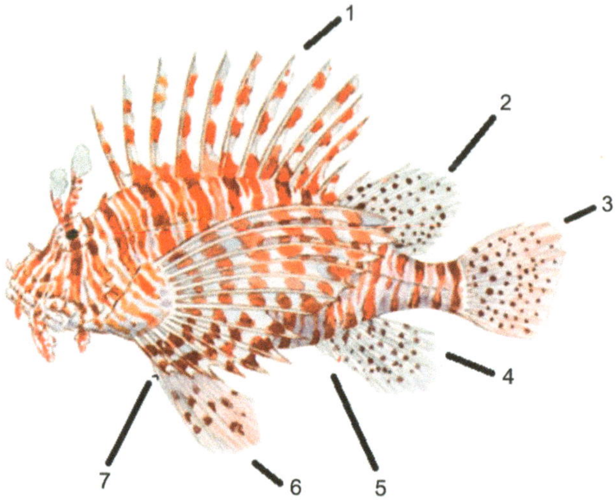

Lionfish with venomous and nonvenomous fins; (1) dorsal fin with 13 venomous spines, (2) nonvenomous 10 dorsal fin rays, (3) nonvenomous caudal fin, (4) anal fin with 6 rays, (5) venomous spines (3) of anal fin, (6) pelvic fin rays, (7) venomous spine (1) of pelvic fin. Source: *Author.*

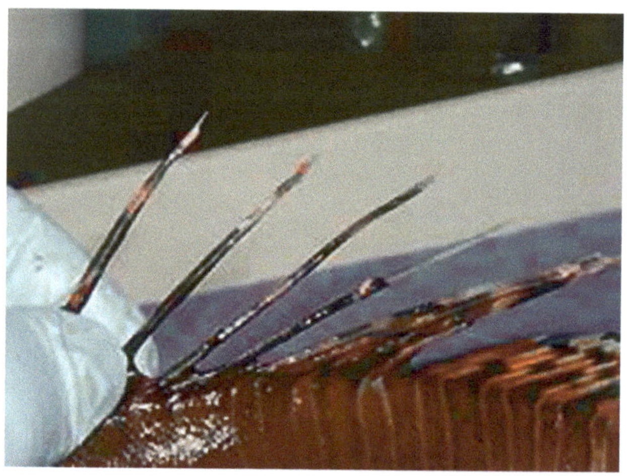

Venomous, dorsal fin spines of a lionfish. *Photo credit: Stephen Vives, National Ocean Service, NOAA.*

4.1.3.1 Pterois volitans

This species is venomous with its spines containing apocrine-type venom glands. Each spine of the lionfish (except caudal spines) is venomous including 13 dorsal spines, 3 anal spines, and 2 pelvic spines. The spines are encased in an integumentary sheath or skin and contain two grooves of glandular epithelium that comprises the venom-producing tissue. Spine glandular tissue extends approximately three quarters the distance from the base of the spine toward the tip (Halstead et al., 1955a,b).

4.1.3.2 *Pterois lunulata*

The crude toxin of this species has been reported to show mouse-lethal, hemolytic, edema-forming, and nociceptive activities (Kiriake et al., 2013).

4.1.3.3 *Pterois radiata*

The top spines of the dorsal fin of this species are venomous and are capable of inflicting a painful sting. The reactions caused are very much similar to that of a bee sting.

4.1.3.4 *Dendrochirus spp.*

All the six species of dwarf lionfishes of the genus *Dendrochirus* are small but venomous. Their size bears little relation to the power of their sting. Like small rattlesnakes, small lionfishes are just as venomous as their larger brethren. These fishes may inflict serious injuries with other spines during handling (http://feelingkeep.com/article/dwarf-lionfish-in-the-aquarium).

4.2 Characteristics of Scorpaenid Fish Venoms

4.2.1 Characteristics of *Scorpaena guttata* Venom

Fractionation studies on the venom of this species indicated that the lethal property of venom is associated with protein(s) having a molecular weight of more than 50,000 and less than 800,000. The semi-purified lethal fraction had an intravenous LD50 in mice of 0.9 mg protein/kg body weight, while the LD50 of the crude venom was 2.6 mg protein/kg. Further, this venom is more lethal than the crude venom (Schaeffer et al., 1971).

4.2.2 Characteristics of Scorpaena plumieri Venom

The crude venom of this species has been reported to induce a decrease in blood pressure, cardiac, and respiratory frequency and exhibited hemorrhagic, hemolytic, and proteolytic activities. The enzyme, protease, which hydrolyzes casein and gelatin, has been isolated for the first time from the venom of this species. The LD(50) (i.e., mouse) was 0.28 mg/kg (Carrijo et al., 2005).

Andrich et al. (2015) reported on the isolation of new vasoactive cytolytic toxin, referred to as Sp-CTx from the venom of this species. This protein was found to possess a potent hemolytic activity on washed rabbit erythrocytes [EC(50) 0.46 nM] (Andrich et al., 2010). The hemagglutinating fraction of the venom of this species also revealed the presence of five C-type isolectins.

4.2.3 Characteristics of Pterois volitans Venom

Intravenous injection of the extracts prepared from the venomous dorsal spines of this species into mice produced death from less than a minute up to about 1.5 hours. In the case of rabbits, the primary action was a fall in blood pressure, accompanied by increase in respiratory rate; and with larger doses, there was evidence of myocardial ischemia or injury. In both the animals, after injection of fatal doses, a variety of electrocardiographic changes occurred and the blood pressure fell to zero and respiratory arrest occurred terminally. The active material was nondialyzable and the extracts contained considerable amounts of protein. The mean lethal dose (LD50) following intravenous injection into mice was about 1 mg of protein/kg (Saunders and Taylor, http://www.physiology.org/doi/abs/10.1152/ajplegacy.1959.197.2.437).

A soluble toxic extract derived from the spine tissue of *Pterois volitans* decreased heart rate in clam and frog hearts. These actions were largely due to the presence of micromolar concentrations of acetylcholine in the extract (Cohen and Olek, 1989). The toxin in the venom of *P. volitans* contains acetylcholine and a neurotoxin that affects neuromuscular transmission (Anon., https://www.fws.gov/fisheries/ans/erss/highrisk/Pterois-volitans-WEB-7-28-2014.pdf).

The venom of lionfish is a combination of protein, a neuromuscular toxin, and a neurotransmitter called acetylcholine (Anon., https://oceanservice.noaa.gov/facts/lionfish-facts.html). Further, it is heat-labile and antigenic (Brown, 2012).

4.2.4 Health-Associated Properties of Scorpaenid Fish Venoms in Humans

Certain properties of scorpaenid venoms (*Scorpaena* and *Pterois*) identified at lethal dose are given below.

Parameter	*Scorpaena*[a]	*Pterois*[b]
Pressure	Irreversible fall in arterial pressure	Irreversible fall in systemic arterial pressure
Electrocardiogram	Extensive changes	Extensive changes
Respiratory rate	Marked decrease	Marked decrease
Paralysis (in mice)	Some paralysis of legs	Complete paralysis of legs

[a]Intravenous LD50 mice 2.0 mg protein/kg body weight.
[b]Intravenous LD50 mice 1.1 mg protein/kg body weight.
Source: Mariscal (1974).

4.3 Envenomation of Scorpaenid Fishes

Fishes of the family Scorpaenidae are responsible for severe injuries and occasionally deaths in humans around the world. The more venomous fishes on the Brazilian coast and in the Southwestern Atlantic region belong to the genus *Scorpaena*. Haddad et al. (2003) reported on 23 accidents caused by *Scorpaena* spp. among fishermen of East Atlantic Ocean.

4.3.1 Envenomation of Scorpionfishes

Wound severity and systemic symptoms were found to be invariably greater in *Scorpaena* envenomation as compared to *Pterois* envenomations.

4.3.1.1 Scorpaena guttata

Envenomation occurs through mechanical pressure on the spine which ruptures the integumentary sheath. The venom is then emptied mechanically into the wound produced by the fin spine (Mariscal, 1974).

4.3.1.2 Scorpaena plumieri

Scorpaena plumieri is the most venomous fish in the Brazilian fauna and is responsible for relatively frequent accidents involving anglers and bathers. In humans, its venom causes edema, erythema, ecchymoses, anxiety, nausea, vomiting, and syncope (Hde et al., 2016). The first spines of the dorsal fin of this fish have a conduit through which the venom passes. Fishermen, divers, and bathers become victims of this species when they handle the fish and have their skin perforated by the spines. The clinical manifestations of accidents include local and systemic effects. The first symptom is always excruciating pain, followed by edema, erythema, occasional skin necrosis, adenopathy, nausea, vomiting, agitation, malaise, sweating, diarrhea, tachycardia, and arrhythmias, culminating, in some cases, in severe hypotension (Campos et al., 2016). The venom extract of this species has also been reported to perform lethal, hemolytic, cardiovascular, inflammatory, integrin-binding-inhibitory, and proteolytic activities. Intravenous injection of the extract in mice has induced loss of muscular coordination, paralysis, urination, hypersalivation, convulsions, and respiratory failure, followed by death (Carrijo et al., 2005).

Harm to humans (case study) by *S. plumieri*: A total of 36 patients were apparently envenomed by this species at Venezuela during 2006–07. Clinical manifestations observed were intense and irradiated pain (100%), edema (27.78%), and erythema (22.22%). One case showed systemic manifestations: hypotension and faintness. Treatment in this case consisted of infiltration with lidocaine (1%), systemic antiinflammatory analgesia (100 mg, orally every 8 hours for 5 days), and antibiotic therapy per os (500 mg, 2 daily dose for 10 days) (Loyo et al., 2008).

4.3.1.3 Scorpaena brasiliensis

These spines can be erected with venom passing into the wound from the spines that penetrate the flesh of the victim. This potent venom is harmful and on occasion deadly to human. Haddad et al. (2003) reported on the injuries caused by this species in the Southwestern Atlantic Ocean (Brazilian coast).

4.3.1.4 Scorpaena elongata

Its venom is believed to be neurotoxic–cardiotoxic in Mediterranean (Bayhan and Kaya, 2015).

4.3.1.5 *Scorpaena loppei*

This venomous fish is believed to be neurotoxic–cardiotoxic in Mediterranean region (Bayhan and Kaya, 2015).

4.3.1.6 *Scorpaena maderensis*

The venom of this species is believed to be neurotoxic and cardiotoxic in the Mediterranean region (Bayhan and Kaya, 2015).

4.3.1.7 *Scorpaena notata*

The venom of this species is dangerous to humans who may step on a camouflaged scorpionfish. Further, it is also neurotoxic and cardiotoxic in the Mediterranean region (Bayhan and Kaya, 2015).

4.3.1.8 *Scorpaena porcus*

The venom of this species is also said to be neurotoxic and cardiotoxic in the Mediterranean region (Bayhan and Kaya, 2015).

4.3.1.9 *Scorpaena scrofa*

The sting of this species is very painful. In some cases, there may be hypotension with fainting or dizziness. The poison is thermolabile and therefore a first remedy may consist in soaking the affected part into the warm water. It is also believed to be neurotoxic–cardiotoxic in the Mediterranean (Bayhan and Kaya, 2015).

4.3.1.10 *Sebastapistes coniorta*

In spite of its small size, the venom of this species is to be considered. In one case study, a person received a slight prick of a dorsal spine on one thumb. It was surprisingly painful and caused the thumb to swell significantly (Randall and Earle, 2009).

4.3.2 Envenomation of Rockfishes

Though about 30 species of *Sebastes* have been reported to be venomous, no detailed research has been made on the venomation of rockfishes.

The tiger rockfish, *Sebastes nigrocinctus*, has been reported to be venomous and cytotoxic in North America. The vasoactive peptides (hypotensive) and calcium-modulating effects on neuromuscular transmission have been reported with three (local envenomation only) hazard levels (https://www.aafp.org/afp/2009/1015/afp20091015p793-s1.pdf).

4.3.3 Envenomation of Lionfish

The lionfish poses a serious threat to divers, as well as aquarists who encounter lionfish on a daily basis. The spines of this species deliver a venomous sting that can last for days and cause extreme pain, sweating, respiratory distress, and even paralysis. After the spine punctures the skin, the venom enters the wound when exposed to the venom glands within the grooves of the spine (Anon., https://oceanservice.noaa.gov/facts/lionfish-facts.html). It is very important to note that even after death, the 18 venomous spines of the lionfish can produce wounds if not handled properly. Although most sources list the lionfish as nonlethal, some have suggested otherwise (Ruiz-Carus et al., 2006). The lionfish venom has also been found to cause cardiovascular, neuromuscular, and cytolytic effects.

4.3.3.1 Pterois volitans Envenomation

Lionfish (*P. volitans*) envenomation occurs when the spine's integumentary sheath is depressed as it enters the victim. This process tears the glandular tissue allowing the venom to diffuse into the puncture wound. The toxin in lionfish venom contains acetylcholine and a neurotoxin that affects neuromuscular transmission. Lionfish venom has been found to cause cardiovascular, neuromuscular, and cytolytic effects ranging from mild reactions such as swelling to extreme pain and paralysis in upper and lower extremities. The sting causes intense pain, redness, and swelling around the wound site. A painful sting may also cause headache, vomiting, and respiratory distress. Although the worst of the pain is over after an hour or two, some people report pain and tingling sensations around the wound for several days or weeks. On rare occasions, when the venom spreads to other parts of the body, people may experience headaches, chills, cramps, nausea, and even paralysis and seizures. The severity of sting reactions in humans is dependent upon such factors as the amount of lionfish venom delivered, the immune system of the victim, and the location of the sting (Morris et al., 2008; https://oceanservice.noaa.gov/education/stories/lionfish/lion04_biology.html).

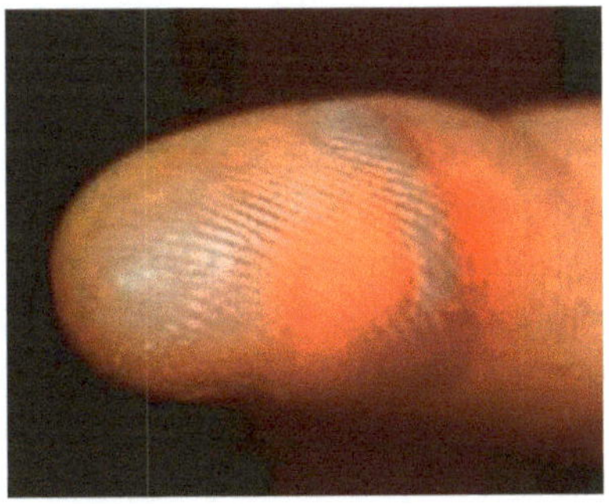

Injury caused by *Pterois volitans*. Source: *Author.*

Other health-associated problems in humans due to lionfish envenomations

1. Severe pain associated with lionfish envenomations can cause shock which may involve shortness of breath, weakness, fainting, and cardiac arrest.
2. Diabetics and those with compromised immune systems may react very badly to the venom and its systemic effects.
3. People who are allergic to the venom may go into anaphylactic shock. Many people die every year from anaphylaxis.
4. There is a chance that any injury caused by lionfish envenomation can become terribly infected.
5. Lionfish envenomations may also lead to paralysis in both arms and legs of a victim who is stuck in the finger.

6. Lionfish venom can cause tissue necrosis (tissue death) that has the ability to spread if not treated immediately when identified.
7. *Pterois* sting-associated blister characteristics: Fluid aspirated from blisters following the sting of this species was analyzed for its qualities. Analysis for prostaglandin F2 alpha demonstrated 16.9 ng/mL, for prostaglandin E2 0.143 ng/mL, for 6-keto-prostaglandin F1 alpha less than 0.1 ng/mL (nondetectable), and for thromboxane B2 1.65 ng/mL. Platelet aggregation studies showed that blister fluid caused aggregation of isolated platelets only, which was inhibited by heat treatment or by the presence of normal donor plasma (Auerbach et al., 1987).

Ciguatera fish poisoning (CFP): *P. volitans* can also cause CFP. Ciguatera is a foodborne illness caused by eating this reef fish whose flesh is contaminated with a toxin (ciguatoxin) made by dinoflagellates such as *Gambierdiscus toxicus*. This acute poisoning syndrome is characterized by a variety of severe gastrointestinal, neurological, and occasionally cardiovascular symptoms that can occur within 4 hours and last up to 6 weeks. A chronic phase of CFP lasting many years has also been reported in up to 20% of those acutely exposed (https://www.fws.gov/fisheries/ans/erss/highrisk/Pterois-volitans-WEB-7-28-2014.pdf).

4.3.3.2 Lionfish Envenomation Case Studies

Eastern Atlantic, Gulf of Mexico, and Caribbean Sea: The invasion of lionfish (native range, Indo-Pacific) into the shallow reef systems in the Eastern Atlantic, Gulf of Mexico, and Caribbean Sea which are popular travel destinations for beachcombing, fishing, swimming, and scuba diving—recreational activities have increase risks of lionfish envenomation. Envenomation case studies conducted in these regions showed the following: (1) most cases occurred in young adult male vacationers visiting endemic regions; (2) victims sought medical attention for pain control within 2 hours of injury and presented with intense pain, edema, and erythema in affected extremities; (3) systemic manifestations and surgical interventions were relatively uncommon following initial management with hot water soaks and parenteral analgesics; and (4) all cases required tetanus prophylaxis and deeply penetrating, lacerated, and necrotic wounds required antibiotic prophylaxis (Diaz, 2015).

Case report from aquarists: Lionfish envenomations are also becoming common among aquarists as these fishes are popular items in home aquariums. The victim (an aquarist) reported that he felt a sting followed by immediate pain and bleeding from the puncture site. He became diaphoretic with continued pain and experienced minimal relief after removing the barbs from the puncture site, taking aspirin and ibuprofen, and soaking his arm for 2 hours in a warm water bath. On initial examination, the patient had erythema, edema, and urticaria extending up the right arm from his fingertips to the lower portion of the humerus, with a visible puncture wound at the volar aspect of the forearm. The extremity was submerged in a hot water bath for 30 minutes. After 10 minutes of the water bath, the patient also received parenteral hydromorphone 1 mg every hour for 4 hours for continued pain. The following morning, 6 hours after the last dose of analgesic medication, the patient's pain score was reported as 0/10 (from 10/10 pain) and the swelling and erythema were also improved. He received a total of 4 mg hydromorphone and 10 mg oxycodone during his hospitalization (Schult et al., 2017).

Envenomation by the invasive *Pterois volitans* in the French West Indies—a 2-year prospective study: The invasion of the lionfish (*P. volitans*) in the French West Indies represented one of the most important marine invasions by alien species. Since its first recognition in Martinique, the lionfish appearance has strongly progressed, resulting in increasing envenomation cases. A total of 117 patients were included in this case report. Localization was mainly at one upper (67%) or lower limb (32%). The stings were single (81%) or multiple (19%). Envenomation resulted in marked pain and local edema (100%), paresthesia (90%), abdominal cramps (62%), extensive edema (53%), tachycardia (34%), skin rash (32%), gastrointestinal disorders (28%), fainting (27%), transient weakness (24%), hypertension (21%), hypotension (18%), hyperthermia (9%), bradycardia (3%), hypophosphatemia (12%), elevated aspartate aminotransferase (10%), and thrombocytopenia (3%). The sting was complicated by local infection (18%) including skin abscess (5%), cellulitis (3%), skin necrosis (3%), and septic arthritis (2%). Twenty-six patients (22%) were hospitalized requiring surgery (8%). Immersion in hot water (44%, performed less than 3 hours after the sting in 36% of the cases) significantly reduced pain duration and local infection. All patients actually improved (Resire et al., 2016).

Case report from Texas poison control centers: Forrester (2008) reported on the lionfish envenomations from Texas poison control centers (PCCs) during 1998–2006. A total of 188 envenomation cases were identified. Of these cases, 94% involved the finger or hand. The most adverse clinical effects were dermal pain (89%) and a puncture or wound (66%). The treatments given were immersion in hot water (84%) and tetanus antitoxin/status (48%).

A 5-year poison center experience: Kizer et al. (1985) reported on 45 lionfish envenomation involving lionfish stings to aquarists or fish handlers. The primary manifestation of envenomation in all the observed victims was intense local pain. Various other local and systemic symptoms were also noted. He also reported a simple and effective treatment which was generally unknown to emergency personnel treating these patients. This exercise included the immersion of affected pats in hot water which produced either complete (80%) or moderate (14%) symptomatic relief in 94% of the victims.

Case report (1) of PCC: The crude venom from *P. volitans* has been reported to increase intracellular Ca^{2+} and muscle contracture in vivo in the victims. A PCC case series from 1979 to 1988 identified 23 cases of *P. volitans* envenomation. It reported on the symptoms of the victims of *P. volitans*. The affected patients had pain, swelling, nausea, numbness, joint pain, anxiety, headache, dizziness, and cellulitis (Nelson et al., 2011).

Case report (2) of PCC: Another PCC series identified 51 Scorpaenidae stings (45 *P. pterois*, 6 *S. guttata*). The victims showed with intense pain were reported in 98%, extension of pain to the limb in 22%, swelling in 58%, and systemic signs (nausea, diaphoresis, dyspnea, chest pain, abdominal pain, weakness, hypotension, and syncope) in 13%. Thirteen percent of patients in the series developed wound infection; one patient's wound healing was delayed several weeks (Nelson et al., 2011).

Lionfish (*Pterois* spp.) envenomations in the United States: In one series of 101 described cases of captive lionfish (genus *Pterois*) envenomations in the United States, 92%

of patients experienced local pain, 60% experienced edema, and 13% experienced systemic symptoms. There were no fatalities. Ninety-five percent of the wounds were found to be grade I (erythema), 4% were found to be grade II (vesicle formation), and 1% were found to be grade III (tissue necrosis). Pain was relieved with hot-water immersion therapy in 97% of the patients, and 0% of the patients required antivenom administration. One patient required intravenous antibiotics, one hypotensive patient responded well to intravenous fluids, and 13% of patients had variable, less severe, systemic symptoms (Anon., http://misc.medscape.com/pi/iphone/medscapeapp/html/A770764-business.html).

Case report from Chicago Poison Control Center: This report deals with 33 cases of envenomations by captive lionfish (*P. volitans*) called to a regional Chicago PCC over a 2-year period. All stings were accidental, and 10 (30%) were treated in an emergency department. The wounds were uniformly on the hand, and all presented with local, intense pain. The majority of envenomations were responsive to prompt immersion in nonscalding water within 90 minutes, and all were advised on tetanus prophylaxis and local wound care. Two patients (6%) required hospitalization. All patients recovered without permanent sequelae. As a result of increasing encounters with lionfish as pets, it is advised that health-care providers, regardless of their locale, should be made familiar with the current treatment recommendations (Aldred et al., 1996).

4.4 Treatment and Management of Scorpaenid Fish Envenomations

The common treatment for scorpaenid fish envenomations is to place the injured area of the victim in warm water for 30–90 minutes, and this process is repeated if the pain recurs. By this method, about 80% pain relief has been reported. Alternatively, the sting of this fish can be treated by heating the afflicted part and application of corticoids.

Treatment for the injury of *Scorpaena plumieri*: Immersing the affected area in hot water (45–50°C) for about 15 minutes is the immediate treatment for the affected victim. If pain persists, infiltrating the area with lignocaine eases the pain and decreases the complications. Tetanus booster is indicated and any resulting infection should be treated with antibiotics. Caution also has to be taken when the fish is dead, as the venom will remain active even if the fish is frozen. Cutting off the dangerous spines immediately when the fish is dead is therefore recommended.

4.4.1 Treatment and Management of Scorpionfish Sting

Symptoms: A scorpion fish sting causes intense pain and swelling at the site of the sting. Swelling can spread and affect an entire arm or leg within minutes. The various symptoms of a scorpion fish sting in different parts of the human body are given below:

Skin: Severe pain at the site of the sting, lighter color of the area around the site of the sting, color of the area changes as the amount of oxygen supplying the area decreases, bleeding

Airways and lungs: Difficulty in breathing
Heart and blood: Collapse
Stomach and intestines: Abdominal pain, diarrhea, nausea, and vomiting
Nervous system: Fainting, fever, headache, mental confusion, muscle twitching, seizures, paralysis

Home care: The affected area is washed with salt water. Foreign materials, such as sand or dirt, if any found around the wound, are to be removed. Then, the wound is soaked in the hottest water for about 30–90 minutes. Subsequently, local emergency services may be contacted of medical help may be sought (Anon., https://medlineplus.gov/ency/article/002849.htm).

4.4.2 First Aid and Treatment for Lionfish Sting (Anon., https://lionfish.co/sting-first-aid/)

For most people affected by lionfish sting, the throbbing, intense pain is going to last for a few hours and will remain less painful for 12 hours or longer. The following aspects of first aid and treatment may safeguard the victim from further hospitalization.

Removal of remaining spines: The wound may be inspected very closely for any signs that pieces of the spine might have broken off and are still in the skin. Such spines may be removed gently.

Cleaning and disinfecting wound: The injured area may be cleaned by flushing with freshwater. Using the first aid kit, the wound may be disinfected with antiseptic towelettes and triple antibiotic ointment may be applied if possible.

Control of bleeding: If necessary, direct pressure may be applied to the wound in order to stop the bleeding.

Application of nonscalding hot water or other heat: Ice should not be applied to the injured area even if there is a swelling. On the other hand, heat may be required to ease the condition. The venom is protein based and begins to breakdown with the application of heat. Further heat will bring a noticeable relief to the pain and shorten the effectiveness of the venom. For this reason, hot water treatment is advised. The wound is immersed in the hot water (i.e., as hot as the victim can stand without scalding or burning the skin) for at least 30 minutes. Physicians may be called or the patient may be hospitalized for further treatment to overcome complications.

4.4.3 Treatment and Management of Lionfish Envenomation (Gallagher, http://emedicine.medscape.com/article/770764-treatment)

Prehospital care: This relates to the recognition of the injury, gentle removal of visible spines, controlling bleeding, administration of analgesia, and transporting the victim for medical evaluation.

Hot water immersion technique: Heat treatment is widely recommended as effective initial treatment for envenomations by the fishes of Scorpaenidae. By this technique, the affected limb is immersed in water no warmer than 114°F or 45°C.

Analgesia: Initial therapy consists of immersion in nonscalding hot water after removal of visible spines and sheath, in order to inactivate the thermolabile components of the venom which may otherwise cause a severe systemic reaction.

Stonefish antivenom: The lionfish venom shows similarities with the stonefish venom. Further, the stonefish (*Synanceia* spp.) antivenom has actually been shown to neutralize the effects of the lionfish venom. There appears to be crossreactivity between the two. The stonefish antivenom is recommended only for predilution intramuscular usage. However, for serious envenomations, the use of this antivenom may not be ideal because of erratic absorption. Following dilution, a slow intravenous administration may be preferable: one ampule (2000 U) for every one to two punctures, up to three ampules for more than four punctures. It should be diluted in 50–100 mL of isotonic sodium chloride solution and run through at least 20 minutes (Church et al., 2003).

References

Abellán, L.J.L., Santamaría, M.T.G., Conesa, P., 2001. Age and growth of *Pontinus kuhlii* (Bowdich, 1825) in the Canary Islands. Sci. Mar. 65, 259–267.

Aissaoui, N., Marzouki, M.N., Abidi, F., 2017. Purification and biochemical characterization of a novel intestinal protease from *Scorpaena notata*. Int. J. Food Prop. 20, 2151–2165.

Aissaoui, N., Abidi, F., Hardouin, J., Abdelkafi, Z., Marrakchi, N., Jouenne, T., et al., 2017. ACE inhibitory and antioxidant activities of novel peptides from *Scorpaena notata* by-product protein hydrolysate. Int. J. Pept. Res. Ther. 23, 13–23.

Aldred, B., Erickson, T., Lipscomb, J., 1996. Lionfish envenomations in an urban wilderness. Wilderness Environ. Med. 4, 291–296.

Allen, G.R., Erdmann, M.V., 2008. *Pterois andover*, a new species of scorpionfish (Pisces: Scorpaenidae) from Indonesia and Papua New Guinea. Aqua Int. J. Ichthyol. 13, 128–138.

Andrich, F., Carnielli, J.B., Cassoli, J.S., Lautner, R.Q., Santos, R.A., Pimenta, A.M., et al., 2010. A potent vasoactive cytolysin isolated from *Scorpaena plumieri* scorpionfish venom. Toxicon 56, 487–496.

Andrich, F., Richardson, M., Naumann, G., Cordeiro, M., Santos, A., Santos, D., et al., 2015. Identification of C-type isolectins in the venom of the scorpionfish *Scorpaena plumieri*. Toxicon 95, 67–71.

Angelini, G., Bonamonte, D., 2016. *Aquatic Dermatology*. Springer International Publishing, Switzerland.

Anon. A–Z Animals, Scorpion Fish. ⟨https://a-z-animals.com/animals/scorpion-fish⟩.

Anon. Lionfish Hunting. ⟨https://lionfish.co/sting-first-aid/⟩.

Anon. What is a Lionfish? National Ocean Service, NOAA. ⟨https://oceanservice.noaa.gov/facts/lionfish-facts.html⟩.

Anon. The Status of California Scorpionfish off Southern California in 2004. California Department of Fish and Wildlife. ⟨https://www.wildlife.ca.gov/Conservation/Marine/Scorpionfish⟩.

Anon. Status of Fisheries Resources in NSW, 2008/09. Eastern Red Scorpionfish (*Scorpaena cardinalis*). Wild Fisheries Research Program. ⟨https://www.dpi.nsw.gov.au/__data/assets/pdf_file/0007/375883/Eastern-Red-Scorpionfish.pdf⟩.

Anon. Family Scorpaenidae, Scorpionfishes/Rockfishes. ⟨http://www.burkemuseum.org/static/FishKey/scorp.html⟩.

Anon. Lionfish and Stonefish Envenomation. ⟨http://misc.medscape.com/pi/iphone/medscapeapp/html/A770764-business.html⟩.

Anon. Rockfish and Scorpionfish Family Photos and Information—Scorpaenidae. Mexico-Fish, Marine Life, Birds and Terrestrial Life. ⟨https://www.mexican-fish.com/scorpionfish-family⟩.

Anon. Introduction to Dive Medicine-Dangerous Creatures of the Sea. ⟨http://web.utah.edu/umed/students/clubs/international/presentations/dangers.html⟩.

Anon. A Study on the Venomous Fishes and Sea Snakes of Southeast Asia. ⟨http://oai.dtic.mil/oai/oai?verb=getRecord&metadataPrefix=html&identifier=AD0761209⟩.

Anon. *Amyloodinium ocellatum/Scorpaena brasiliensis*. ⟨http://www.aquasymbio.fr/en/amyloodinium-ocellatum-scorpaena-brasiliensis⟩.

Anon. Canary Scorpionfish (*Scorpaena canariensis*). ⟨http://www.whatsthatfish.com/fish/canary-scorpionfish/2441⟩.

Anon. Devil Firefish (*Pterois miles*) Ecological Risk Screening Summary. U.S. Fish and Wildlife Service. ⟨https://www.fws.gov/fisheries/ans/erss/highrisk/Pterois-miles-WEB-7-28-2014.pdf⟩.

Anon. Eschmeyer's Scorpionfish (*Rhinopias eschmeyeri*). ⟨https://reefapp.net/en/lex/details/rhinopias-eschmeyeri⟩.

Anon. Family Scorpaenidae—Scorpionfishes or rockfishes. FishBase. ⟨http://www.fishbase.org/summary/FamilySummary.php?ID=264⟩.

Anon. Identificazione e distribuzione nei mari italiani di specie non indigene, *Scorpaena stephanica*. ⟨http://www.marinealien.sinanet.isprambiente.it/uploads/Scorpaena%20stephanica.pdf⟩.

Anon. Key anatomical features of the lionfish. VAPAGUIDE. ⟨http://www.vapaguide.info/catalogue/FIS-138⟩.

Anon. Leaf Fish (*Taenianotus triacanthus*). ⟨http://www.theaquariumwiki.com/wiki/Taenianotus_triacanthus⟩.

Anon. *Pterois volitans* (Red Lionfish), The Online Guide to the Animals of Trinidad and Tobago, ⟨https://sta.uwi.edu/ffa/geography/fst/lifesciences/documents/Pterois_volitans.pdf⟩.

Anon. Red Lion Fish (*Pterois volitans*) Ecological Risk Screening Summary. ⟨https://www.fws.gov/fisheries/ans/erss/highrisk/Pterois-volitans-WEB-7-28-2014.pdf⟩.

Anon. Scorpionfish, Deep Water (*Idiastion kyphos*). ⟨http://portal.gulfcouncil.org/CreaturesOfTheDeep/ScorpionFish.html⟩.

Anon. Scorpionfishes, rockfishes, rosefishes, stingfishes, stonefishes, turkeyfishes and waspfishes. FAO Species Identification Sheets. Scorpaenidae. Fishing Area 51 (W. Indian Ocean). FAO, 1983. ⟨http://www.fao.org/tempref/docrep/fao/009/ad468e/AD468eKV.pdf⟩.

Anon. Species: *Scorpaena brachyptera*, Shortfin Scorpionfish. ⟨http://biogeodb.stri.si.edu/caribbean/en/the-fishes/species/3421⟩.

Anon. Species: *Scorpaena brasiliensis*, the Barbfish. ⟨http://biogeodb.stri.si.edu/caribbean/en/thefishes/species/3422⟩.

Anon. Species: *Scorpaena cocosensis*, Cocos Scorpionfish. ⟨http://biogeodb.stri.si.edu/sftep/en/thefishes/species/4634⟩.

Anon. Species: *Scorpaena histrio*, Bandfin Scorpionfish, Player Scorpionfish. ⟨http://biogeodb.stri.si.edu/sftep/en/thefishes/species/1015⟩.

Anon. Species: *Scorpaena russula*, Reddish Scorpionfish, Short-Spine Red Scorpionfish. ⟨http://biogeodb.stri.si.edu/sftep/en/thefishes/species/1017⟩.

Anon. Species: *Scorpaenodes tredecimspinosus*, Deepreef Scorpionfish. ⟨http://biogeodb.stri.si.edu/caribbean/en/thefishes/species/3431⟩.

Anon. Species: *Scorpaenodes xyris*, Rainbow Scorpionfish. ⟨http://biogeodb.stri.si.edu/sftep/en/thefishes/species/1020⟩.

Anon. Sport Fishing. ⟨https://www.sportfishingmag.com/gallery/2014/05/mysterious-fish-deep-may-fish-facts#page-3⟩.

Anon. Scorpionfishes: Venomous, but Not Poisonous. ⟨http://www.wetwebmedia.com/scorps_dang_mars_art.htm⟩.

Anon. Scorpion Fish Sting. MedlinePlus, US National Library of Medicine. ⟨https://medlineplus.gov/ency/article/002849.htm⟩.

Anon. Lionfish Sting First Aid and Treatment. ⟨https://lionfish.co/sting-first-aid/⟩.

Anon. Status of the Fisheries Report 2008, 12 Black Rockfish, *Sebastes melanops*. ⟨file:///E:/My%20Downloads/status2008black.pdf⟩.

Anon. The Water Zoo. ⟨https://www.waterzoo.co.uk/shop/FUZZY-DWARF-LION-(VENOMOUS)-M3951⟩.

Anon. Scorpion & Lion Fish—Scorpaenida. ⟨http://www.underwaterasia.info/reef-spotter/scorpionfish-lion-fish.php⟩.

Anon. Scorpionfishes—Family Scorpaenidae. ⟨http://www.wildsingapore.com/wildfacts/vertebrates/fish/scorpaenidae/scorpaenidae.htm⟩.

Anon. Species: *Pseudocarcharias kamoharai*, Crocodile Shark. ⟨http://biogeodb.stri.si.edu/caribbean/en/thefishes/species/34⟩.

Anon. Species: *Scorpaena dispar*, Hunchback Scorpionfish. ⟨http://biogeodb.stri.si.edu/caribbean/en/thefishes/species/3424⟩.

Anon. Species: *Scorpaena elachys*, Dwarf Scorpionfish. ⟨http://biogeodb.stri.si.edu/caribbean/en/thefishes/species/3425⟩.

Anon. Species: *Scorpaena grandicornis*, Plumed sScorpionfish. ⟨http://biogeodb.stri.si.edu/caribbean/en/thefishes/species/3426⟩.

Anon. Species: *Scorpaena isthmensis*, Smoothcheek Scorpionfish. ⟨http://biogeodb.stri.si.edu/caribbean/en/thefishes/species/3428⟩.

Anon. The Lionfish Invasion. ⟨https://oceanservice.noaa.gov/education/stories/lionfish/lion04_biology.html⟩.

Anon. Calico Rockfish, *Sebastes dalli* (Eigenmann and Beeson). ⟨http://content.cdlib.org/view?docId=kt8z09n9rg;NAAN=13030&doc.view=frames&chunk.id=d0e1220&toc.depth=1&toc.id=d0e847&brand=cali⟩.

Anon. (1983). FAO Species Identification Sheets. Fishing Area 51 (W. Indian Ocean), Scorpaenidae. ⟨http://www.fao.org/tempref/docrep/fao/009/ad468e/AD468eKV.pdf⟩.

Arai, H.P., Cathryn, H., Koski, C.H., 1964. A new species of *Trochopus* (Monogenea: Capsalidae) from *Scorpaena guttata* Girard. Can. J. Zool. 42, 1007–1010.

Auerbach, P.S., McKinney, H.E., Rees, R.S., Heggers, J.P., 1987. Analysis of vesicle fluid following the sting of the lionfish *Pterois volitans*. Toxicon 25, 1350–1353.

Baldwin, C.C., Pitassy, D.E., Robertson, D.R., 2016. A new deep-reef scorpionfish (Teleostei, Scorpaenidae, Scorpaenodes) from the southern Caribbean with comments on depth distributions and relationships of western Atlantic members of the genus. ZooKeys 606, 141–158.

Barnhart, P.S., Hubbs, C.L., 1946. *Pontinus vaughani*, a new scorpaenid fish from Baja California. Bull. Scripps Inst. Oceanogr. 5, 371–390.

Başçınar, N.S., Sağlam, H., 2009. Feeding habits of black scorpionfish *Scorpaena porcus*, in the southeastern Black Sea. Turk. J. Fish. Aquat. Sci. 9, 99–103.

Bayhan, B., Kaya, M., 2015. An updated checklist of poisonous fishes of Turkish Aegean Sea. J. Coastal Life Med. 3, 579–581.

Béarez, P., Motomura, H., 2009. Description of a new scorpionfish (Scorpaenoidei, Sebastolobinae) from the tropical eastern Pacific. Zootaxa 2277, 61–68.

Bo, J., Zheng, R., Kuang, W., Hong, F., Xie, Q., Zhang, Y., 2017. The use of rockfish *Sebastiscus marmoratus* as a sentinel species to assess petroleum hydrocarbons pollution: a case study in Quanzhou Bay, China. Mar. Pollut. Bull. 124, 984–992.

Bonecker, A.C.T., Pereira Namiki, C.A.P., de Castro, M.S., Campos, P.N., 2014. Ordem *Scorpaeniformes*. Catalogo dos estágios iniciais de desenvolvimento dos peixes da bacia de Campos [online]. *Sociedade Brasileira de Zoologia, Curitiba*, pp. 175–179.

Brown, E.M., Hovasse, R., 1946. *Amyloodinium ocellatum* (Brown), a peridinian parasitic on marine fishes. A complementary study. Proc. Zool. Soc., London 116, 32–46.

Brown, J.W. (2012). Lionfish: Kudzu of the Caribbean. ALS 398. 11p. ⟨http://www.mbio.ncsu.edu/lionfish/LionfishPaper.pdf⟩.

Butler, J.L., Jacobson, L.D., Barnes, J.T., Moser, H.G., 2003. Biology and population dynamics of cowcod (*Sebastes levis*) in the southern California Bight. Fish. Bull. 101, 260–280.

Carrijo, L.C., Andrich, F., de Lima, M.E., Cordeiro, M.N., Richardson, M., Figueiredo, S.G., 2005. Biological properties of the venom from the scorpionfish (*Scorpaena plumieri*) and purification of a gelatinolytic protease. Toxicon. 45, 843–850.

Carrijo, L.C., Andrich, F., de Lima, M.E., Cordeiro, M.N., Richardson, M., Figueiredo, S.G., 2005. Biological properties of the venom from the scorpionfish *Scorpaena plumieri* and purification of a gelatinolytic protease. Toxicon 45, 843–850.

Castro, R., Gonzalez, M.T., 2005. *Clavellotis sebastidis* sp. nov. (Copepoda, Lernaeopodidae) parasitic on *Sebastes oculatus* Valenciennes, 1833 from Argentina. Acta Parasitol. 50, 74–79.

Catarino, D., Stefanni, S., Menezes, G.M., 2013. Size distribution and genetic diversity of the offshore rockfish (*Pontinus kuhlii*) from three Atlantic archipelagos and seamounts. Deep Sea Res. 98, 160–169.

Cohen, A.S., Olek, A.J., 1989. An extract of lionfish (*Pterois volitans*) spine tissue contains acetylcholine and a toxin that affects neuromuscular transmission. Toxicon 27, 1367–1376.

Çulha, S.T., Yabanlı, M., Baki, B., Yozukmaz, A., 2016. Heavy metals in tissues of scorpionfish (*Scorpaena porcus*) caught from Black Sea (Turkey) and potential risks to human health. Environ. Sci. Pollut. Res. Int. 23, 20882–20892.

Diaz, J.H., 2015. Marine Scorpaenidae envenomation in Travelers: epidemiology, management, and prevention. J. Travel Med. 22, 251–258.

Dulcic, J., Jug-Dujakovic, J., Bartulovic, V., Glamuzina, B., Haskovic, E., Skaramuca, B., 2007. Embryonic and larval development of largescaled scorpionfish *Scorpaena scrofa* (Scorpaenidae). Cybium 31, 465–470.

Else, P., Haldorson, L., Krieger, K., 2002. Shortspine thornyhead (*Sebastolobus alascanus*) abundance and habitat associations in the Gulf of Alaska. Fish. Bull. 100, 193–199.

Eschmeyer, W.N., 1965. Three new scorpionfishes of the genera *Pontinus, Phenacoscorpius* and *Idiastion* from the Western Atlantic Ocean. Bull. Mar. Sci. 15, 521–534.

Eschmeyer, W.N., 1969. A new scorpionfish of the genus *Scorpaenodes* and *S. muciparus* (Alcock) from the Indian Ocean, with comments on the limits of the genus. Occas. Pap. Calif. Acad. Sci. 76, 1–11.

Eschmeyer, W.N., 1998. Catalog of fishes, Vol. 3. Calif. Acad. Sci., San Francisco, 2905 p.

Fenner, P.J., 2004. Venomous marine animals (review articles). South Pac. Underwater Med. Soc. (SPUMS) J. 34, 196–202.

Ferri, J., Petrić, M., Matić-Skoko, S., Dulčić, J., 2008. New host record, black scorpionfish *Scorpaena porcus* (Pisces, Scorpaenidae) for *Nerocila orbigny* and *Ceratothoa parallela* (Crustacea, Isopoda, Cymothoidae). Acta Adriat. 49, 255–258.

Ferri, J., Petrić, M., Matić-Skoko, S., 2010. Biometry analysis of the black scorpionfish, *Scorpaena porcus* (Linnaeus, 1758) from the eastern Adriatic Sea. Acta Adriat. 51, 45–53.

Field, C. (2014). Eat More Fish—Bourbetto (Scorpion Fish Stew). ⟨http://www.loverofcreatingflavours.co.uk/2014/07/eat-fish-bourbetto-scorpion-fish-stew/⟩.

Forrester, M.B., 2008. Pattern of lionfish envenomations reported to Texas poison control centers, 1998–2006. Toxicol. Environ. Chem. 90, 385–391.

Fricke, R., Durville, P., Mulochau, T., 2013. *Scorpaenopsis rubrimarginatus*, a new species of scorpionfish from Réunion, southwestern Indian Ocean (Teleostei: Scorpaenidae). Cybium 37, 207–215.

Gallagher, S.A. Lionfish and Stonefish Envenomation Treatment & Management. MedScape. ⟨http://emedicine.medscape.com/article/770764-treatment⟩.

Gardner, P.G., Frazer, T.K., Charles, A., Jacoby, C.A., Yanong, R.P.E., 2015. Reproductive biology of invasive lionfish (*Pterois* spp.). Front. Mar. Sci. Available from: https://doi.org/10.3389/fmars.2015.00007.

Gastoni, A., Mulas, A., Porcu, C., Cabiddu, S., 2011. Trophic habits of the blackbelly rockfish, *Helicolenus dactylopterus* (Osteichthyes, Scorpaeniformes) in the Sardinian Channel. Biol. Mar. Mediterr. 18, 346–347.

Gautam, K., Nazar, A.R., Begum, A.H.N., Marisamy, P., Ganesh, E.A., 2014. Variation in the fatty acid profile of lion fishes *Pterois volitans* (Linnaeus, 1758) and *Pterois antennata* (Bloch, 1787) of Parangipettai coast, Tamil Nadu, India. Adv. Appl. Sci. Res. 5, 166–172.

Gilbert, T.H., 2006. Prey Swimming Behavior and Culture Techniques for Copper, *Sebastes caurinus*, and Quillback Rockfish, *S. maliger*. Unpublished Master of Science Dissertation. Oregon State University.

Haddad Jr, V., Martins, I.A., Makyama, H.M., 2003. Injuries caused by scorpionfishes (*Scorpaena plumieri* Bloch, 1789 and *Scorpaena brasiliensis* Cuvier, 1829) in the southwestern Atlantic Ocean (Brazilian coast): epidemiologic, clinic and therapeutic aspects of 23 stings in humans. Toxicon 42, 79–83.

Halstead, B.W., 1970. Poisonous and venomous marine animals of the world, Vol. III. U.S. Gov. Printing Office, Washington, DC, p. 1006.

Halstead, B.W., Chitwood, M.J., Modglin, F.R., 1955a. The venom apparatus of the California scorpionfish, *Scorpaena guttata* Girard. Trans. Am. Microsc. Soc. 74, 145–158.

Halstead, B.W., Chitwood, M.J., Modglin, F.R., 1955b. The anatomy of the venom apparatus of the zebrafish, *Pterois volitans* (Linnaeus). Anat. Rec. 122, 317–333.

Hamza, F., Boxshall, G., Kechemir-Issad, N., 2007. A new species of *Prohatschekia* Nunes-Ruivo, 1954 (Copepoda: Hatschekiidae) parasitic on *Scorpaena elongata* (Cadenat) off Algeria. Syst. Parasitol. 67, 119–124.

Harmelin-Vivien, M.L., Kaim-Malka, R.A., Ledoyer, M., Jacob-Abraham, S.S., 1989. Food partitioning among scorpaenid fishes in Mediterranean seagrass beds. J. Fish Biol. 34, 715–734.

Hde, T.A., Costa, R.B., Marques, C.M.E., Santos, V.D.C.W., Gomes, F.S., Pereira, V.H.J., 2016. Angiotensins processing activities in the venom and epidermic mucus of *Scorpaena plumieri*. Toxicon 119, 92–98.

Higgs, N.D. The Feeding Habits of the Indo-Pacific Lionfish *Pterois volitans* at Artificial Lobster Habitats in the Bahamas. ⟨http://www.nickhiggs.com/uploads/1/5/7/5/15756104/higgs_lionfish_09.pdf⟩.

Huge, D.H., Schofield, P.J., Jacoby, C.A., Thomas, K., Frazer, T.K., 2014. Total mercury concentrations in lionfish (*Pterois volitans/miles*) from the Florida Keys National Marine Sanctuary, USA. Mar. Pollut. Bull. 78, 51–55.

Imamura, H., Yabe, M., 1996. Larval record of a red firefish, *Pterois volitans*, from Northwestern Australia (Pisces: Scorpaeniformes). Bull. Fac. Fish., Hokkaido Univ. 47, 41–46.

Ishida, M., Amaoka, K., 1992. A new species of the fish genus *Idiastion* (Pisces: Scorpaenidae) from the Kyushu-Palau Ridge, western Pacific. Jpn. J. Ichthyol. 38, 357–360.

Jones, O.A., Endean, R., 1976. Biology and Geology of Coral Reefs V3: Biology 2. Academic Press, New York-San Francisco-London, 435 p.

Johnson, E.G., Swenarton, M.K., 2016. Age, growth and population structure of invasive lionfish (*Pterois volitans/miles)* in northeast Florida using a length-based, age-structured population model. Peer J 4, e2730.

Kaya, Y., Kocatepe, D., 2014. Chemical composition and nutritional quality of scorpion fish (*Scorpaena porcus*, Linnaeus 1758) muscle. Indian J. Anim. Res. 48, 83–87.

Khalil, M.Y. Insights Into the Venomology of the Red Sea lionfish (*Pterois miles*, Scorpaenidae): Venom Protein Identification and Anti-cancer Efficacy. ⟨http://dar.aucegypt.edu/handle/10526/3344⟩.

Kiriake, A., Suzuki, Y., Nagashima, Y., Shiomi, K., 2013. Proteinaceous toxins from three species of scorpaeniform fish (lionfish *Pterois lunulata*, devil stinger *Inimicus japonicus* and waspfish *Hypodytes rubripinnis*): close similarity in properties and primary structures to stonefish toxins. Toxicon 70, 184–193.

Kizer, K.W., McKinney, H.E., Auerbach, P.S., 1985. Scorpaenidae envenomation. A five-year poison center experience. J. Am. Med. Assoc. 283, 807–810.

Krishna, N.M., 2016. First record of raggy scorpion fish *Scorpaenopsis venosa* (Cuvier, 1829) (Family Scorpaenidae) from Visakhapatnam, central eastern coast of India. Int. J. Zool. Stud. 1, 25–27.

Krishna, N.M., Sujatha, K., 2016. First record of the endemic cocks comb fire fish species *Ebosia falcata* Eschmeyer and Rama-Rao in the Visakhapatnam coastal waters, east coast of India and its distribuiontion range, with notes on some aspects of length frequency and length weight relationship studies. NeBIO 7, 146–150.

Laidig, T.E., Sakuma, K.M., Nishimoto, M.M., 1996. Description of pelagic larval and juvenile stripetail rockfish, *Sebastes saxicola* (family Scorpaenidae), with an examination of larval growth. Fish. Bull. 94, 289–299.

Luca, C., Manuela, F., Maria, F., Pierpaolo, C., Cristina, P., Valentina, E., et al., 2012. Trophic relationships among scorpaeniform fishes associated with gas platforms. Helgoland Mar. Res. 66, 401–411.

Mariscal, R.N., 1974. ExperimentalMarine Biology. Academic Press, N.Y.

Matsunuma, M., Motomura, H., 2013. A new lionfish of the genus *Dendrochirus* (Scorpaenidae: Pteroinae) from the Tuamotu Archipelago, South Pacific Ocean. Species Diversity 18, 1–7.

Matsunuma, M., Motomura, H., 2013. Newly recognized diagnostic characters of the poorly known lionfish *Pterois brevipectoralis* (Scorpaenidae: Pteroinae), with notes on fresh coloration. Species Diversity 18, 163–173.

Matsunuma, M., Motomura, H., 2015. A new species of scorpionfish, *Ebosia saya* (Scorpaenidae: Pteroinae), from the western Indian Ocean and notes on fresh coloration of *Ebosia falcata*. Ichthyol. Res. 62, 293–312.

Matsunuma, M., Motomura, H., 2016. A new species of scorpionfish, *Ebosia vespertina* (Scorpaenidae: Pteroinae), from the southwestern Indian Ocean. Ichthyol. Res. 63, 110–120.

Matsunuma, M., Sakura, M., Motomura, H., 2013. Revision of the Indo-West Pacific genus *Brachypterois* (Scorpaenidae: Pteroinae), with description of a new species from northeastern Australia. Zootaxa 3693, 401–440.

McCain, B.B., Miller, S.D. and Wakefield, W.W., II (2005). Life history, geographical distribution, and habitat associations of 82 West Coast groundfish species: a literature review. Pacific Coast Groundfish Fishery Management Plan for the California, Oregon, and Washington Groundfish Fishery, Appendix B Part 2, Groundfish Life History Descriptions. Pacific Fishery Management Council, Portland, Oregon.

Mendez, E., Jachmanian, I., Grompone, M.A., 1993. Lipid distribution in blackbelly rosefish (*Helicolenus dactylopterus lahillei)* in relation to its possible functions as hydrostatic agent and energy reserve. Comp. Biochem. Physiol., B: Biochem. Mol. Biol. 105, 193–198.

Mendonça, A., Isidro, E., Menezes, G., Pinho, M.R., Melo, O., Estácio, S., 2006. New contribution to the reproductive features of bluemouth *Helicolenus dactylopterus dactylopterus* from the northeast Atlantic (Azores Archipelago). Sci. Mar. 70, 679–688.

Mesa, G.L., Mesa, M.L., Tomassetti, P., 2007. Feeding habits of the Madeira rockfish *Scorpaena maderensis* from central Mediterranean Sea. Mar. Biol. 150, 1313–1320.

Mesaa, M.L., LaMesa, G., Micalizzic, M., 2005. Age and growth of Madeira scorpionfish, *Scorpaena maderensis* Valenciennes, 1833, in the central Mediterranean. Fish. Res. 74, 265–272.

Monteiro, L.R., Isidro, E.J., Lopes, H.D., 1991. Mercury content in relation to sex, size, age and growth in two scorpionfish (*Helicolenus dactylopterus* and *Pontinus kuhlii*) from Azorean waters. Water Air Soil Pollut. 56, 359–367.

Moon, S.Y., Youn, S., Venmathi Maran, B.A., 2016. A new species of *Prohatschekia* Nunes-Ruivo, 1954 (Siphonostomatoida: Hatschekiidae) from the scorpaenid fish, *Scorpaena neglecta* Temminck & Schlegel, 1843 (Scorpaeniformes: Scorpaenidae), in Korean waters. Zootaxa 4174, 153–160.

Moreno, G. (1990). Description of Tile Larval Stages of Five Northern California Species of Rockfishes (Family Scorpaenidae) from Rearing Studies. A Thesis Presented to the Faculty of California State University,

Stanislaus and Moss Landing Marine Laboratories in Partial Fulfillment of the Requirements for the Degree Master of Science in Marine Science.

Morris, J.A., Jr., Akins, J., Barse, A., Cerino, D., Freshwater, D.W., Green, S.J., et al. (2008). Biology and ecology of the invasive lionfishes, *Pterois miles* and *Pterois volitans*. In: Proceedings of the 61st Gulf and Caribbean Fisheries Institute. Gosier, Guadeloupe, French West Indies, pp. 409–414.

Morris Jr., J.A., Sullivan, C.V., Govoni, J.J., 2011. Oogenesis and spawn formation in the invasive lionfish, *Pterois miles* and *Pterois volitans*. Sci. Mar. 75, 147–154.

Moser, H.G., Ahlstrom, A.H., 1978. Larvae and pelagic juveniles of blackgill rockfish, *Sebastes melanostomus*, taken in midwatet trawls off southern California and Baja California. J. Fish. Res. Board Can. 35, 981–996.

Motomura, H., 2004. New species of scorpionfish, *Scorpaena cocosensis* (Scorpaeniformes: Scorpaenidae) from the Cocos Islands, Costa Rica, Eastern Pacific Ocean. Copeia 2004, 818–824.

Motomura, H., 2008. *Scorpaenopsis stigma* Fowler, 1938, a junior synonym of *Phenacoscorpius megalops* Fowler, 1938, with comments on the type series of *P. megalops* (Teleostei: Scorpaenidae). Zool. Stud. 47, 774–780.

Motomura, H., 2009. *Sebastapistes taeniophrys* (Fowler 1943): a valid scorpionfish (Scorpaenidae) from the Philippines. Ichthyol. Res. 56, 62–68.

Motomura, H., Causse, R., 2011. A new deepwater scorpionfish of the genus *Scorpaenopsis* (Scorpaenidae) from Wallis and Futuna Islands, southwestern Pacific Ocean. Bull. Mar. Sci. 87, 45–53.

Motomura, H., Johnson, J.W., 2006. Validity of the poorly known scorpionfish, *Rhinopias eschmeyeri*, with redescriptions of *R. frondosa* and *R. aphanes* (Scorpaeniformes: Scorpaenidae). Copeia 2006, 500–515.

Motomura, H., Kanade, Y., 2015. Review of the scorpionfish genus *Pteroidichthys* (Scorpaenidae), with descriptions of two new species. Zootaxa 4057, 490–510.

Motomura, H., Shinohara, G., 2005. Assessment of taxonomic characters of *Scorpaenopsis obtusa* and *S. gibbosa* (Scorpaenidae), with first records of *S. obtusa* from Japan and Australia and comments on the synonymy of *S. gibbosa*. Cybium 29, 295–301.

Motomura, H., Causse, R., Béarez, P., Mishra, S.S., 2002. Redescription of the Indo-West Pacific scorpionfish (Scorpaenidae), *Neomerinthe erostris* (Alcock 1896), a senior synonym of *Scorpaena gibbifrons* Fowler 1938, *N. rotunda* Chen 1981, and *N. bathyperimensis* Zajonz & Klausewitz. Zootaxa 4021, 529–540.

Motomura, H., Paulin, C.D., Stewart, A.L., 2005a. First records of *Scorpaena onaria* (Scorpaeniformes: Scorpaenidae) from the southwestern Pacific Ocean, and comparisons with the Northern Hemisphere population. N.Z. J. Mar. Freshwater Res. 39, 865–880.

Motomura, H., Fricke, R., Eschmeyer, W.N., 2005b. Redescription of a poorly known scorpionfish, *Scorpaena canariensis* (Sauvage), and a first record of *Pontinus leda* Eschmeyer from the Northern Hemisphere (Scorpaeniformes: Scorpaenidae). Stuttgarter Beitr. Naturk. 674, 15.

Motomura, H., Last, P.R., Yearsley, G.K., 2005c. *Scorpaena bulacephala*, a new species of scorpionfish (Scorpaeniformes: Scorpaenidae) from the northern Tasman Sea. Zootaxa 1043, 17–32.

Motomura, H., Last, P.R., Gordon, K., Yearsley, G.R., 2006. New species of shallow water scorpionfish (Scorpaenidae: Scorpaena) from the Central Coast of Western Australia. Copeia 360–369.

Motomura, H., Poss, S.G., Shao, K., 2007. *Scorpaena pepo*, a new species of scorpionfish (Scorpaeniformes: Scorpaenidae) from northeastern Taiwan, with a review of *S. onaria* Jordan and Snyder. Zool. Stud. 46, 35–45.

Motomura, H., Sakurai, Y., Shinohara, G., 2009. First records of a scorpionfish, *Scorpaenodes albaiensis*, from East Asia, with a synopsis of *S. minor* (Actinopterygii: Scorpaeniformes: Scorpaenidae). Species Diversity 14, 75–87.

Motomura, H., Arbsuwan, S., Musikasinthorn, P., 2010. *Thysanichthys evides*, a senior synonym of *Sebastella littoralis*, and a valid species of *Scorpaenodes* (Actinopterygii: Scorpaenidae). Species Diversity 15, 71–81.

Motomura, H., Matsunuma, M., Ho, H., 2011a. New records of three scorpaenid fishes (Teleostei: Scorpaeniformes) from Taiwan. J. Fish. Soc. Taiwan 38, 97–107.

Motomura, H., Béarez, P., Causse, R., 2011b. Review of Indo-Pacific specimens of the subfamily Scorpaeninae (Scorpaenidae), deposited in the Muséum national d'Histoire naturelle, Paris, with description of a new species of *Neomerinthe*. Cybium 35, 55–73.

Motomura, H., Kanehira, N., Imamura, H., 2012. Redescription of a poorly known southeastern Pacific scorpionfish (Scorpaenidae), *Phenacoscorpius eschmeyeri* Parin and Mandrytsa. Species Diversity 17, 145–150.

Motomura, H., Aizawa, M., Endo, H., 2014. *Sebastapistes perplexa*, a new species of scorpionfish (Teleostei: Scorpaenidae) from Japan. Species diversity 19, 133–139.

Motomura, H., Causse, R., Struthers, C.D., 2016. Redescription of the Indo-Pacific scorpionfish *Scorpaenodes guamensis* (Quoy & Gaimard 1824) (Scorpaenidae), a senior synonym of seven nominal species. Zootaxa 4067, 345–360.

Munoz, M., Sabat, M., Vila, S., Casadevall, M., 2005. Annual reproductive cycle of *Scorpaena notata* (Teleostei: Scorpaenidae). Sci. Mar. 69, 555–562.

Muñoz, M., Dimitriadis, C., Casadevall, M., Vila, S., Delgado, E., Lloret, J., et al., 2010. Female reproductive biology of the bluemouth *Helicolenus dactylopterus dactylopterus*: spawning and fecundity. J. Fish Biol. 77, 2423–2442.

Nakabo, T., Yamada, U., 1996. *Ursinoscorpaenopsis kitai*, a new genus and species of Scorpaenidae from the East China Sea. Ichthyol. Res. 43, 141–145.

Naranji, M.K., Kandula, S., 2017. Australian sawcheek scorpionfish, *Brachypterois curvispina* Matsunuma, Sakurai & Motomura 2013 (Scorpaeniformes: Scorpaenidae) new record from Indian waters. FishTaxa 2, 71–75.

Naranji, M.K., Velamala, G.R., 2016. First record of Rama Rao's scorpionfish, *Scorpaenopsis ramaraoi* Randall and Eschmeyer, 2001 (Scorapenidae) from Visakhapatnam, Andhra Pradesh Coast, India. FishTaxa 1, 84–88.

Nelson, L.S., Lewin, N.A., Howland, M.N., Hoffman, R.S., Lewis, R., Goldfrank, L.R., et al., 2011. Goldfrank's Toxicologic Emergencies. The McGraw-Hill, China, pp. 1629–1642.

Neves, A., Sequeira, V., Vieira, A.R., Paiva, R.B., Gordo, L.S., 2015. Age and growth of small red scorpionfish, *Scorpaena notata* (Actinopterygii: Scorpaeniformes: Scorpaenidae), a common discard species from the Portuguese fishery. Acta Ichthyol. Piscatoria 45, 13–20.

Ordines, F., Valls, M., Gouraguine, A., 2012. Biology, feeding, and habitat preferences of Cadenat's rockfish, *Scorpaena loppei* (Actinopterygii: Scorpaeniformes: Sscorpaenidae), in the Balearic Islands (western Mediterranean). Acta Ichthyol. Piscatoria 42 (1), 21–30.

Padate, V.P., Rivonker, C.U., Anil, A.C., Sawant, S.S., Krishnamurthy, V., 2017. New records of marine fishes from the coral reefs and deep waters of Gulf of Mannar, India. Acta Ichthyol. Piscatoria 47, 145–161.

Parker, S.J., Berkeley, S.A., Golden, J.T., Gunderson, D.R., Heifetz, J., Hixon, M.A., et al., 2000. Management of Pacific rockfish, Am. Fish. Soc. (AFS) Policy Statement, 25. pp. 22–29.

Parker, S.J., Rankin, P.S., Olson, J.M., Hannah, R.W., 2007. Movement Patterns of Black Rockfish (Sebastes melanops) in Oregon Coastal Waters. In: Heifetz, J., DiCosimo, J., Gharrett, A.J., Love, M.S., O'Connell, V.M., Stanley, R.D. (Eds.), Biology, Assessment, and Management of North Pacific Rockfishes. Alaska Sea Grant, University of Alaska, Fairbanks, Alaska, pp. 39–47.

Pavlov, D.A., Emel'yanova, N.G., 2007. Features of reproductive biology in two tropical fish species from the family Scorpaenidae. J. Ichthyol. 47, 353–365.

Pearson, K.E., Gunderson, D.R., 2003. Reproductive biology and ecology of shortspine thornyhead rockfish, *Sebastolobus alascanus*, and longspine thornyhead rockfish, *S. altivelis*, from the northeastern Pacific Ocean. Environ. Biol. Fishes 67, 117–136.

Poss, S.G., Scorpaenidae, Scorpionfishes (Also, Lionfishes, Rockfishes, Stingfishes, Stonefishes, and Waspfishes). FAO. ⟨http://www.fao.org/tempref/docrep/fao/009/x2400e/x2400e20.pdf⟩.

Poss, S.G. Scorpionfishes (Also Rosefishes, Rockfishes, Stingfishes). ⟨http://www.fao.org/tempref/docrep/fao/009/ad468e/AD468eKV.pdf⟩.

Poss, S.G., Collette, B.B., 1990. *Scorpaenodes immaculatus*, a new species of scorpionfish (Osteichthyes: Scorpaenidae) from Walters Shoals, Madagascar Ridge. Proc. Biol. Soc. Wash. 103, 543–549.

Poss, S.G., Duhamel, G., 1991. Neomerinthe bauchotae, a new scorpionfish (Scorpaenidae) from Saint Paul and Amsterdam Islands (southern Indian Ocean), with comments on the limits of the genus. Cybium 15, 93–102.

Poss, S.G., McCosker, J.E., Baldwin, C.C., 2010. A new species of *Scorpaenodes* (Pisces: Scorpaenidae) from the Galápagos and Cocos Islands with discussions of the limits of *Scorpaenodes* and *Thysanichthys*. Proc. Calif. Acad. Sci. 61, 235–267.

Ragonese, S., Gancitano, S., Norrito, G., Bono, R.G., 2002. Life history traits of the slender rockfish, *Scorpaena elongata* Cadenat, 1943 (Pisces—Scorpaenidae), of the Strait of Sicily (Mediterranean Sea). Biol. Mar. Mediterr. 10, 223–232.

Ramos-Ascherl, Z., Williams Jr., E.H., Bunkley-Williams, L., Lillian, J., Tuttle, L.J., Sikkel, P.C., et al., 2015. Parasitism in *Pterois volitans* (Scorpaenidae) from coastal waters of Puerto Rico, The Cayman Islands, and the Bahamas. J. Parasitol. 101, 50–56.

Randall, J.E., Earle, J.L., 2009. Variation in habitat and color of the Hawaiian scorpionfish *Sebastapistes coniorta*. Aqua (Int. J. Ichthyol.) 15, 61–68.

Resire, D., Cerland, L., De Haro, L., Valentino, R., Criquet-Hayot, A., Chabartier, C., et al., 2016. Envenomation by the invasive *Pterois volitans* species (lionfish) in the French West Indies—a two-year prospective study in Martinique. Clin. Toxicol. 54, 313–318.

Roche, E.T., Halstead, B.W., 1972. The venom apparatus of California rockfishes (family Scorpaenidae, Fish Bulletin. Department of Fish and Game, University of California, California, p. 156.

Rogers, B.L., Lowe, C.G., Fernández-Juricic, E., 2011. Recovery of visual performance in rosy rockfish (*Sebastes rosaceus*) following exophthalmia resulting from barotrauma. Fish. Res. 112, 1–7.

Sasikala, R., Ravindran, V.S., 2013. Antimicrobial activity of *Scorpaenopsis venosa* toxic extracts and its structural elucidation. J. Chem. Pharm. Res. 5, 254–259.

Saunders, P.R., Taylor, P.B., 1959. Venom of the lionfish *Pterois volitans*. ⟨http://www.physiology.org/doi/abs/10.1152/ajplegacy.1959.197.2.437⟩.

Schaeffer, R.C., Richard, W., Carlson, R.W., Russell, F.E., 1971. Some chemical properties of the venom of the scorpionfish *Scorpaena guttata*. Toxicon 9, 69–78.

Schofield, P.J., Huge, D.H., Rezek, T.C., Slone, D.H., Morris Jr, J.A., 2015. Survival and growth of invasive Indo-Pacific lionfish at low salinities. Aquat. Invasions 10, 333–337.

Schult, R.F., Acquisto, N.M., Stair, C.K., Wiegand, T.J., 2017. A case of lionfish envenomation presenting to an Inland Emergency Department. Case Rep. Emergency Med. 3. Article ID5893563.

Schwartz, S., Meinking, T., 1997. Venomous marine animals of Florida: morphology, behavior, health hazards. J. Fla. Med. Assoc. 84, 433–440.

Shaw, F.R., 2008. Notes on the life history traits of the rosethorn rockfish, Sebastes helvomaculatus. Calif. Fish Game 94, 123–136.

Simmons, K.R., Kerstetter, D.W., Blanar, C., 2014. Parasites of the lionfish complex (*Pterois volitans* and *P. miles*) in the Western North Atlantic, Gulf of Mexico, and Caribbean Sea: evidence of the enemy release hypothesis. Oceanography Faculty Proceedings, Presentations, Speeches, Lectures. 230. ⟨http://nsuworks.nova.edu/occ_facpresentations/230⟩.

Stewart, J., Hughes, J.M., 2010. Life-history traits of the southern hemisphere eastern red scorpionfish, Scorpaena cardinalis (Scorpaenidae: Scorpaeninae). Marine and Freshwater Research 61, 1290–1297.

Takano, K., Takemura, A., Mitsuru Furihata, M., Teruyuki Nakanishi, T., Akihiko Hara, A., 1991. Annual reproductive and spawning cycles of female *Sebastiscus marmoratus*. Environ. Biol. Fishes 30, 39–48.

Vaske Jr., T., de Lima, K.L., Aurelyanna, C.B., Ribeiro, A.C.B., Lessa, R.P., 2008. Record of the St. Helena deepwater scorpionfish, *Pontinus nigropunctatus* (Günther) (Scorpaeniformes: Scorpaenidae), in the Saint Peter and Saint Paul Archipelago, Brazil. Pan-Am. J. Aquat. Sci. 3, 46–48.

Victor, B.C., 2013. *Scorpaena wellingtoni* n. sp., a new scorpionfish from the Galápagos Islands (Scorpaeniformes: Scorpaenidae). J. Ocean Sci. Found. 8, 30–43.

Washington, B.B., Moser, H.G., Laroche, W.A., Richards, W.J., 1984. Scorpaeniformes: development. In: Moser, H.G., Richards, W.J., Cohen, D.M., Fahay, M.P., Kendall Jr., A.W., Richardson, S.L. (Eds.), Ontogeny and systematics of fishes, 1. Am. Soc. Ichthyol. Herpetol. Spec. Publ., pp. 405–428.

Wibowo, K., Motomura, H., 2017. A new species of the deepwater scorpionfish genus *Phenacoscorpius* (Teleostei: Scorpaenidae) from the Galápagos Islands. Zootaxa 4323, 261–268.

Zajonz, U., Klausewitz, W., 2002. *Neomerinthe bathypevimensis* sp. nov. from deep waters of the southern Red Sea. J. Fish Biol. 61, 1481–1488.

Zhang, X., Guo, H., Zhang, S., Song, J., 2015. Sound production in marbled rockfish (*Sebastiscus marmoratus*) and implications for fisheries. Integr. Zool. 10, 152–158.

Further Reading

Alexander, Q., Fogg, A.Q., Ruiz, C.F., Curran, S.S., Bullard, S.A., 2016. Parasites from the red lionfish, *Pterois volitans*, from the Gulf of Mexico. Gulf Caribb. Res. 27, SC1–SC5.

Anon, 2001. Fishery Management Plan for Coral Reef Ecosystems of the Western Pacific Region: Environmental Impact Statement,, vol. 3. National Oceanic and Atmospheric Administration, United States, p. 6.

Anon, 2009. COSEWIC assessment and status report on the Darkblotched Rockfish Sebastes crameri in Canada. Committee on the Status of Endangered Wildlife in Canada, Ottawa, p. 48.

Anon. (2014). Red Lion Fish (*Pterois volitans*) Ecological Risk Screening Summary. U.S. Fish and Wildlife Service, Web Version. ⟨https://www.fws.gov/fisheries/ans/erss/highrisk/Pterois-volitans-WEB-7-28-2014.pdf⟩.

Anon. (2016). Wildlife Protection (Amendment) Ordinance, 2016. ⟨http://www.ascension-island.gov.ac/wp-content/uploads/2017/03/ORD-8-Wildlife-Protection-Ord-Schedule-Amendment-2016-Asc.pdf⟩.

Cole, K.S., 2003. Hermaphroditic characteristics of gonad morphology and inferences regarding reproductive biology in *Caracanthus* (Teleostei, Scorpaeniformes). Copeia 2003, 68–80.

Field, J.C. Status of the Chilipepper Rockfish, Sebastes Goodei, in 2007. ⟨https://www.pcouncil.org/wp-content/uploads/ChilipepperJan2009final.pdf⟩.

Gomes, H.L., Menezes, T.N., Carnielli, J.B.T., Andrich, F., Evangelista, K.S., Chávez-Olórteguid, C., et al., 2011. Stonefish antivenom neutralises the inflammatory and cardiovascular effects induced by scorpionfish *Scorpaena plumieri* venom. Toxicon 57, 992–999.

Green, S.J., Akins, J.L. and Morris, J.A., Jr. (2012). Lionfish dissection: techniques and applications. NOAA Technical Memorandum NOS NCCOS 139. 24 pp.

Hayden-Spear, J., 2006. Nearshore Habitat Associations of Young-of-Year Copper (*Sebastes caurinus*) and quillback (*S. maliger*) rockfish in the San Juan Channel, Washington. Unpublished Master of Sciecne Dissertation. University of Washington.

Jones, O.A., Endean, R., 1976. Biology and Geology of Coral Reefs V3: Biology 2. Academic Press.

Kwik, J.T.B. (2011). The biology and ecology of small tropical scorpaenoids inhabiting shallow coastal habitats in Singapore. ScholarBank@NUS. Repository. ⟨https://scholarbank.nus.edu.sg/bitstream/10635/31654/1/KwikTBJ%20PhD%20Thesis%2029%20Feb%202012.pdf⟩.

Laan, V.D.R., Eschmeyer, W.N., Fricke, R., 2014. Family-group names of Recent fishes. Zootaxa 3882, 001–230.

Love, M.S., Axell, B., Morris, P., Collins, R.S., Brooks, A., 1987. Life history and fishery of the california scorpionfish, *Scorpaena guttata*, within the Southern California Bight. Fish. Bull. 85, 99–116.

Mancuso, M., 2015. *Scorpaena scrofa*: a promising aquaculture candidate for sicilian aquaculture. J. Aquacult. Res. Dev. 6, 11.

Morris, J.A., Jr., Whitfield, P.E., 2009. Biology, ecology, control and management of the invasive Indo-Pacific lionfish: an updated integrated assessment. NOAA Technical Memorandum NOS NCCOS 99. 57 pp.

Motomura, H., Senou, H., 2009. New records of the dwarf scorpionfish, *Sebastapistes fowleri* (Actinopterygii: Scorpaeniformes: Scorpaenidae), from East Asia, and notes on Australian records of the species. Species Diversity 14, 1–8.

Nichol, D.G. (1990). Life History Examination of Darkblotched Rockfish (*Sebastes crameri*) off the Oregon Coast. Unpublished Master of Science Dissertation. Oregon State University.

Ouanounou, G., Malo, M., Stinnakre, J., Kreger, A.S., Molgo, J., 2002. Trachynilysin, a neurosecretory protein isolated from stonefish (*Synanceia trachynis*) venom, forms nonselective pores in the membrane of NG108-15 cells. J. Biol. Chem. 277, 39119–39127.

Prithiviraj, N., Annadurai, D., Kumaresan, S., 2014. Length-weight relationship of lionfish *Pterois mombasae* (Smith, 1957) and spine used as biomedical agent from Parangipettai coast, Tamilnadu, India. Adv. Appl. Sci. Res. 5, 7–12.

Priyadharsini, S., Manoharan, J., Varadharajan, D., Subramaniyan, A., 2013. Reproductive biology and histological study of red lionfish *Pterois Volitans* from Cuddalore, South East Coast of India. J. Aquacult. Res. Dev. 4, 201. Available from: https://doi.org/10.4172/2155-9546.1000201.

Washington, B.B., Moser, H.G., Laroche, W.A., Richard, W.J., 2013. Scorpaeniformes: development. Ontogeny and Systematics of Fishes. Bibliogov, pp. 405–428.

Wong, M.Y.L., Munday, P.L., Jones, G.P., 2005. Habitat patch size, facultative monogamy and sex change in a coral-dwelling fish, Caracanthus unipinna. Environ. Biol. Fishes 74, 141–150.

Zem, A., Gey, D., Justine, J.L., Tazerouti, F., 2017. A new species of *Microcotyle* (Monogenea: Microcotylidae) from *Scorpaena notata* (Teleostei: Scorpaenidae) in the Mediterranean Sea. Parasitol. Int. 66, 37–42.

Index

FSC
www.fsc.org
MIX
Papier | Fördert gute Waldnutzung
FSC® C083411

Kolibri 360 GmbH
Ferdinand-Jühlke-Straße 7
99095 Erfurt, Deutschland
produktsicherheit@kolibri360.de